9th Seminar on Magnetic Materials 2015

Journal of Physics: Conference Series
Volume 1091

Palembang, Indonesia
19 – 21 October 2015

ISBN: 978-1-5108-7433-6
ISSN: 1742-6588

Printed from e-media with permission by:

Curran Associates, Inc.
57 Morehouse Lane
Red Hook, NY 12571

Some format issues inherent in the e-media version may also appear in this print version.

Printed by Curran Associates, Inc. (2018)

For permission requests, please contact the Institute of Physics
at the address below.

Institute of Physics
Dirac House, Temple Back
Bristol BS1 6BE UK

Phone: 44 1 17 929 7481
Fax: 44 1 17 920 0979

techtracking@iop.org

Additional copies of this publication are available from:

Curran Associates, Inc.
57 Morehouse Lane
Red Hook, NY 12571 USA
Phone: 845-758-0400
Fax: 845-758-2633
Email: curran@proceedings.com
Web: www.proceedings.com

Table of contents

Volume 1091

The 9th Seminar on Magnetic Materials (9th SMM)

19–21 October 2015, Palembang, Indonesia

Accepted papers received: 22 August 2018
Published online: 20 September 2018

IOP Conf. Series: Journal of Physics: Conf. Series **1091** (2018) 011001 doi:10.1088/1742-6596/1091/1/011001

Preface

We are very pleased to present the Proceedings of the 9th Seminar on Magnetic Materials 2015 held in Palembang, South Sumatera, Indonesia on October 19th – 21st, 2015. The seminar aimed to be a platform on academic exchange for materials scientists, especially in magnetic materials through invited talks and distinguish lectures series. In total, we received **40 manuscripts** and all of them have been reviewed by the reviewers. Those manuscripts are covering 7 topics on magnetic materials and applications meeting: magnetism and characterizations; magnetic materials applications on clean energy materials, transportation and vehicles; aerospace and defense; health care; environment and oil refining; electronics; and natural resources of magnetic materials.

Among those manuscripts, after a very intensive reviewing process, **3 papers** have been accepted by the Physica B's Editors and then published **in Physica B: Condensed Matter 2017, Vol. 526** of Elsevier and the other **30 manuscripts** are published in the Journal of Physics Conference Series 2018, Institute of Physics (IoP).

We are really honored as the selected seminar papers were published in the regular issues of high impact factor international journals, i.e. Physica B of Elsevier. This achievement is showing the level and quality of the regional/international meeting in Indonesia, organised by the Center for Science and Technology of Advanced Materials (PSTBM) - National Nuclear Energy Agency (BATAN) Indonesia in collaboration with the Faculty of Mathematics and Natural Sciences (FMIPA) - University of Sriwijaya.

Here, we are delighted to acknowledge the incredible support from the Ministry of Industry of the Republic Indonesia (Kemenperin), the Indonesian Institute of Sciences (LIPI), the University of Indonesia (UI), the Institute Technology of Sepuluh Nopember (ITS) Surabaya, the University of Gadjah Mada (UGM) Yogyakarta, the University of Padjadjaran (Unpad) Bandung, the Institute Technology of Bandung (ITB), the Agency of the Assessment and Application of Technology (BPPT).

We hope that the present proceedings provide the valuable information to the readers in the field of magnetic materials and its applications.

Dr. Poedji Loekitowati Hariani (Unsri) and Edy Giri Rachman Putra, Ph.D (BATAN)

Chairpersons of the 9th Seminar on Magnetic Materials

IOP Conf. Series: Journal of Physics: Conf. Series **1091** (2018) 011002 doi:10.1088/1742-6596/1091/1/011002

Peer review statement

All papers published in this volume of *Journal of Physics: Conference Series* have been peer reviewed through processes administered by the proceedings Editors. Reviews were conducted by expert referees to the professional and scientific standards expected of a proceedings journal published by IOP Publishing.

Magnetic Properties and Electrical Conductivity of $NiFe_2O_4$-MWNT/ PVA Nanocomposite Films

Ade Mulyawan*, Setyo Purwanto, Mashadi
Center for Science and Technology of Advanced Materials, National Nuclear Energy Agency of Indonesia (BATAN), Indonesia.

Corresponding author : ademulyawan@batan.go.id

Abstract. Synthesis and characterization of $NiFe_2O_4$-MWNT/PVA nanocomposite films have been performed. $NiFe_2O_4$-MWNT as a filler of nanocomposite films was gained using a simple mixing method from an aqueous solution containing $Fe(NO_3)_3.9H_2O$, $Ni(NO_3)_2.6H_2O$ and MWNT. A wet paste-like filler was first characterized by Raman spectroscopy to confirm the presence of $NiFe_2O_4$ phase. Refering to the result of Raman spectra, it was known that nanocrystalline nickel ferrite ($NiFe_2O_4$) have been successfully synthesized.The morphological investigations using scanning electron microscopy (SEM) and energy dispersive spectroscopic (EDS) revealed that the filler has hair-like structure which correspond to the carbon nanotube with the results of atomic percentage ratio of Fe/Ni correspond to the phase formation of $NiFe_2O_4$. Nanocomposite films were made in 4 various concentrations (5, 10, 20, and 50 %vol filler) have been characterized in order to investigate their magnetic properties and electrical conductivity by using vibrating sample magnetometer (VSM) and Inductance-Capacitance-Resistance (LCR) meter. Based on magnetization hysteresis curves, the variation of filler addition has increased the magnetic properties of nanocomposite samples from diamagnetic into paramagnetic. It was also known that the electrical conductivity enhanced align with the increasing consentration of the filler in nanocomposite films.

1. Introduction

Magnetoelectronic polymer composite in which magnetic nanoparticles are insulated from each other in conductive materials are used in a wide array of applications such as electromagnetic interference (EMI) shielding, chemical sensors, sensing components, etc as written by Bartholome et al. and Hudziak et al [1,2]. The conventional method for preparing magnetoelectronic polymer composite is by admixing magnetic nanoparticles and conductive materials as a fillers into the polymer [3]. Carbon nanotubes (CNT) filled by magnetic nanoparticles as fillers in polymer matrix composites have captivated much interest from many research groups due to the magnetic properties and fascinating electrical conductivities. CNT as a filler offering a good conducting network in polymer matrix due to their chain-like aggregate structures. Magnetic nanoparticles especially cubic nanosized spinel ferrites, with the general formula MFe_2O_4, hold significant potential to the polymer composite because of their unique magnetic, chemical and mechanical properties [4]. One of the most interesting spinel ferrite group is nickel ferrite ($NiFe_2O_4$). Based on the formation and particle size, $NiFe_2O_4$ shows several magnetic properties from paramagnetic to ferrimagnetic behaviour [5].

In the past few years, many research on fabrication of CNT filled by magnetic nanoparticles/polymer composites have been reported, polymers such as polyacrylonitrile by Nilmoung et al [6], polymethyl methacrylate by Sundaray et al [7] , and polyethylene oxide by Naebe et al [8] have been widely used as matrix. However, the development in polymer/CNT composites has been strongly limited by the problems associated the dispersion of CNT fillers and load transfer across the polymer-CNT interface. In this study, to overcome these barriers and in order to further improve not only the magnetic properties and electrical conductivity but also to prevent the agglomeration of the nanocomposite films, polyvinyl alcohol (PVA) was used as a matrix in nanocomposite films. PVA not only has high solubility in various solvent like water and alcohol but also chemically soft bonding with acid treated MWNT so it can be made to possess an unique capacity to solubilize CNT in aquaeous solution. Another advantage of using PVA as matrix is because among the conducting conjugated polymers, PVA is highly stable in air, good electrical conductivity, and easy processibility as written by G.Chakraborty,et al. and Massoti et al [9,10].

A $NiFe_2O_4$-MWNT/PVA nanocomposite film is expected to have diverse properties because each component contributes different properties to the nanocomposite films. The aims of this study is to investigate the effect of the filler addition ,which has been made by using simple mixing method, in matrix PVA to the magnetic properties and electrical conductivity of the nanocomposite films. The phase and morphology of the filler was characterized by using Raman spectroscopy and scanning electron microscopy (SEM). Magnetic properties of the nanocomposite films were measured by vibrating sample magnetometer (VSM) and electrical conductivity were measured by Inductance-Capacitance-Resistance (LCR) meter.

2. Experimental procedure

2.1. Starting materials

The chemicals were employed in this study as the starting materials included : Iron nitrate ($Fe(NO_3)_3.9H_2O$) in crystal form with purity ≥ 98.5 wt%, Nickel nitrate ($Ni(NO_3)_2.6H_2O$) in crystal form with purity ≥ 98.5 wt%, Sodium Dodecyl Sulfate (SDS) with purity ≥ 98.5 wt%, Polyvinyl alcohol with purity ≥ 99+ %wt (Sigma Aldrich), and MWNT black powder with outer diameter 13-18 nm and purity 99 wt% (cheaptubes).
Some text.

2.2. Synthesis of NiFe2O4-MWNT/PVA nanocomposites

Simple mixing method is very suitable for the synthesis of inorganic precursors. It consists of two main process which are dispersion process of the compound materials under vigorous stirring and sonication process using ultrasonicator. $NiFe_2O_4$-MWNT as a filler of nanocomposite films were gained using simple mixing method. The filler was synthesized form an aqueous solution containing $Fe(NO_3)_3.9H_2O$ and $Ni(NO_3)_2.6H_2O$ (Fe/Ni = 2:1 mmol) after that MWNT black powder was added and constantly stirred by magnetic stirrer for 20 hours at 100 ^{0}C then followed by heating treatment at 160 ^{0}C. This filler was then dispersed in SDS with ratio 9:12 and homogenized by ultrasonicator for 15 minutes at 40 ^{0}C and 360 rpm. The matrix nanocomposite was synthesized by using 10 gram of PVA in 100 ml aquades and stirred by magnetic stirrer for 15 minutes at 80 ^{0}C and 360 rpm. In this study, there were 4 different variations filler in nanocomposite films; 5, 10, 20, and 50 %vol. filler. The mixed solution ($NiFe_2O_4$-MWNT/PVA) was poured into a petri dish glass and dried overnight in room temperature to evaporate the water.

2.3. Characterization Technique

In order to confirm and examine the functionalization of the $NiFe_2O_4$-MWNT as a filler, Raman spectra carried out at room temperature by using Senterra Raman spectroscopy. The Raman spectra were measured in the range 100 – 1700 cm^{-1} with an excitation length of 785 nm. The microstructure micrograph and compound composition of the filler product were observed using Scanning Electron Microscope (SEM) on a JEOL JSM-6510 LA and energy dispersive spectroscopic (EDS) operating at 20kV. The observation of morphologies and nanostructures on the sample was performed by using Transmission Electron Microscope (TEM) at Gadjah Mada University. The observation condition are

800,000 times for magnification at a voltage of 120kV. The magnetization–magnetic field hysteresis curves of the filler and nanocomposite samples were measured using a vibrating sample magnetometer (Oxford, 1.2 Tesla) at room temperature. The conductivity of the nanocomposite samples were measured by using LCR meter (Hioki 3532-50) varying from 1 to 100 kHz.

3. RESULTS AND DISCUSSION

3.1. $NiFe_2O_4$-MWNT phase confirmation and formation

The phase confirmation of the $NiFe_2O_4$–MWNT as a filler was determined by Raman spectroscopy. It is a powerful tool for $NiFe_2O_4$-MWNT phase identification and further investigation since it provides into their electronic and vibration properties [11]. There are several dominant peaks of $NiFe_2O_4$–MWNT Raman spectra at 327, 478, 566, 700, 1043, 1300, and 1595 cm^{-1}. A strong peak in the range of 670–710 cm^{-1} is a common feature of the Raman spectra of the inverse spinels otherwise peaks at at 327, 478, 566 $cm^{-1}$are assigned to the $NiFe_2O_4$ crystal phase [12].

A similar Raman spectra pattern of $NiFe_2O_4$ phase was also observed by Zhou et al who studied $NiFe_2O_4$ nanoparticles in silica matrix made by mechanical activation method with calcination process at 900 ^{o}C [13].The Raman peaks appearing in the 1300, and 1595 cm^{-1} regions correspond to the scretching mode of graphite and poor crystallization of graphite, respectively as reported by Xiong et al. The result of this Raman spectra demonstrated good combination of $NiFe_2O_4$ and MWNT as a filler [14].

The SEM micrograph of $NiFe_2O_4$-MWNT is shown in Fig.2.(a). From this figure it can be seen that the surface of the micrograph consists of some hair-like nanostructure which correspond to the well spread MWNT material. Fig.2.(b) shows the energy-dispersive-spectroscopy (EDS) confirming the presence of the elemental compound of the filler. The result clearly exhibits elemental compound such as C, O, Fe, and Ni. The result of the mass and atomic percentage ratio of Fe/Ni in the filler is more or less correspond to $NiFe_2O_4$ phase. This result is equal with the confirmation of $NiFe_2O_4$ phase using Raman spectra (Fig.1).

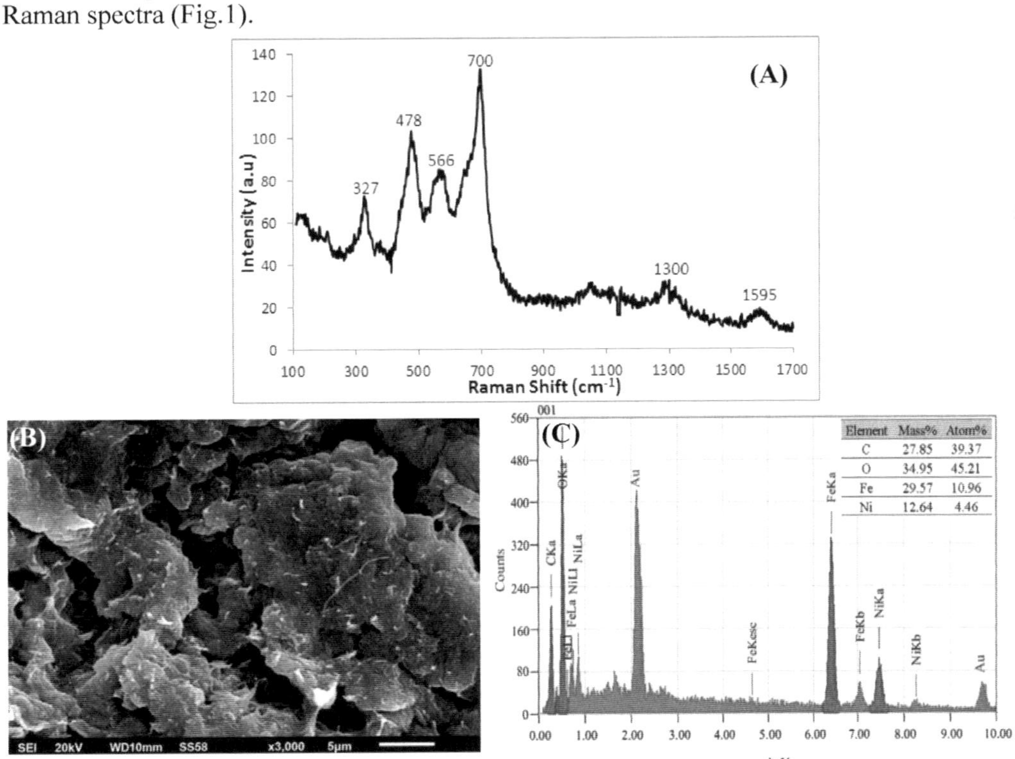

Figure 1. (a) the Raman spectra , (b) the SEM image, and (c) The EDS pattern of $NiFe_2O_4$-MWNT as a filler in the composite.

Table 1. The EDS results of $NiFe_2O_4$-MWNT/PVA nanocomposite films

Nanocomposite Samples	Elemental Compound							
	Atomic Percentage (%)				Mass Percentage (%)			
	C	O	Fe	Ni	C	O	Fe	Ni
5%.vol Filler	76.88	23.64	0.05	0.04	71.19	28.40	0.21	0.20
10%.vol Filler	58.90	40.77	0.18	0.15	51.33	47.32	0.72	0.64
20%.vol Filler	63.36	36.38	0.17	0.09	56.03	42.85	0.71	0.41
50%.vol Filler	59.63	38.96	0.99	0.42	50.46	43.91	3.90	1.73

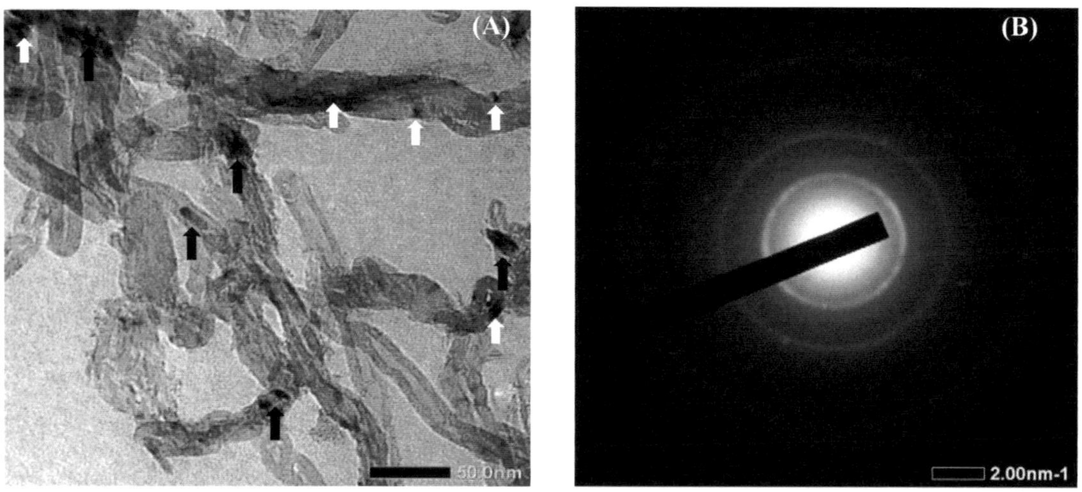

Figure 2. (a)The TEM image of $NiFe_2O_4$-MWNT (b) The Selected area electron diffraction (SAED) pattern of $NiFe_2O_4$-MWNT

The EDS of the nanocomposite films are shown in Table 1. Referring to these results it can be confirmed that the amount of Fe and Ni are increased align with increasing filler in the nanocomposite films. The amount of Ni and Fe (%mass and %atomic) in 10%.vol and 20%.vol filler nanocomposite are increased comparing with the 5%.vol filler nanocomposite. The amount of Ni and Fe in 50%.vol filler nanocomposite is the biggest among of all $NiFe_2O_4$-MWNT/PVA nanocomposite samples. The amount of Ni and Fe plays an important role to determine the characteristic of $NiFe_2O_4$-MWNT/PVA nanocomposite films such as magnetic properties and electrical conductivity of the nanocomposite films. It also can be seen in table 1 that along with the increasing volume filler the result of both atomic and mass percentage of Ni and Fe become clearer to confirm the $NiFe_2O_4$ phase.

The formation of the $NiFe_2O_4$-MWNT as a filler of all nanocomposite films were carried out by using TEM. It is a suitable tools to further study the morphologies and microstructures of nanosized materials. The TEM image is shown that the hollow tube of the MWNT are partially filled with the $NiFe_2O_4$ nanoparticles and at the same time it also decorating the outer surface of the MWNT. As shown in Fig.2.(a), the image indicated the $NiFe_2O_4$ phase could partially penetrate into the hollow tube of the nanotube which represented by black arrows, it happened because the ends of some tubular caps were opened as the result of the acid treatment and mixing process. The white arrows in Fig.2.(a) represented the $NiFe_2O_4$ phase which decorating the outer surface of the nanotube, it can be seen that the $NiFe_2O_4$ phase attach to each other strongly and form black line into the wall of the nanotube. The SAED pattern of $NiFe_2O_4$-MWNT are shown in Fig .2.(b), it can be seen a spotty ring which correspond to the crystal of the $NiFe_2O_4$.

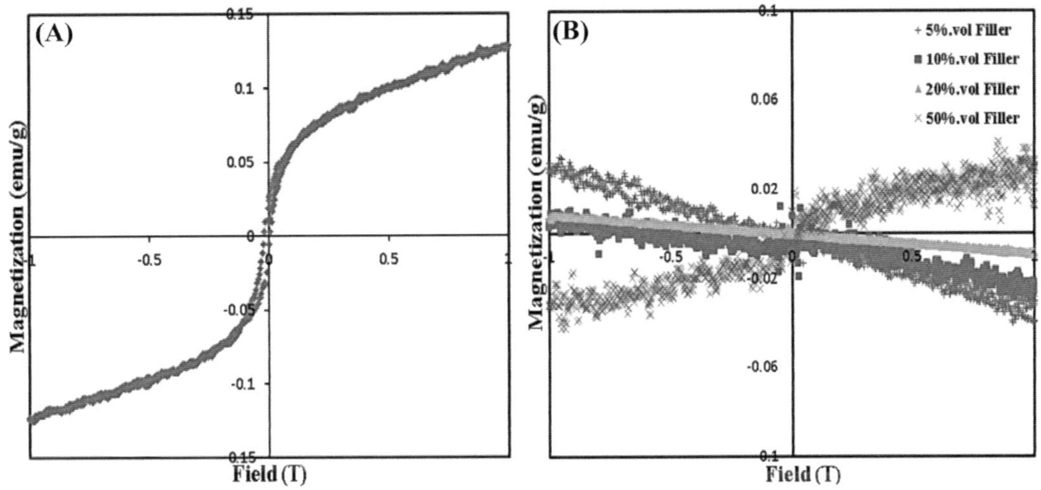

Figure 3. (a) Magnetization curves of wet solid paste $NiFe_2O_4$-MWNT (b) $NiFe_2O_4$-MWNT/PVA nanocomposite films

3.2. Magnetic Properties

The magnetization hysteresis curve of the $NiFe_2O_4$-MWNT in wet solid paste form is shown in Fig. 3.(a). The hystereresis curve is typical for a soft magnetic material and demonstrate the superparamagnetic behaviour with specific saturation magnetization (Ms) values of 0.13 emu/g and coercive forces (Hc) 119.5 Oe. The magnetization hysteresis curve shows the existence of hysteresis indicating the presence of magnetic materials which correspond to the attachment of the $NiFe_2O_4$ in MWNT reffering to the result of EDS and Raman spectra (Fig. 2.). Superparamagnetic behavior which appeared in this experiment can be a clear evidence that simple mixing method is succeed to synthesize $NiFe_2O_4$. Referring to the result of the Ms value, in the process to synthesize the filler still needs a further optimization especially in optimizing heat treatment process so that the $NiFe_2O_4$ can be improved to have high ordered structure crystal. The Ms and Hc of the $NiFe_2O_4$-MWNT as a filler value was comparable with the report by Nilmoung,et al. (2011) who studied the magnetic properties of the carbon/$NiFe_2O_4$ nanofibres stabilized at 280 ^{0}C for 0.5 hour with polyacryonitrile as the polymer source, he reported the Ms value of 0.36 emu/g and Hc 98.97 Oe.

The magnetization hysteresis curve of all $NiFe_2O_4$-MWNT/PVA nanocomposite films are given in Fig. 3.(b), It exhibits a lower Ms value than the Ms of $NiFe_2O_4$-MWNT as a filler. It can be explained due to the existence of diamagnetic phase of PVA in the nanocomposites; Interaction between $NiFe_2O_4$-MWNT and PVA as matrix in nanocomposite samples may affect the uniformity of magnetization by extinguishing the surface magnetic moment [15]. The improvement of magnetic properties of the nanocomposite films because the presence of the filler in the matrix is expected. From the shape of the magnetization hysteresis curves in Fig.3.(b) is known that the addition of $NiFe_2O_4$-MWNT as a filler can improve the magnetic properties of the nanocompsite film. The 5, 10, and 20%.vol filler nanocomposite films exhibit diamagnetic behavior which indicated that the role of the PVA as the matrix is too strong but there is a clear improvement of the magnetic properties from 5%.vol to 20%.vol filler nanocomposite film, respectively. Unlike the magnetic properties of other nanocompsite films, the 50%.vol filler nanocomposite film exhibits paramagnetic behavior which indicating a random spin interaction of the $NiFe_2O_4$ magnetic nanoparticle inside the nanocomposite films. The analysis results of the magnetic properties showed that the magnetic properties of nanocomposite materials are dependent on the content of $NiFe_2O_4$ as a compound of the filler. It can be confirmed from the result of the atomic and mass percentage of Ni and Fe (table 1), the 50%.vol filler nanocomposite film exhibited the biggest amount of Ni and Fe which also had the highest Ms value of 0.04 emu/g whereas the lowest value of the Ms was obtained in 5%.vol filler nanocomposite film which also has least amount of Ni and Fe among all nanocomposite films.

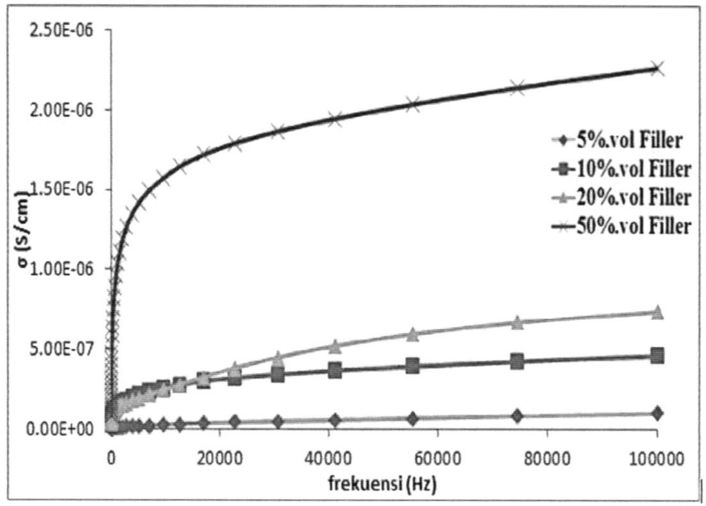

Figure 4. The conducivity of $NiFe_2O_4$-MWNT/PVA nanocomposite films (Mulyawan,2014)

3.3. Electrical Conductivity

Fig.4. shows the effect of $NiFe_2O_4$-MWNT loading as a filler to the conductivity values of the nanocomposite films, varying from 5%.vol to 50%.vol filler nanocomposite film, respectively [16]. The $NiFe_2O_4$-MWNT/PVA nanocomposite samples were characterized by using LCR meter to define the relation of the volume filler to the electrical conductivity values. A universal relationship between the addition of the filler and the electrical conductivity values can be described using the equation as follows;

$$\sigma \propto (V - V_p)^t \quad (1)$$

The conductivity behavior can be described by using the equation (1) when the volume fractions greater than the percolation threshold. V is the volume fraction of the filler, Vp is the percolation threshold, and σ and t are the conductivity and a critical exponent respectively, Oskouyi et al.(2014). Refering to the quantum mechanical tunneling theory known that electrical current can flow through the insulator matrix based, so it can be assumed that the filler as a conductive materials are connected by resistor formed of the matrix which allowed the electrons to interact with the neighbor electrons [17]. Refering to the equation and the explanation above, it can be noticed that the conductivity values are equal to the volume filler contain in nanocomposite films.

In accordance with those theory the results of all nanocomposite films are increased gradually along with the increasing filler. The electrical conductivity of 5, 10, and 20%.vol filler nanocomposite films were 9.87 x 10^{-8} S/cm, 4.60 x 10^{-7} S/cm, and 7.38 x 10^{-7} S/cm, respectively. The low conductivity value of the 5%.vol filler nanocomposite film was because the role of PVA as the insulator matrix still dominant rather than the filler materials. Along with the increasing volume filler composition, it can be seen from 10% and 20%.vol filler nanocomposite films that the influence of PVA to the conductivity of the films were being replaced by the filler. The 50%.vol filler nanocomposite film has the highest conductivity among all nanocomposite films with the value of 1.92 x 10^{-6} S/cm. As shown in fig.4 the conductivity value was increased sharply in the low range frequency (less than 5kHz), it happened because a greater conductive materials in the nanocomposite films could increase the contact of the filler to filler junction to create some electron tunneling which allowed for the electron hopping under the insulator condition. The results of our research to investigate the relationship between volume filler concentration and the conductivity values is in accordance with the research that has been done by Mohd Faiz Muaz,et al. (2011) who reported that the addition of the filler of MnO_2-MWNT/PVA could enhance the electrical conductivity of polymer composites in the range of 10^{-6} S/cm from its neat conductivity.

4. CONCLUSION

Referring to the result of the Raman spectra and the EDS results $NiFe_2O_4$-MWNT were successfully synthesized by using simple mixing method from the strarting compound of $Ni(NO_3)_2.6H_2O$, $Fe(NO_3)_3.9H_2O$, and MWNT black powder. $NiFe_2O_4$-MWNT/PVA nanocomposite samples were successfully prepared by using a simple casting method. Magnetic measurement by using VSM revealed that magnetic properties of nanocomposite samples were improved with the increasing of filler content. The synergistic effect of the filler volume to the nanocomposite films was observed, where the electrical conductivity gradually increased with the increasing filler content.If this progress continues apace, we can expect a continuing bright future in this fascinating and potentially very useful area in some new applications as electronic material for energy and nanodevices.

Acknowledgements
The authors would like to thank to the Center for Science and Technology of Advanced Materials for providing Raman spectrometer, SEM/EDS, VSM, and LCR meter facilities. In addition the authors would like to thank to Dr.Siti Nikmatin,S.Si from department of Physics, Faculty of Mathematics and Science, Bogor Agricultural University for a fruitful discussion during the progress to create this work.

References

[1] Bartholome C, Miaudet P, Derré A, Maugey M, Roubeau O, Zakri C. nanotube-PVA composites Reference : Influence of surface functionalization on the thermal and electrical properties of nanotube-PVA composites. *Composites Science and Technology*. 2009;68: 2568-2573.
[2] Hudziak S, Darfeuille A, Zhang R, Peijs T, Mountjoy G. Magnetoresistive phenomena in an Fe-filled carbon nanotube / elastomer composite. *Nanotechnology*. 2010;125505.
[3] Faiz M, Ahmad M, Hussein S, Zein S, Abdullah AZ, Irwin N. Improved Electrical Conductivity of Polyvinyl Alcohol / Multiwalled Carbon Nanotube Nanofibre Composite Films with MnO_2 as Filler Synthesised using the Electrospinning Process. *International Journal of Engineering & Technology IJET-IJENS*. 2011;11(6):15-21.
[4] Lazarević ZŽ, Jovalekić Č, Sekulić D, Slankamenac M, Romčević M, Characterization of nanostructured spinel NiFe2O4 obtained by soft mechanochemical synthesis. *Science of Sintering*, 2012;44(3):331–339.
[5] Sivakumar P, Ramesh R, Ramanand A, Ponnusamy S, Muthamizhchelvan C. Synthesis and characterization of $NiFe_2O_4$ nanoparticles and nanorods. *Journal Alloys and Compound*. 2013;563:6–11.
[6] Nilmoung S, Rugmai S, Maensiri S. Fabrication, Structure, and Magnetic Properties of Carbon / Nickel Ferrite ($Nife_2O_4$) Composite Nanofibers. *Presented in the 18th international conference on composite materials*.
[7] Sundaray B, Subramanian V, Natarajan ST, Krishnamurthy K. Electrical Conductivity of A Single Electrospun Fiber of Poly (Methyl Methacrylate) and Multiwalled Carbon Nanotube Nanocomposite. *Journal Of Applied Physics*.2006;88:143114.
[8] Naebe M, Lin T, Tian W, Dai L. Effects of MWNT Nanofillers on Structures and Properties of PVA Electrospun Nanofibres. *Nanotechnology*. 2007;15:225605.
[9] Chakraborty G, Gupta K, Meikap AK, Babu R, Blau WJ. Synthesis , Electrical and Magnetotransport Properties Of Polypyrrole-MWCNT Nanocomposite. *Solid State Communications*. 2011;
[10] Masotti A, Caporali A. Preparation of Magnetic Carbon Nanotubes (Mag-CNTs) for Biomedical and Biotechnological Applications. *International Journal of Molecular Sciences*. 2013;14:24619–42.

[11] Chen I H., Wang C C. Fabrication And Characterization Of Magnetic Cobalt Ferrite/Polyacrylonitrile And Cobalt Ferrite/Carbon Nanofibers By Electrospinning.*Carbon.* 2010;48:604-611.
[12] Ivanov VG, Abrashev M V, Iliev MN, Gospodinov MM, Meen J, Aroyo MI. Short-Range B-site Ordering in Inverse Spinel Ferrite $NiFe_2O_4$. 2010;1–10
[13] Zhou ZH, Xue JM, Wang J, Chan HSO, Yu T, Shen ZX, et al. $NiFe_2O_4$ Nanoparticles Formed in situ in Silica Matrix by Mechanical Activation $NiFe_2O_4$ Nanoparticles Formed in situ in Silica Matrix by Mechanical Activation. *Journal of Applied Physiscs.*2002;91(9):6015.
[14] Xiong P, Fu Y, Wang L, Wang X. Multi-walled carbon nanotubes supported nickel ferrite : A magnetically recyclable photocatalyst with high photocatalytic activity on degradation of phenols. *Chemical Engineering Journal.*2012;195-196:149–157.
[15] Poddar A, Bhowmik RN, De A, Sen P. Magnetic Response of $NiFe_2O_4$ nanoparticles in polymer matrix. *Journal of Magnetism and Magnetic Materials.* 2009;
[16] Mulyawan,A, "Fabrication and Optimization Physical Properties of $NiFe_2O_4$-Multiwalled Nanotube/PVA Nanocomposite Film," Undergraduate Thesis, Bogor Agricultural University, 2014.
[17] Oskouyi AB, Sundararaj U, Mertiny P. Tunneling Conductivity and Piezoresistivity of Composites Containing Randomly Dispersed Conductive Nano-Platelets. *Materials.* 2014;7:2501–21.

IOP Conf. Series: Journal of Physics: Conf. Series **1091** (2018) 012002 doi:10.1088/1742-6596/1091/1/012002

The Surface Functionalization of Fe_3O_4 Nanoparticles by CTAB as Adsorbent for Methyl Orange Elimination in Water

Adel Fisli, Didin S. Winatapura and Alfian

Center for Science and Technology of Advanced Material
National Nuclear Energy Agency, Kawasan PUSPIPTEK Serpong,Tangerang Selatan (15314) Indonesia

Corresponding author : adel@batan.go.id

Abstract. Adsorption is the most extensively used technique for elimination of dye in contaminant water. Magnetically separation of toxic pollutant is becoming a potential method for water purification and obtained more effective and simple compared to conventional method of treatment. The surface-functionalized magnetic iron oxide nanoparticles with cetyl trimethyl ammonium bromide (CTAB-coated CTAB) has been investigated. CTAB-coated Fe_3O_4 were synthesized by precipitation of ferrous/ferric chloride (mole ratio 2:1) in CTAB/ammonium hydroxide solution mixture. The samples were characterized by X-ray diffraction (XRD), transmission electron microscopy (TEM), nitrogen adsorption-desorption isothermal and Fourier transforms infrared spectroscopy. The CTAB-coated Fe_3O_4 nanoparticles exhibited ability significantly to eliminate methyl orange in water compared to uncoated Fe_3O_4. Dye adsorption equilibrium state data were fitted well to the Langmuir isotherm were obtained the favorable of constant separation factor value (RL) (0.005-0.03). The CTAB coated Fe_3O_4 followed the pseudo-second-order kinetic model, resulting in the rate constant adsorption value for the elimination of methyl orange was 0.0694 $g.mg^{-1}.h^{-1}$. CTAB coated Fe_3O_4 nanoparticles have relatively high saturation magnetization, which allowed their highly-efficient magnetic separation from wastewater. This investigation was concluded than CTAB-coated Fe_3O_4 nanoparticles was efficient to eliminate dye from wastewater

1. Introduction

Iron oxide magnetic nanoparticles with novel properties and functions have many unique magnetic properties such as superparamagnetic, high coercivity, high surface area to volume ratios, low Currie temperature and high magnetic susceptibility. It is great potential in electronics, magnetic storage, biomedical, optoelectronic, ferrofluid, separation and environmental application [1.2]. The superparamagnetic material can response external magnetic field but retain no residual magnetism after the field has been removed. This property is to be used as possible utilized as carrier adsorbent to remove pollutant from wastewater. Magnetically separation of magnetic particles is considered as a quick and effective technique for many application in biochemistry, analytical chemistry, mining ores and environment [3]. Adsorption procedure combined with magnetic separation has been used extensively in water treatment and environmental cleanup which considerable advantages with respect to fast separation of adsorbent via employing a strong external magnetic field and cost savings for decontamination of wastewater without centrifugation or filtration treatment (versus non-magnetic) [4].

Iron oxide magnetite, Fe_3O_4, $Fe^{II}Fe^{III}{}_2O_4$, ferrimagnetic possesses superparamagnetic behavior in size less than 15 nm. Accordingly experimental, saturation magnetization value of magnetite nanoparticles have been reported 30-80 emu/g, lower than that bulk magnetite of 100 emu/gr.

However, this value is still sufficient for magnetic separation with a magnet. On the other hand, Fe_3O_4 exhibits adsorption characteristics, possessing the capability to treat a large volume of wastewater and being convenient for magnetic separation [2]. Several experiments show that Fe_3O_4 nanoparticles have been undertaken to examine the removal efficiency of inorganic and organic contaminants [5]. It recommended that Fe_3O_4 nano sorbent were effective and economical adsorbents for rapid removal and recovery of metal ions from wastewater effluents.

To increase its performance, surface modification of Fe_3O_4 with inorganic shell or organic molecules have been reported [5]. It is in order to not only stabilize the nanoparticles and eventually prevents the oxidation but also enhance the capacity or selective for uptake contaminants in water treatment. The surface functionalized magnetite for environmental application have been utilized such as Fe_3O_4/silica for methylene blue and Zinc removal [6, 7], biopolymer functionalized iron Oxide for antimicrobial activity [8].

Cetyltrimethylammonium bromide (CTAB), a cationic surfactant has been widely used for the extraction of DNA and synthesis of nanoparticles (sphere, rods, bipyramid, mesoporous). Previous studies found that modification of some adsorbents with CTAB could greatly enhance the sorption of pollutants. For example, dextrin-oxalic acid/CTAB-montmorillonite for heavy metal remediation [9], CTAB-modified graphite for adsorption bisphenol A [10].

Previous studies, we found that modification of magnetite surface with cetyltrimethylammonium bromide (Fe3O4@CTAB). In general, there are two step synthesis of Fe_3O_4@CTAB. In the first step, preparation of Fe_3O_4 nanoparticles by co-precipitation. In the second step, coating prepared Fe_3O_4 with CTAB. For example, Jin et al. (2012) [11] reported that Fe_3O_4 nanoparticles carried out by precipitation of a mix Fe(II)/Fe(III) solution with a Fe(II)/Fe(III) of 0.5 in alkali condition. The obtained Fe_3O_4 was then transferred into CTAB solution under sonification. The founded Fe_3O_4@CTAB was used for removal of arsenate. Wang et al. (2012) [12] reported that Fe_3O_4-CTAB was prepared by a facile one-pot solvothermal method in using an autoclave. Series of Fe_3O_4 morphologies are obtained by controlling the concentration of CTAB for removal of dyes.

In this study, Fe_3O_4-CTAB was prepared simply by one step (in-situ) process where the mixture of Fe(II)/Fe(III) was dropped in alkali-containing CTAB. This method has a shorter time in the synthesis process. The structural properties and morphology of obtained material were investigated in detail. Adsorption properties of Fe_3O_4-CTAB for the elimination of methyl orange in water were evaluated using Langmuir isotherms and kinetic adsorption equation.

2. Experiment

2.1. Materials and Reagents

All reagents used were of analytical reagent grade, i.e., ferric chloride ($FeCl_3.6H_2O$), ferrous chloride ($FeCl_2.4H_2O$), cetyltrimethylammonium bromide (CTAB), ammonium hydroxide (NH_4OH), hydrochloride acid, methyl orange. All reagents were purchased from Merck and used without further purification. Deionized water was used in all experiments.

2.2. Synthesis of Fe_3O_4-CTAB

Amount 5.2 g $FeCl_3.6H_2O$ and 2 g $FeCl_2.4H_2O$ (mol ratio of Fe(III):Fe(II) = 2:1) were dissolved in 10.3 mL of 1 N HCl and diluted to 25 mL with deionized water under vigorous stirring. On the other hand, 0.36445 g CTAB was dissolved in 250 mL of 1.5 M ammonium hydroxide solution under vigorous stirring. The iron salt solution was dropped into CTAB/ammonia solution. The mixture was subsequently shaken for 8 hours. The formed black-colored precipitates were separated by helping magnetic bar and washed with deionized water. The obtained black-solid were dried in the oven at 60^oC overnight and then 100^oC for 2 hours.

2.3. Characterization

The synthesized samples were subsequently characterized by X-ray diffractometer (XRD, Empyrean Panalytical with Cu-Kα radiation, λ=1.154 Å), vibrating sample magnetometer (VSM, Oxford type 1.2 T at room temperature), transmission electron microscope (TEM, JEM-14000 JOEL), Fourier transform infrared-attenuated total reflection (FTIR-ATR Bruker TENSOR 27), N_2 adsorption-desorption isotherms (Qauantasorb SI-4-Kr/MP, Quantachrome Instruments at the liquid nitrogen temperature (77K)). Before measurement, the samples were degassed at 150°C for 8 hours.

2.4. Adsorption evaluation

All studies were conducted on the batch adsorption. The amount of Fe_3O_4-CTAB was inserted into methyl orange solution and stirred by shaker at room temperature for 4 hours. The effects of contact time, pH solution, adsorbent dose and concentration dye on adsorption were investigated. The adsorbed amount of methyl orange was calculated by the difference between the initial and final concentration in the equilibrium solution. The concentration of methyl orange was determined in the UV-Lambda 25 Spectrometer Perkin Elmer at 469 nm.

3. Result and Discussion

In order to determine crystal phase of iron oxide particles, the samples were analyzed by powder X-ray diffractometer (XRD). The XRD patterns of the samples were measured from 20 to 90° (2θ) and revealed the natural of the sample. Figure 1 showed XRD patterns of Fe_3O_4 without and coated CTAB. The peaks of Fe_3O_4 was clearly observed in diffractograms which similar to the standard diffraction pattern of magnetite in JCPDS file no. 03-0863 with 2θ at 30.167, 35.451, 43.253, 56.781, 62.726. Thus, the particles can be indexed to the pure phase of magnetite (Fe_3O_4) structure. The XRD pattern of Fe_3O_4-CTAB is identical with Fe_3O_4 pattern. This result indicated that Fe_3O_4 structure phase was no change after coated with CTAB.

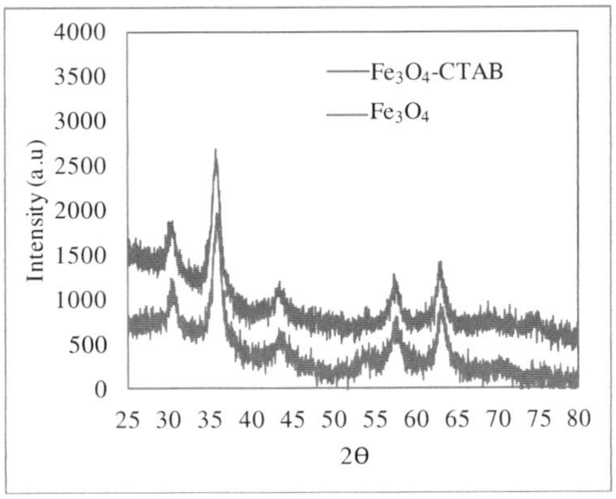

Figure 1. XRD pattern of Fe_3O_4-CTAB and Fe_3O_4

The morphology and microstructure of Fe_3O_4-CTAB were investigated by transmission electron microscopy (TEM). Fig 2 (a) shown that the Fe_3O_4-CTAB nanoparticles tend to aggregate form around with size 200 nm. The particle of Fe_3O_4 is observed possess particles size diameter less than 20 nm. (Fig. 2 (b). In this case, beneficial for magnetic separation due to retaining the superparamagnetic nature of individual nanoparticles, while aggregations of magnetic nanoparticle have relatively large volume enables their effective high-gradient magnetic separation (HGMS).

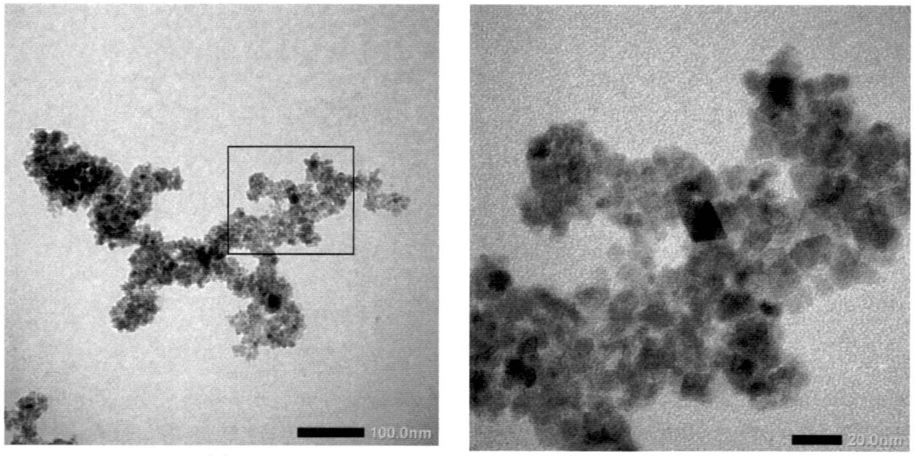

(a) (b)

Figure 2. Transmission electron microscope (TEM) of Fe_3O_4-CTAB (a) image on bar scale 100 nm (b) image on bar scale 20 nm.

The isotherms of nitrogen adsorption and desorption in sample Fe_3O_4 and Fe_3O_4-CTAB are presented in Fig. 3. The shape of the isotherms are attributed to type –IV curves according to the IUPAC classification. The pattern of the isotherms suggests that the founded Fe_3O_4 possess characteristic mesoporous materials. Mesoporous samples generally show hysteresis in the adsorption and desorption branches above P/Po = 0.4, attributed to multilayer formation and especially, capillary condensation in mesopores [13]. Mesoporous was occurred on Fe_3O_4 nanoparticles because of formation of inter-particles spaces of Fe_3O_4 agglomeration. There was no significant difference the total pore volume value for both Fe_3O_4 and Fe_3O_4-CTAB. However, the average pore diameter of Fe_3O_4-CTAB is higher than that Fe_3O_4 (Tabel 1). It is caused by the inter-particles space is rather broad due to the existence of CTAB attached on the surface of Fe_3O_4.

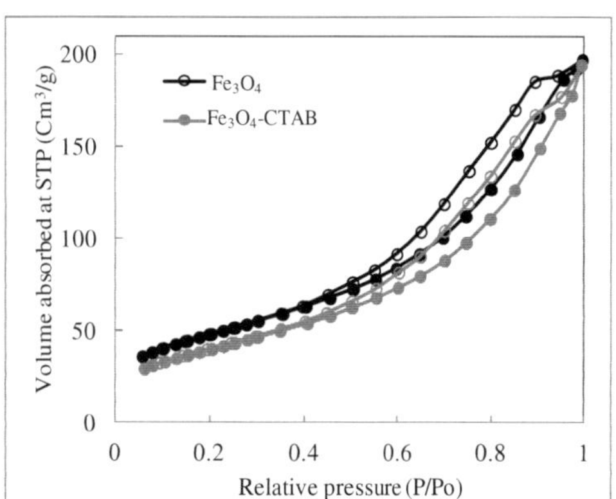

Figure 3. Nitrogen adsorption-desorption isothermal of Fe_3O_4 and Fe_3O_4-CTAB

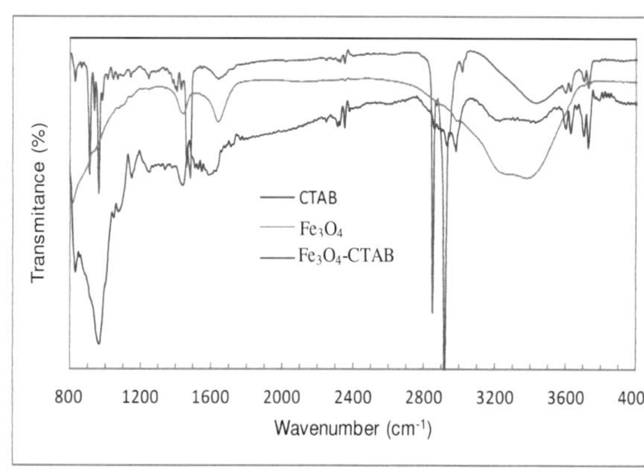

Figure 4. FTIR spectra of Fe_3O_4 and Fe_3O_4-CTAB

Table1. Surface area and porosity analysis of Fe_3O_4 and Fe_3O_4-CTAB

Sampel	BET surface area	Total Pore volume	Avarage pore diameter
	(m^2/g)	cc/g	nm
Fe_3O_4	172	0.3050	7.086
Fe_3O_4-CTAB	146	0.3008	8.204

The binding of CTAB to Fe_3O_4 nanoparticles surface was investigated by FTIR-ATR analysis as shown in Fig. 4. On CTAB curve, the asymmetric and symmetric vibration of CH_2 frequencies observed in 2914 and 2850 cm^{-1}, respectively. The asymmetric vibration of (CH_3-N^+) and C-N stretching observed around at 3015 and 960 [14]. The bending vibration bands of the head [$N(CH_3)_3$] methyl group at 1489 and 1429 cm^{-1} and (CN^+) stretching vibrations at 904 [15]. On Fe_3O_4 curve is appear a peak 3400- 3400 cm^{-1}. It is not from water, but it is attributed to the stretching vibrations of -OH groups which are assigned to surface OH group of Fe_3O_4 nanoparticles. This peak is wide due to the formation of hydrogen bond on hydroxyl group. While the peak at 1462 and 1627 is the plane and out of plane bonds of O-H groups [16]. On the Fe_3O_4/CTAB curve, the binding of CTAB to Fe_3O_4 nanoparticles surface result shift the spectrum IR bands peaks of CTAB vibration from 2850 to 2933 cm^{-1}, 2914 to 2980 $cm^{-1}$and 1489 to 1452 cm^{-1} suggesting an interaction between the Fe_3O_4 nanoparticles surface and the chemisorbed CTAB molecule [17]. In addition, the spectrum of CTAB-coated nanoparticles indicated a small shift in the positions of the main peaks in comparison with the bare nanoparticles. This could be ascribed to the alterations in the surface environment of the particles after modification.

Room-temperature magnetic hysteresis of Fe_3O_4 and Fe_3O_4-CTAB were measured by VSM (Fig. 5). It has a typical ferromagnetic behavior. The value of saturation magnetization (Ms) is 62.3 emu/ for Fe_3O_4. According, experimental value for Ms in magnetic iron oxide nanoparticles have been range the 30-80 emu/g. The experiments are lower than that bulk magnetic value 100 emu/g due to Fe_3O_4 nanoparticles are not very stable under ambient condition and easily change to Fe_2O_3 [2]. The saturation magnetization (Ms) of Fe_3O_4-CTAB is 55.9 emu/gr, lower than that Fe_3O_4 pure. This result is reasonable due to the present of non-magnetic material containing (CTAB) on samples. The magnetic remanence (Mr) and coercivity (Hc) value are 11.7 emu/gr and 138 Oe for Fe_3O_4 and 8.49 emu/gr and 116.7 Oe for Fe_3O_4-CTAB. The magnetic remanence and coercivity value of Fe_3O_4-CTAB is lower than that Fe_3O_4. It is favorable due to the degree of its superparamagnetic properties is increase. This alteration of magnetic properties may be caused by several factors, including particles size, structure, morphologies, etc.

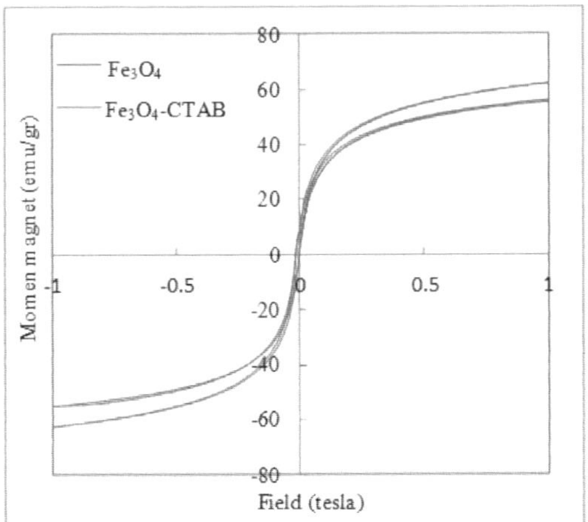

Figure 5. M-H hysteresis curves of Fe_3O_4 and Fe_3O_4-CTAB

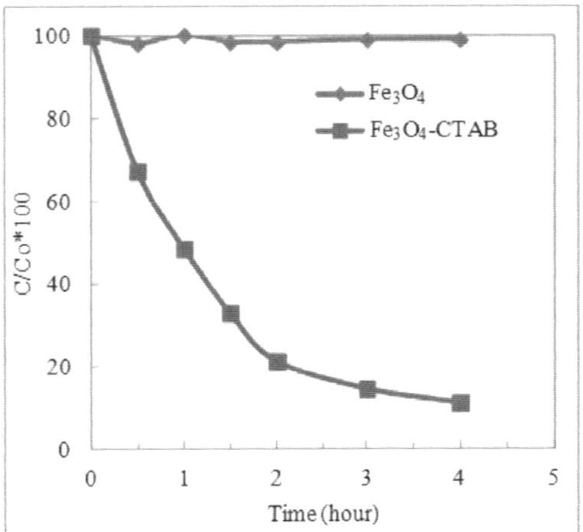

Figure 6. The adsorption of methyl orange on Fe_3O_4 and Fe_3O_4-CTAB

Fig. 6 shows the experiment result of methyl orange (MO) adsorption on Fe_3O_4 and Fe_3O_4-CTAB. The data presented a plot of C/Co (%) MO versus contact time, where C= concentration of MO in sampled water at t time, Co= concentration of MO initial. As shown in this figure, methyl orange was no adsorbed by Fe_3O_4 surface. The surface of Fe_3O_4 nanoparticles is negatively charged and the methyl orange containing $-SO_3^-$ group are negative too. So, the dyes can not directly interaction with Fe_3O_4 nanoparticles surface. Next curve, the methyl orange can be adsorbed drastically on CTAB coated Fe_3O_4 (Fe_3O_4/CTAB) which are adsorbed over 90% for 4 hours processes. CTAB containing

≡S^+ group is the type of cationic surfactant. The functionalization of Fe_3O_4 surface with CTAB has altered into a positively charged. Since the Fe_3O_4-CTAB surface and methyl orange contain positive group and negative group ≡S^+ and –SO_3^- respectively, electrostatic interactions will occur both materials. This result indicates that CTAB plays an important role in the dye adsorption mechanism. The methyl orange is adsorbed via CTAB on the surface of Fe_3O_4 nanoparticles. In addition, Sun et al. (2014) [18] reported that methyl orange could interact with alkyl chain of CTAB through hydrophobic intermolecular interaction which can contribute methyl orange adsorbed on the surface of Fe_3O_4-CTAB.

Initial pH solution is an important parameter to control the adsorption process. The pH of the solution can change adsorption capacity of adsorbent due to (1) the surface change of the adsorbent, (2) the degree of ionization of the adsorptive molecule and (3) extent of dissociation of functional groups on the active sites of the adsorbent [19]. The adsorption capacity (*q*) of adsorbent was calculated by the following equation:

$$q = \frac{(C_o - C_e)V}{m} \quad (1)$$

where C_0 and Ce are the initial and equilibrium concentration of methyl orange (mgL^{-1}), *m* is the mass of composite (g) and *V* is the volume of solution (L). The adsorption of methyl orange onto Fe_3O_4-CTAB was investigated in range pH from 4 to 12 at initial methyl orange concentration of 15 mg/L and the contact time of 4 hours. As shown in figure 7, the methyl orange uptake tends to decrease with increasing pH solution. As described previously, the methyl orange is a dye anionic. The decrease in the adsorption of methyl orange with an increase of pH value is due to the competition between the anionic dye and excess OH- ion in the solution.

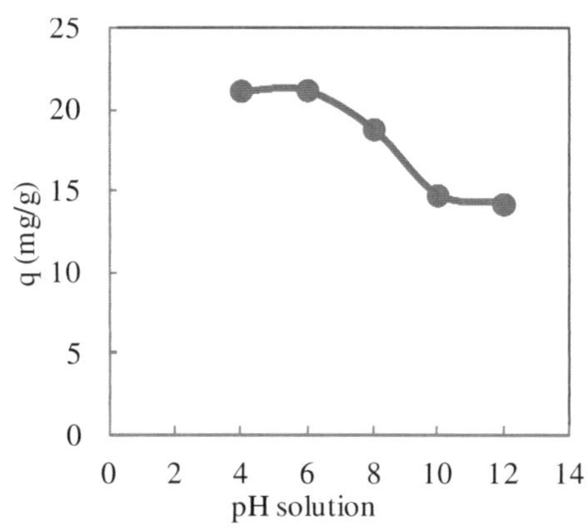

Figure 7. pH effect of adsorption methyl orange on Fe_3O_4-CTAB

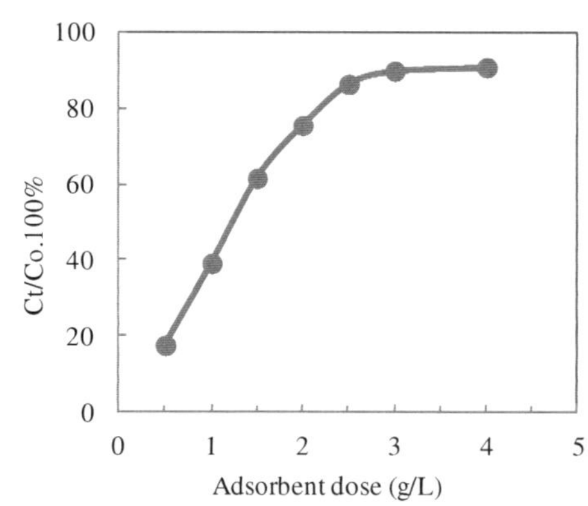

Figure 8. Effect of adsorbent dose on percentage removal of methyl orange

The effect of the adsorbents dose was necessary in order to observe the effective of adsorbent used and also shows the maximum adsorption stochiometric. The effect of adsorbent dosage gives an idea for the ability of a dye adsorption to be adsorbed with the smallest amount of adsorbent. The influence of Fe_3O_4-CTAB dose on the adsorption capacity of methyl orange was shown in figure 8. The amount of Fe_3O_4-CTAB in the solution was varied from 0.5 to 2.5 g/L in solution with the constant of the initial methyl orange concentration of 20 mg/L. It can see that an adsorbent dose of 1.5 g/L is sufficient for optimal removal of methyl orange. Increasing the dose further did not affect, instead of decrease the adsorption capacity. Generally, increase in the solid concentration increase the surface area of the adsorbent, which in turn increase the number of binding sites. Thus the total amount of dye is removed increase. The adsorption capacity was low at high dose rate and vice versa. The increasing adsorbent concentration can result in a low specific site uptake, due to interference between binding sites. When the mass of the adsorbent in the liquid phase is higher, the interaction between adsorbent particles become more important. It may cause physical blockage of some adsorption sites, decreasing the adsorption efficiency [20].

The interactive behavior between solutes and adsorbent was described in the equilibrium adsorption isotherm. The equilibrium adsorption data are analyzed using Langmuir isotherms model the following formula:

$$\frac{C_e}{q_e} = \frac{1}{bq_m} + \frac{C_e}{q_m} \quad (2)$$

where qe is the amount adsorbed at equilibrium ($mg.g^{-1}$), Ce is the equilibrium concentration of the methyl orange ($mg.L^{-1}$), constant b is related to the energy of adsorption ($L.mg^{-1}$), qm is the Langmuir monolayer adsorption capacity ($mg.g^{-1}$). If Ce/qe is plotted against Ce, one will get a straight line with a slope of 1/qm. figure 9 shows adsorption isotherm and of methyl orange adsorption onto Fe_3O_4-CTAB, plotting Ce/qe against Ce, one would then obtain a high regression correlation coefficient with R2 value of 0.9941. This indicates that the Langmuir model is suitable for describing the adsorption equilibrium of methyl orange by the Fe_3O_4-CTAB. The maximum monolayer capacity value which is determined from the Langmuir isotherm was found to have a value of 7.58 $mg.g^{-1}$. The characteristic of the Langmuir isotherm can also be illustrated in a dimensionless constant separation factor RL that is given by

$$R_L = \frac{1}{1+bC_0} \quad (3)$$

where C_0 is the initial concentration in the liquid phase ($mg.L^{-1}$). The values of calculated Langmuir constants b, q_m and R_L are listed in Table 2. The value of R_L indicates the shape of the isotherm to be either unfavorable ($R_L>1$), linear ($R_L=1$), favorable ($0<R_L<1$) or irreversible ($R_L=0$). The obtained R_L value is in the range of 0.005-0.03, indicating that the adsorption is a favorable process.

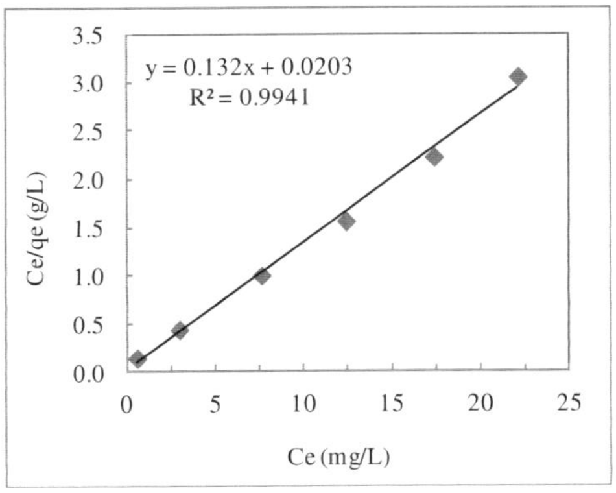

Figure 9. Langmuir isotherms of methyl orange adsorption onto Fe_3O_4-CTAB

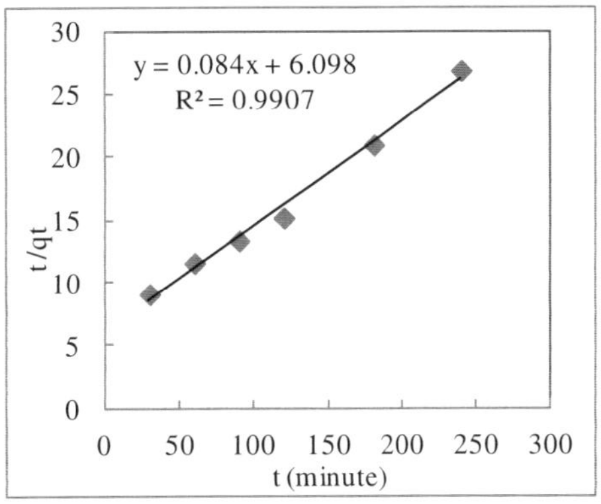

Figure 10. Pseudo-second-order kinetics of adsorption methyl orange onto the Fe_3O_4-CTAB

Kinetic parameters, which are useful for the prediction of the adsorption rate, give an important information for designing and modeling the adsorption processes. The adsorption mechanism of methyl orange onto the adsorbent was investigated by a pseudo-second-order kinetic model which was expressed in followed equation.

$$\frac{t}{q_t} = \frac{1}{k_2 q_e^2} + \frac{t}{q_e} \quad (4)$$

where q_e and q_t are the amounts of dye adsorbed ($mg.g^{-1}$) at equilibrium and at time t (h) and k_2 ($g.mg^{-1}.h^{-1}$) is the rate constant of the pseudo-second-order adsorption. Kinetic constants obtained by linear regression for a pseudo-second-order kinetic model with R^2 value of 0.9907 (figure 10). The rate constant of pseudo-second-order adsorption (k_2) is 0.0694 $g.mg^{-1}.h^{-1}$ as showed in Table 2. The $q_{e,(cal)}$

value close to the $q_{e.exp}$ (7.21 mg/g), which indicates the applicability of the second-order model to describe the adsorption process of methyl orange onto the Fe_3O_4-CTAB.

Tabel 2. Adsorption isothermal and kinetic parameters of methyl orange adsorption onto Fe_3O_4-CTAB

Sampel	q_m ($mg.g^{-1}$)	b ($L.mg^{-1}$)	R^2	R_L	qe ($mg.g^{-1}$)	k_2 $g.mg^{-1}.h^{-1}$	R^2
Fe_3O_4-CTAB	7.58	6.5024	0.9941	0.005-0.03	11.905	0.0694	0.9907

Tabel 3. Comparison of the maximum monolayer adsorption (q_m) of methyl orange onto various adsorbents

Sampel	q_m(mg/g)	reference
Fe_3O_4-CTAB	7.6	This study
Goethite	55	[21]
MoSe2 Microsphere	36.9	[22]
Stratum Landform-like composite	41.65	[23]
Wheat Bran	12	[24]
Cork powder	17	[25]

The removal of methyl orange by various adsorbent has been studied extensively, and dye adsorption capacity was reported in the literature [21-25]. Tabel 3 compares the adsorption capacities of composite obtained in this work with different adsorbents previously used for removal methyl orange. It can be seen that the adsorption capacities of the Fe_3O_4-CTAB for methyl orange are lower slightly than that of many other previously reported adsorbents, indicating that the prepared composite has potential application in dye removal from aqueous solution

4. Conclusion

We have demonstrated the synthesizing of the Fe_3O_4-CTAB on a simple one step (in-situ) process. The founded material has good adsorbent for methyl orange removal. The sorption capacity for methyl orange was 7.58 $mg.g^{-1}$. The optimal sorption pH was 4-6. The adsorbent dose of 1.5 g/L is sufficient for optimal removal of methyl orange. The dimensionless constant separation factor (R_L) was in the range 0.005-0.03, meaning that the adsorption is a favorable process. The kinetic parameter was applicable for pseudo-second-order adsorption with rate constants of 0.0694 $g.mg^{-1}.h^{-1}$. The synthesized Fe_3O_4-CTAB could easily recollect from water by the magnetic bar. So, they could be utilized as a convenient and effective adsorbent for methyl orange elimination from the water.

Acknowledgements

This work was supported by Research and Development of Smart Magnetic Programs for DIPA 2015 grand of Center for Science and Technology of Advanced Materials, National Nuclear Energy Agency from Indonesia Government Research of the Republic of Indonesia.

5. References

[1] Xu P, Zeng G M, Huang D L, Feng C.L, Hu S, Zhao M H, Lai C, Wei Z, Huang C, Xie G X, Liu Z F 2012 *Sci.Tot. Environ.* **424** 1-10.
[2] Issa B, Obaidat I. M, Albiss B.A and Haik Y 2013 *Int. J. Mol. Sci.* **14** 21266-21305
[3] Ai L, Zhou Y and Jiang, J 2011. *Desalination* **266** 72–77.
[4] Ambashta R.D and Sillanpää M 2012 J. Haz. Mater. **180** 38–49.
[5] Xu P, Zeng G M, Huang D L, Feng C L, Hu S, Zhao M H, Lai C, Wei Z, Huang C, Xie G X, Liu Z F 2012 *Sci.Tot. Environ.* **424** 1-10
[6] Fisli A, Yusuf S, Ridwan, Gunlazuardi J 2014 *Adv. Mater. Res.* **896** 525–531.

[7] Emadi M, Shams E, Amini M K 2013 *J. Chem.* **2013**, ID 787682.
[8] Dhivya S M, Sathiya S M, Manivannan G, Rajan M A J. 2016 Mater. Today: *Proceeding 3(B)* 3866-3871.
[9] Ahmad R and Mirza A 2017 *Groundwater Sustain. Develop.* **4** 2017.
[10] Wang L, Ni X, Cao Y and Cao G 2018 Appl. Surf. Sci. 428 165-170.
[11] Jin Y, Liu F, Tong M and Hou Y 2012 J. Haz. Mater. **227-228** 46–468.
[12] Wang Y, Cheng R, Wen Zi and Zhao L 2012 *Chem. Engineer. J.* **181-182** 823–827.
[13] Thommes M, Cychosz K.A 2014 Adsorption **20** 233-250.
[14] Guivar J A R, Sanches E A, Magon C J and Fernandes E G R 2015 *J. Electroanal. Chem.* **755** 158–166.
[15] Viana R B, Silva A B F and Pimentel A S 2012 *Adv.Phys. Chem.* 2012 ID 903272.
[16] Lopez J, González F, and Bonilla F 2010. Rev. Latin. Metal. Mater 30 60-66
[17] Guivar J.A.R, Sanches E.A, Magon C.J and Fernandes E.G.R 2015 *J. Electroanal. Chem.* **755** 158–166.
[18] Sun T, Gao G, Chen Q and Shen X 2014 *Coll. Sur. A: Physicochem. Eng. Aspects* **456** 18-25.
[19] Ahangaran F, Hassanzadeh A and Nouri S 2013. *Inter. Nano Lett.* **3** 23.
[20] Malamis S and Katsou E 2013 *J. Hazard. Mater.* **252-253** 428-461.
[21] Munagapati V S, Yarramuthi V and Kim D S 2017 *J. Mole.Liq* **240** 329-339.
[22] Tang Hua, Huang H, Wang X, Wu K, Tang G and Li C 2016 App. Surf. Sci. 379 296-303
[23] Chen J, Shi Xiaowen, Zhan Y, Qiu X, Du Y, Deng H 2017 *Appl. Surf. Sci.* **397** 133-143
[24] Alzaydien A S 2015 *Orient. J. Chem.* **31** 643-65.
[25] Krika F, Benlahbib O F 2015 *Des. Wat. Treat.* **53** 3711-3723

IOP Conf. Series: Journal of Physics: Conf. Series **1091** (2018) 012003 doi:10.1088/1742-6596/1091/1/012003

Crystal Structure and Magnetic Properties of Magnesium Ferrite ($MgFe_2O_4$) Nanoparticles Synthesized by Coprecipitation Method

E Suharyadi*, A Hermawan, and D L Puspitarum

Department of Physics, UniversitasGadjahMada, Sekip Utara BLS 21 Yogyakarta, Indonesia

*Corresponding author: esuharyadi@ugm.ac.id

Abstract.$MgFe_2O_4$has been successfully synthesized by coprecipitation method. The Precursors used are $MgCl_2.6H_2O$ (magnesium chloride hexa-hydrate, M_r= 203.205 g/mol) and $FeCl_3$. $6H_2O$ (ferric chloride hexa-hydrate, Mr= 270.19 g/mol) and NaOH (sodium hydroxide, Mr = 39.99 g/mol) as coprecipitant. This synthesis process was done by varying the NaOH concentration, the synthesis temperature and the stirring duration. XRD analysis results showed that the grain size increases with synthesis temperature and decreases with the increase in the concentration of NaOH and the duration of stirring. The smallest grain size is 2.1 nm and the largest grain size is 10 nm. Analysis of magnetic properties $MgFe_2O_4$by Vibrating Sample Magnetometer (VSM) showed that nanoparticles $MgFe_2O_4$can be assumed in multidomain region, with relationships. Sample with the smallest particle size 2.1 nm, has highest Hc value 133.9 Oe. The maximum magnetization at H = 15 kOe and remanent magnetization (*M*r) increase with the degree of crystallinity of the sample. Although there is also the effect of the presence of hematite phase (α-Fe_2O_3) and the grain size. Magnesium ferrite magnetic nanoparticles ($MgFe_2O_4$) has great potential to be applied as adsorption, sensors, drug delivery, semiconductor materials and other magnetic technology

1. Introduction

Magnetic nanoparticles have been studied for a variety of applications in the study of technology and materials science, chemistry, physics, biology, and environmental science. Nanoparticles are particles with a nanometer size, which is about 1-100 nm[1]. The nanoparticles have unique physical-, chemical-, mechanical-, magnetic- and optical characteristic which are not present in other type of bulk materials[2]. Magnetic nanoparticles show a variety of unique magnetic phenomenon that differs from their bulk magnetic property, so that the properties may be advantageous for a variety of applications such as magnetic fluids, catalysts, bio-applications, magnetic resonance imaging and data storage media[3]. Magnetic properties of nanoparticles are determined by many factors, including the chemical composition, particle size and shape, morphology, particle interaction with the surrounding matrix and particles [4]. By changing the size, shape, composition, and structure of the nanoparticles, the magnetic properties of the material can be controlled[5].

Presently, much research work and numerous studies have been carried outon spinel ferrite nanoparticlesdue to their very wide field of application in data storage systems, transformers, computer memory, inductors, recording heads, microwave and medical diagnostics [6]. This material has excellent magnetic properties such as high permeability,high resistivity and low coercivity. Spinel ferrite nanoparticles are soft ferrite which have a cubic crystal structure.

Among the researchinterests focusing on ferrite nanoparticles, the study of nanoparticles of magnesium ferrite ($MgFe_2O_4$) seems to stand out. Some of the reasons which led $MgFe_2O_4$to attract the interest of many researchers compared to other ferrites is its wide-range application potential, because $MgFe_2O_4$has a high value of saturation magnetization, Curie temperature and high electrical resistivity. $MgFe_2O_4$is a soft magnetic material and one inverse spinel group are very important. $MgFe_2O_4$is also an n-type semiconductor material that can be applied as adsorption device, sensors, and used in magnetic technology. Another interesting pointis that $MgFe_2O_4$has unique chemical properties and thermal stability, as well as the dependence of the magnetic properties on the particle size. $MgFe_2O_4$nanoparticles can also behave as superparamagnetic nanoparticles. When an external magnetic field is removed, then the sum of the magnetic moment of the magnetic nanoparticles are each pointing in different directions, thus the overall bulk magnetic moment is zero.

Various methods were employed to synthesize magnetic nanoparticles, for example, the thermal decomposition-, microemulsion-, coprecipitation-, sol gel-, hydrothermal-, and sonochemical method[7]. Among these methods, coprecipitation method is an effective method and can work at low temperatures. Moreover, because it can control the particle size, coprecipitation method can be used to evaluate the dependence of the magnetic properties of the particle size[8].

This research investigates in more details related to the influence of the grain size from magnetic properties of nanoparticles $MgFe_2O_4$synthesized by coprecipitation method by varying the synthesis parameters such as NaOH concentration and synthesistemperature. The results of this study will provide information about the influence of parameter variations to magnetic properties as well as the synthesis of the acquired information about the properties of nanoparticles $MgFe_2O_4$, in order to improve its performanceand to make it more effective in its application.

2. Materials and Methods

The material synthesis of $MgFe_2O_4$nanoparticle is $MgCl_2.6H_2O$ and $FeCl_3.6H_2O$ as a provider of Mg^{2+} ions and Fe^{3+} by reaction coefficient ratio 1:2. Coprecipitation synthesis method is applied by mix1.018 grams $MgCl_2.6H_2O$; 2.703 grams $FeCl_3.6H_2O$ and 3.5 ml of HCl (37%) in 50 mL of distilled water until homogeneous. Then put that solutionto 25 mL of NaOH solution by slowlydropwise while stirring use a magnetic stirrer at 1000 rpm for 60 minutes with a temperature variation of 36, 60 and 90°C (Table 1). For samples with varying concentrations of NaOH is applied in the same way at 90°C with various concentrations of NaOH as 3 M; 6 M and 10 M (Table 1).

Those solution are placed over a permanent magnet to accelerate precipitation. In order to minimize the salts dissolved in solution $MgFe_2O_4$, then washing approximately 7 repetitions. After the washing process is completed, then the precipitate is heated in a furnace to dry byarrange the temperatures around 95°C. Finally, black powder is succefully made. $MgFe_2O_4$ samples with temperatures and concentrations variations of NaOH which has formed, then characterized by X-ray Diffraction (Shimadzu models XD-3H) with CuKα tube (wavelength 1.5406 Å) to determine the phase contained in the sample. Calculation of the sample size distribution is calculated by using Scherrer equation, as shown in equation 1 below:

$$t = \frac{k\lambda}{\beta \cos\theta} \qquad (1)$$

where t is the crystal grain size, k is a constant Scherrer (0.89), λ is the wavelength of X-rays and β is the width of half peak (full width at half maximum = FWHM) of the main peak. Particle morphology

IOP Conf. Series: Journal of Physics: Conf. Series **1091** (2018) 012003 doi:10.1088/1742-6596/1091/1/012003

will be investigated by using transmission electron microscopy (JEOL JEM 1400) and characterization of magnetic properties by vibrating sample magnetometer (Riken Denshi Co., Ltd.).

Table 1. Synthesis parameters of $MgFe_2O_4$ nanoparticles

No.	Name of Sample	$MgCl_2.6H_2O$ (g)	$FeCl_3.6H_2O$ (g)	HCL (37%) (mL)	Concentration of NaOH (M)	duration of stirring (minute)	the synthesis temperature (^0C)
1.	A	1.017	2.703	3.37	3	60	90
2.	B	1.017	2.703	3.37	6	60	90
3.	C	1.017	2.703	3.37	10	60	90
4.	D	1.017	2.703	3.37	10	60	50
5.	E	1.017	2.703	3.37	10	60	RT
6.	F	1.017	2.703	3.37	10	30	90
7.	G	1.017	2.703	3.37	10	90	90

3. Results and Discussion

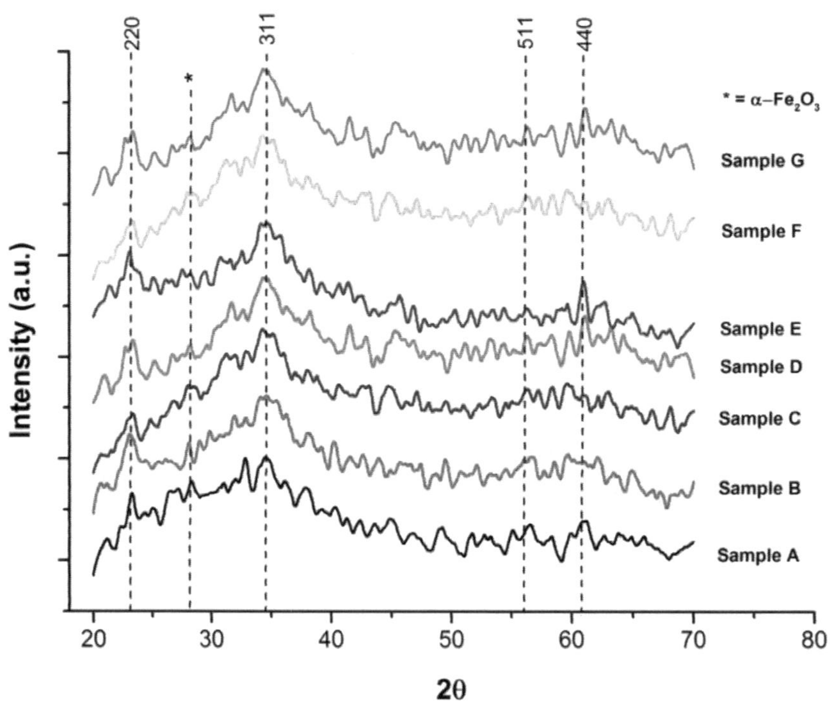

Fig 1. XRD patterns of $MgFe_2O_4$ nanoparticles

Table 2. Crystal structure of$MgFe_2O_4$ nanoparticles

No	Sample	Latticeparameters (Å)	Particles Size (nm)
1.	A	8.57	10.7
2.	B	8.56	7.8
3.	C	8.60	6.4
4.	D	8.58	2.1
5.	E	8.55	6.5
6.	F	8.59	4.4
7.	G	8.67	8.4

Figure 1 provides information related to the peaks of $MgFe_2O_4$and obtained information related to the lattice parameters (a), grain size (t), α-Fe_2O_3ratio and lattice parameter deviations. It is presented in Tables 2.Based on Table 2, the value of $MgFe_2O_4$ lattice parametertend to increase by the increasing concentrations of NaOH. Based on XRD analysis of seven samples which were synthesized by varying the NaOH concentration, synthesis temperature and stirring duration, the lattice parameter values obtained for samples A, B, C, D, E, F and Gare 8.57; 8.56, 8.60, 8.58, 8.55, 8.59 and 8.67 Å. These lattice parameter values are closer to the value of the lattice parameter MgFe2O4 bulk size, 8.4 Å[9].

Table 2shows that the particle size of nanoparticles synthesized $MgFe_2O_4$ decrease by the increasing concentrations of NaOH. The highest nanoparticle size is sample A with 10.7 nm and the smallest size issample Dwith2.1 nm. Particle size variation can be attributed to the influence of the formation and the growth rate of the core particles. Rising concentrations of NaOH causes the rate of deposition (deposition) ion configuration is higher than the rate of soluble (dissolution) resulting in the nucleation stage. The nucleation stage is more dominant than the crystal growth, so that the resulting particle size is small. The more concentration of NaOH causes OH^- higher. The numbers of OH^- accelerate the formation ofcore particles, but reduces the space between the particles to grow so that the diameter is smaller. Moreover, the decreasing particle size with increasing concentrations of NaOH showed that NaOH is a good substance breaker (decomposers).

In Table 2also shows that the particle size increased along with the increase in the synthesis temperature. This is caused by nucleation process and crystal growth that is taking place rapidly and simultaneously as a result of the high degree of supersaturated solution when the temperature in synthesisprocessis increased, supplemental information is that the process of growing grains occurs when the temperature of the synthesis below the critical temperature and will undergo a phase change or the formation of the core when the temperature is above the critical temperature (T_c).

At $MgFe_2O_4$naonoparticle samples which os synthesized by the synthesis temperature variations, the presence of α-Fe_2O_3 (hematite) influences the peak intensity in the sample C at which time the synthesis temperature increases, the peak intensity tends to decrease. This is as a result of the oxidation reaction between the sample ions with oxygen ions. Are shown in Table 2 are generally the ratio of the volume presentation phase presence of impurities has increased along with the increase in the synthesis temperature.

Observation of Morphological $MgFe_2O_4$nanoparticles was acquired by TEM.This morphological observation uses sample A because it has the largest size. TEM results are used to reinforce the results of measurement using XRD. Morphology of the sample A is shown in Figure 2.

Figure 2(a) shows that the grain sample A has a spherical shape, although not perfectly round. Samples were formed seem prone to agglomeration (clumping) resulting in low dispersibility of nanoparticles obtained. This may occur because due to the magnetic interaction between single item with other items. If this trend continues then it is predicted to lead tonanoparticle size becomes larger. So, a treatment to reduce the occurrence of agglomeration is neccessary.

TEM can also be acquired image diffraction patterns that indicate the presence of ring structures in the polycrystalline sample, as shown in Figure 2(b). The pattern of diffraction rings show rings that indicate discrete crystal planes. Diffraction rings innermost to outermost show diffraction peaks with miller index (220), (311), (511) and (440). The miller indices in accordance with *hkl*the diffraction pattern of XRD that are characteristic samples of $MgFe_2O_4$ phase.

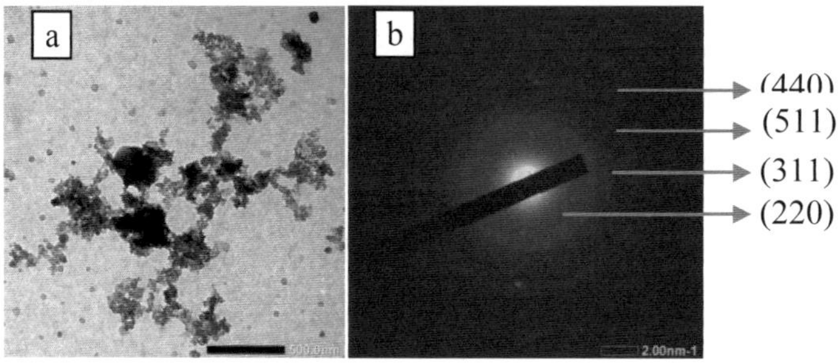

Fig 2. (a) morphologyof$MgFe_2O_4$ nanoparticles; (b) pattern ofdiffractionringsof$MgFe_2O_4$ nanoparticles

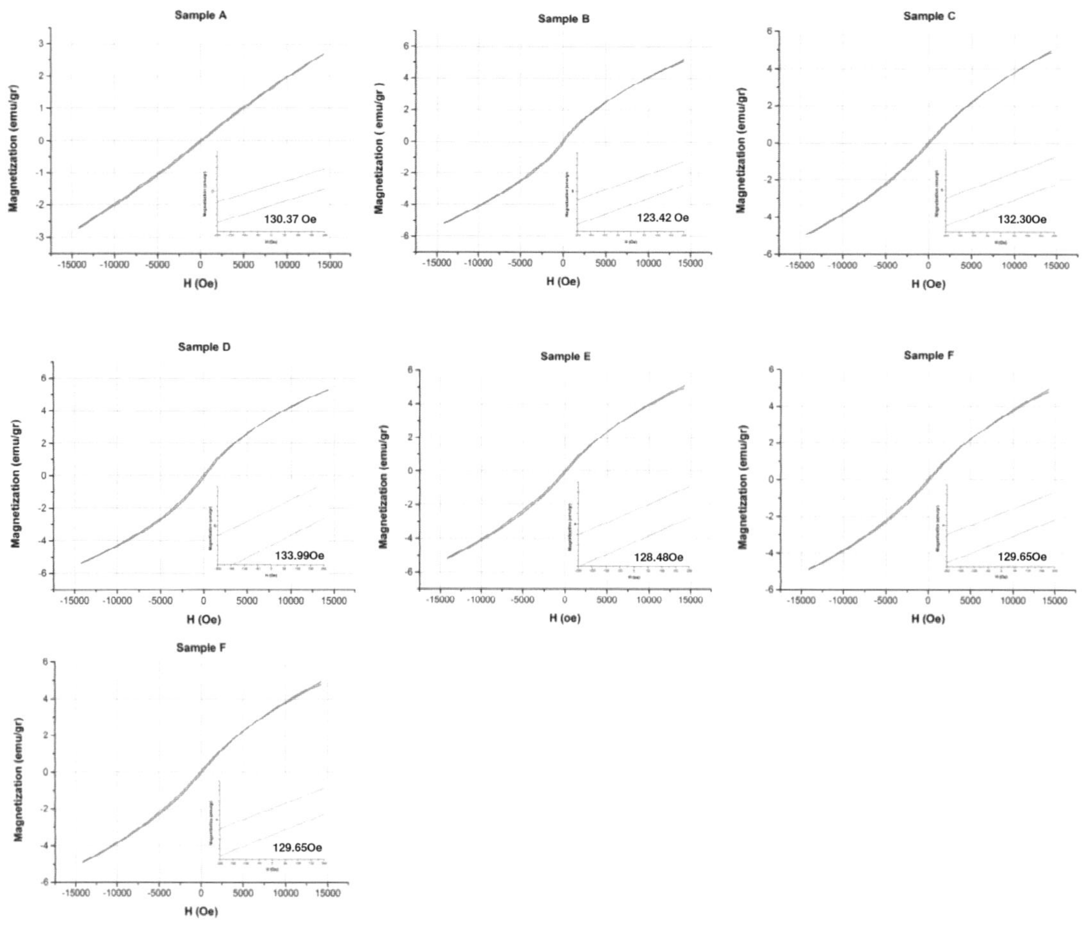

Fig 3. Hysterisis loop of $MgFe_2O_4$ nanoparticles

IOP Conf. Series: Journal of Physics: Conf. Series **1091** (2018) 012003 doi:10.1088/1742-6596/1091/1/012003

Characterization of magnetic properties of nanoparticles $MgFe_2O_4$is investigated by using VSM. The data obtained and plotted in the form of hysteresis curve. The hysteresis curve obtaines relevant maximum magnetization data at H = 15 Koe, remanent magnetization (M_r), and the coercivity (H_c). The results of observations with VSM shown in Figure 3.VSM observations showed that all the samples classified into soft magnetic which arecharacterized by expanses of hysteresis curve that form narrow and relatively small coercivity value. Based onFigure 3, the value of $H_c$shownin Table3.

Table 3. Coersivity value (H_c) of $MgFe_2O_4$ nanoparticles

No	Sample	Particles size (nm)	H_c(Oe)
1.	A	10.7	130.4
2.	B	7.8	123.4
3.	C	6.4	132.3
4.	D	2.1	133.9
5.	E	6.5	128.5
6.	F	4.4	129.7
7.	G	8.4	123.2

Based on Table 3, Sample D with the smallest particle size 2.1 nm, has highest H_c value 133.99 Oe, and sample G with lowest H_c value 123.18 Oe has particle size 8.4 nm.In this study, for variations the concentration of NaOH (Table 3) the value of the highest H_c was 132 Oe owned by the particles with the smallest size (t = 6.4 nm). As well as variations in temperature synthesis the value of the highest H_c was 133 Oe owned by the particles with the smallest size (t = 2.1 nm). Multidomain region H_c value and particle size are inversely. Nanoparticles in this study are assumed in multidomainregion. The influence of nanoparticlessize on the value of coercivity can be described by the equation $H_c = 1/t$, where t is the particle size and H_c is the coercivity [10]. The result shows that the coercivity value will increaseby the smaller size of the nanoparticles. This can be explained that the small size of nanoparticles which will provide extensive surface interaction and string magnetic interaction between the nanoparticles.

Table 4. Magnetization value (M)and remanentmagnetization (M_r) of $MgFe_2O_4$ nanoparticles at$H$=15 kOe

No	Sample	M (emu/g)	M_r (emu/g)
1.	A	2.7	0.014
2.	B	5.1	0.083
3.	C	4.9	0.062
4.	D	5.3	0.082
5.	E	5.1	0.064
6.	F	4.9	0.064
7.	G	6.2	0.094

Study of the relationship of *Hc* value and particle size at multidomain region may also be caused by the movement of domain walls. Magnetization which caused by the movement of the domain wall requires energy is smaller than the magnetization caused by the rotation of the magnetic moment. In the nanoparticles multidomainregion, the particle size is directly proportional to the length of the domain (*l*) and magnetization saturation (M_s) value of each domain.

Variations in the value of the magnetization at each sample can be linked to the presence of α-Fe_2O_3 phase (hematite) in the sample that have different properties with $MgFe_2O_4$ phase. The existence of α-Fe_2O_3 phase that is antiferromagnetic greatly affect the value of the magnetization of the sample $MgFe_2O_4$. When α-Fe_2O_3 phase small amounts, $MgFe_2O_4$nanoparticles towards a single phase. Phase of $MgFe_2O_4$ increasingly towards a single phase to create a domain-domain uniform size, it makes the dipole moments in the growing domain direction. Therefore, the value of M is getting bigger. And bigger ratio of α-Fe_2O_3 phase in the sample, then the value of M will be smaller.

This can be seen in Table 4. the lowest ratio of α-Fe_2O_3 phase obtained on the sample G is 2.59 %. The low value of the ratio of α-Fe_2O_3 phase showed that the presence of α-Fe_2O_3 phase in the sample only in small quantities, so it caused the value of the magnetization of the sample G is the largest value compared with other samples in the amount of 6.15 emu/g.

4. **Conclusions**

$MgFe_2O_4$ nanoparticles which have been synthesized by coprecipitation method,was applied by varying the NaOH concentration, the synthesis temperature and the stirring duration. In this study, the grain size decreased by the increasing concentrations of NaOH and increases with rising temperature synthesis. Analysis of magnetic properties showed that no linear relationship between the coercivity and grain size, so it is assumed that $MgFe_2O_4$ nanoparticles are in multidomain region. Observation of magnetization shows higher magnetization values with increasingly smaller ratio of α-Fe_2O_3 phase. $MgFe_2O_4$ nanoparticles are synthesized with a concentration of 10 M NaOH at room temperature has the smallest particle size of 2.1 nm, the highest *Hc* value of 133.9 Oe, and the highest value of *M*6.1 emu/g at $H = 15$ kOe has by sample G.

Acknowledgments

The authors would like to acknowledge the Competency Research Grant (HIKOM) financial support courtesy of the Directorate General of Higher Education (DIKTI), 2015-2016.

References

[1] Huang J 2006 Pure Appl. Chem. 78 1 15–27,
[2] Abdullah M 2009 Introduction for Nanoscience ITB Bandung Indonesia.
[3] Jun Y W, Seo J W and Cheon J 2008 Acc. Chem. Res. 41 2 179-189.
[4] Mathew D S and Juang R S 2007 Chemical Engineering Journal 129, pp. 51–65.
[5] Gubin S P 2009 Magnetic Nanoparticles John Wiley & Son,
[6] Tartaj P, Morales M P,Veintemillas V S, Gonzalez C T and Serna C 2003 J. Phys. D: Appl. Phys. 36 183-197.
[7] Lu A H, Salabas E L and Schuth F 2007 Angenwadte Chemistry of Sciences 46 1222-1244.
[8] Qu Y, Yang H, Yang N, Fan Y, Zhu H and Zou G 2006 Materials Letters 60 29-30 3548–3552.
[9] Omer M I, Elbadawi A A, Yassin O A 2013 Journal of Applied and industrial Science 1 4 20-23.
[10] Sharifi I, Shokrollahi H and Amiri S 2012 J. Magn. Mag. Mater 325 9003-9015.

IOP Conf. Series: Journal of Physics: Conf. Series **1091** (2018) 012004 doi:10.1088/1742-6596/1091/1/012004

Study on Cold-Drawing in Powder-in-Tube Process in Fabricating Silver-Sheathed (Bi,Pb)-Sr-Ca-Cu-O Tapes

A Imaduddin , Hendrik, S Dwi Yudanto, H Nugraha, P Sebleku, A W Pramono

Research Center for Materials and Metallurgy (P2MM - LIPI) Kawasan Puspiptek, Serpong, Tangerang 15314, Indonesia

Corresponding author: agungi@gmail.com

Abstract. Since Cu oxide superconductors have been discovered, the application for this type of superconductor especially in the form of superconducting tape to be used as electric cable to transport the electricity. The aim to this study is to analyse the manufacture of superconducting wire with superconducting pellets as its precursor. A pellet disc of High Temperature Superconducting (HTS) Bi,Pb-Sr-Ca-Cu-O has been made by using solid state method. This pellet has been prepared by mixing each of oxide powders together, and then sintered at 865°C for 90 hours. We have analysed the samples using X-Ray Diffractometer (XRD), Energy Dispersive x-ray Spectroscopy (EDS), Scanning Electron Microscopy (SEM) and also resistivity measurement to confirm its superconductivity. After confirmed the superconducting properties of the pellet, it was grinded into powder and inserted into silver tube with outer diameter of 10 mm and thickness of 1 mm. And then, drawn to make monofilament wire. It were then cut and inserted into another silver tube, and then re-drawing and rolled into multifilament tapes using Powder-in-Tube (PIT) method. The multifilament tapes was heated at 850°C for 12 hours. We confirmed its phase crystalline and superconductivity at pellet disc. However, when the pellet disc was made into wire and sintered at 850°C, it became amorphous and conductivity behaviour was decline. According to SEM analysis and resistivity measurement result, silver also penetrated inside the sample region of multifilament wire.

1. Introduction

Superconductors have big potential to be applied in practical equipment when it is in the form of wire. Since Cu-O type superconductor was discovered on 1986, Bednorz and Muller [1], research on superconducting wires has attracted much attentions among scientists. These type of superconductors have critical temperature above nitrogen liquid (77K). So, it will increase their possibility to be used in practical application. Also, the great demand for application of superconductivity promotes the need for wire superconductor [2].

In this research, the superconductor used is of Bi,Pb-Sr-Ca-Cu-O (or so-called BPSCCO). In making superconducting wire, solid state reaction is used and raw materials are powders form which a pellet disc of superconducting BSCCO can be fabricated [3]. The crystal structure was investigated when superconducting filaments had been made from its pellet by PIT method.

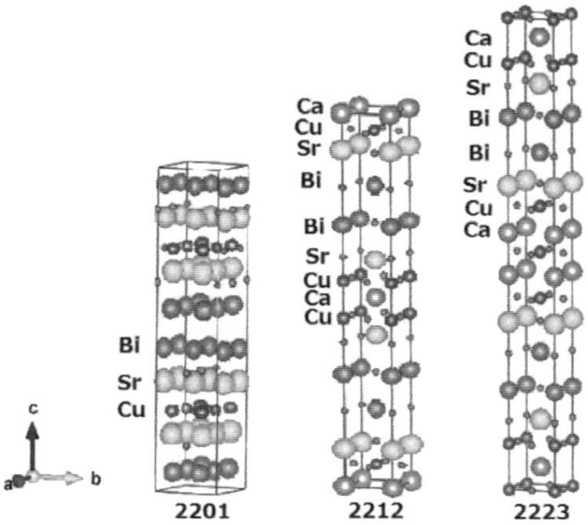

Figure 1. Illustration of crystal structures of three phases of BSCCO crystals.

Figure 1 shows the crystal structure of three phases of Bi-Sr-Ca-Cu-O (BSCCO). Phase 2201 has space group of Ama2 with a = 5.36200 Å, b = 5.37400 Å, and c = 24.6220 Å. Phase 2212 has space group of I4/mmm with a = b = 3.82500 Å, c = 30.8200 Å. Phase 2223 has space group of I4/mmm with a = b= 3.82300 Å, c = 37.3740 Å. Pb substituted Bi on the phase 2223.These were illustrated using crystal drawing application of VESTA. Pb atoms substitute Bi atoms on the phases of 2223 and 2212. Each element has its layer of oxide. These phases were determined by the sum of Cu-O layers of its unit cell [4].

The 2201, 2212 and 2223 phases have transition temperature of 20 K, 95 K and 110 K, respectively. This research aims to study the effect of cold drawing using ground pellet superconductor as precursor in order to make superconducting wire. Superconducting wire is predicted to be used in power devices such as motor, transformer, generator and electric cable. In case of BSCCO superconductor, the application is especially for power electric cable [3].

2. Method of Sample Preparation

A disc pellet of superconducting BSCCO was carried out by solid state reaction. Fabrication in the form of wires required superconductors in powders form as precursor materials. Therefore, superconducting powders was made by grinding pellet of superconductor. The phase 2223 has narrow forming temperature range (840°C to 880°C); therefore, adjustment of temperature in the furnace is critical [5,6].

BPSCCO pellet superconductor was made through solid state reaction. Raw material used were powders of: Bi2O3, PbO, SrCO3, CaCO3, and CuO. Drying was followed by heating 200°C for 2 hours for each element to remove its vapors.

Before calcination, drying was done in temperature of 300°C for 10 minutes. Weighing followed by compaction with a press machine was conducted. Based on stoichiometry results, in order to get the phase of $Bi_{1.6}Pb_{0.4}Sr_2Ca_2Cu_3O_6+\alpha$, the weight ratio of materials should be: Bi_2O_3:PbO : $SrCO_3$: $CaCO_3$: CuO = 9.34 : 2.24 : 7.41 : 5.02 : 5.98

These powders were mixed together for 5 hours [7,8]. Mixed powder was pressed and heated at 820°C for 19 hours. After it has been cooled into room temperature, it was ground into powder, pressed and then sintered. The increase in temperature, the holding time, and the decrease in temperature of sintering can be seen in Figure 2. From room temperature, heating was carried out up to temperature of 865°C reached about 3 hours [6]. After that, temperature was held for about 90 hours. Cooling to room temperature in the furnace was reached in about 24 hours.

IOP Conf. Series: Journal of Physics: Conf. Series **1091** (2018) 012004 doi:10.1088/1742-6596/1091/1/012004

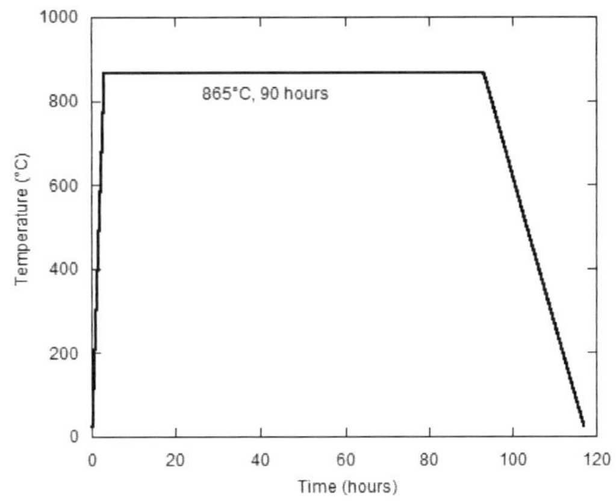

Figure 2. Heating and cooling rates as well as holding time during sintering process

Final pellet sample has diameter of 10 mm and thickness of about 2 mm. Morphological analysis of the sample was performed using Scanning Electron Microscopy (SEM). The equipment allows semi-quantitative analysis of the samples by X-ray microanalysis with an Energy Dispersive detector (EDS). Sample was also analyzed by X-ray diffraction. Resistivity measurement was conducted using four-point probe method of cryogenic magnet system (Oxford Cryotron),

After synthesizing and confirming the superconducting properties of the pellets, a wire was manufactured through Powder-In-Tube (PIT) method.

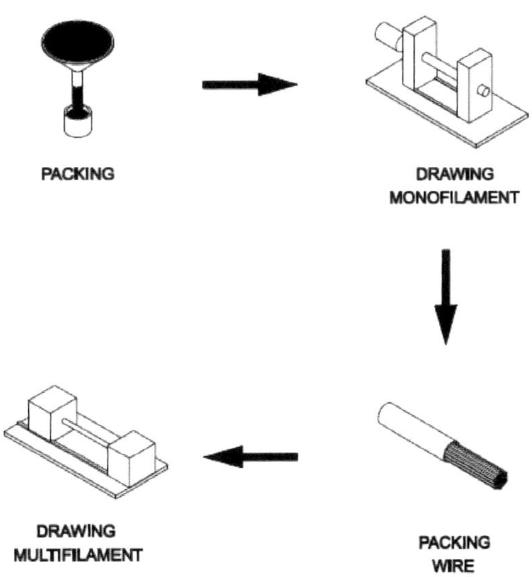

Figure 3. Process flow of making superconducting wire

Figure 3 illustrates theprocess flow of making wire with PIT methods. After the confirmation of superconducting properties, the pellet was ground into powder and inserted into silver tube, drawn and packed into bigger silver tube. The last silver tube was then re-drawn and rolled into wire [9].

The elemental composition, crystal structure, morphology, and electrical properties of the multifilament wire were also investigated with EDS, SEM, XRD and resistivity measurement methods.

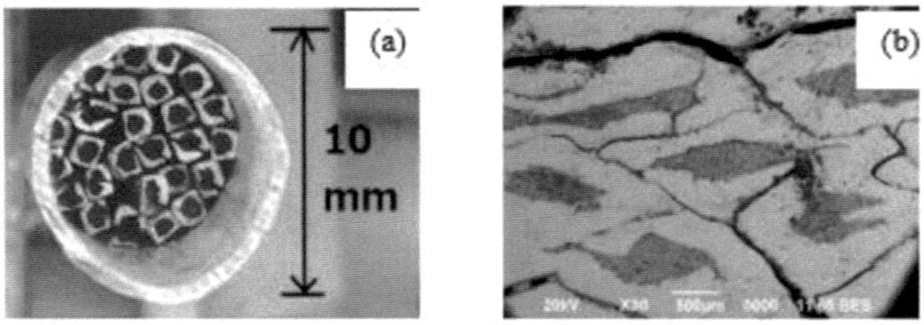

Figure 4. Cross section of Multi filament (a) before been drawn and rolled (photograph), and (b) after been drawn, rolled and sintered (SEM analysis)

Figure 4 shows the cross section of multi filament before and after the drawing, rolling and sintering. Figure 4 (a) shows 27 monofilaments afterinsertion into the silver tube. This multifilament billet was drawn and rolled into multifilament wire. Figure 4 (b) shows cross sectional of multifilament wire analyzed with SEM.

3. Result and Discussion

A disc pellet of superconducting BSCCO was carried out by solid state reaction. Fabrication in the

3.1. XRD Analysis

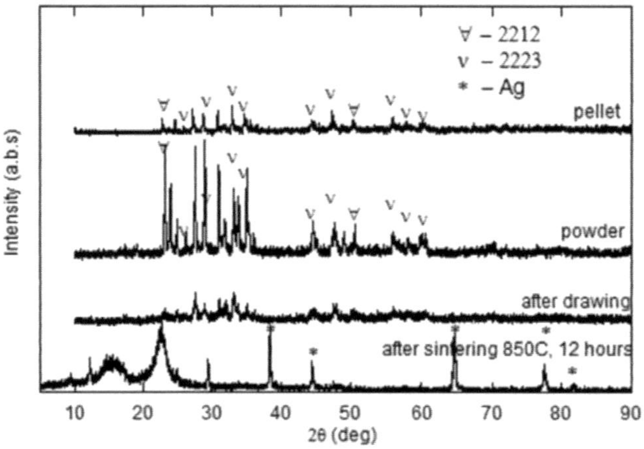

Figure 5. XRD analysis of pellet, powder (ground pellet), monofilament (after drawing/rolling and before sintering) and multifilament (after sintering at 850°C for 12 hours)

Figure 5 shows the XRD analysis of pellet, powder (ground pellet), monofilament (after drawing/rolling and before sintering) and multifilament (after sintering at 850°C for 12 hours).

It was observed that formation of main structure of phase 2212 was detected. This formation was also accompanied by minor structure of Sr2Bi2O5. XRD analysis also shows the phase of 2223 within the sample [10].

These XRD peaks were investigated with Rietveld method to analyze the percentage of its phases in the pellet. The results show that 2223 phase has the portion of 47% and 2212 phase has 36.4%. While in the multifilament, 2223 and 2212 phases both have the portion of 0%. It means that 2223 and 2212 phases in the multifilament have become amorphous. Only the peaks of silver, i.e. silver tubes, were found in the multifilament sample. The peaks of silver are accentuated, as the silver tubes experience elongated deformation during wire-drawing and their corresponding crystallographic textures become more oriented.

IOP Conf. Series: Journal of Physics: Conf. Series **1091** (2018) 012004 doi:10.1088/1742-6596/1091/1/012004

3.2. SEM/EDS Analysis

Figure 6 points out the analysis result of SEM on the sample surface. This SEM analysis observes the non-uniform phase on its surface. But some areas show layer structure estimated belonging to crystal structure of BPSCCO system. Sample also forms a porous material.

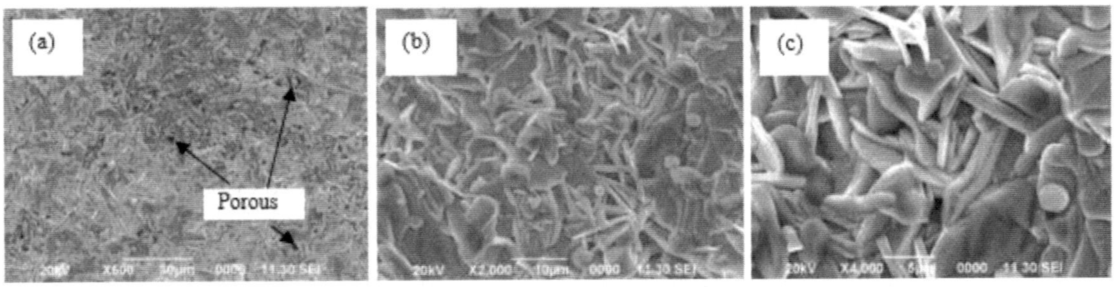

Figure 6. SEM analysis of pellet of (a) 500 times,(b) 2000 times and (c) 4000 times of enlargement. Sample forms porous material.

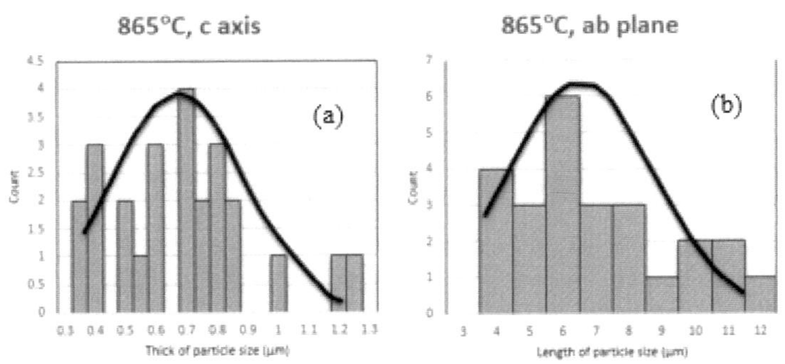

Figure 7. Counted particle size of the pellet (a) the thick of particles (c-axis of crystal) and (b) the width of particles (ab plane of crystal)

Figure 7 shows the counted particle size of the sample. Figure 7(a) shows the thickness of particle, whereas figure 7(b) shows the length of particle. The sample forms layers of particles. According to its crystal structure, the width of layer is correlated with the ab plane, while the thickness of layer is correlated with the c axis of its crystal structure [11].

The particle length in the ab plane (i.e. the particle width) is around 6.46 μm ± 2.43 μm whereas the particle length in the c-axis (i.e. the particle thickness) is around 0.67μm ± 0.23 μm.

IOP Conf. Series: Journal of Physics: Conf. Series **1091** (2018) 012004 doi:10.1088/1742-6596/1091/1/012004

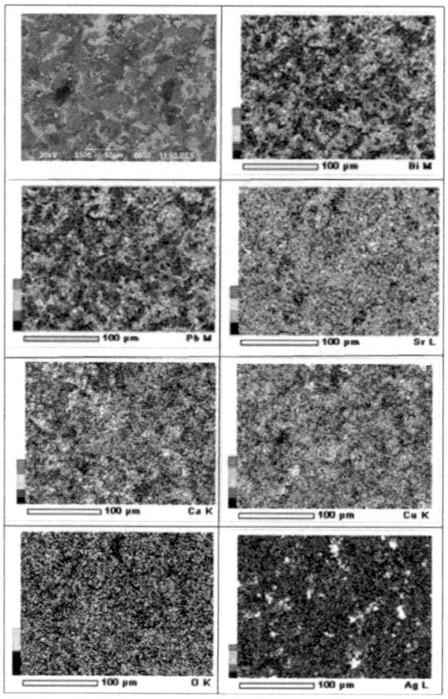

Figure 8. Distribution of elements of Bi, PbSr, Ca, Cu ,O and Ag in multifilament sample.

Figure 8 shows the mapping of Bi, Pb, Sr, Ca, Cu, O and Ag in the multifilament sample. According to this analysis, it is noted that Bi, Pb, Sr, Ca, Cu has similar pattern of uneven distribution. This shows that although sample has formed crystals in the pellet form, the dissociation of compounds in the multifilament sample occurs.

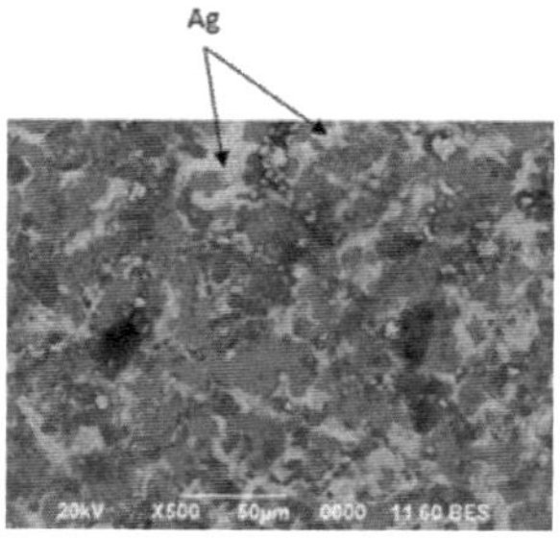

Figure 9. SEM analysis of multifilament sample

Figure 9 shows the SEM analysis of multifilament sample. The crystal like particles that has been observed in pellet sample were not observed in multifilament one. By comparison with figure 8, it is also shown that the silver has infiltrated into the region of BPSCCO.

The BPSCCO crystal was formed in pellet, but such crystal became amorphous after sintered at 850°C for 12 hours. Silver also could be found inside the sample. Silver is supposed to melt and diffuse inside sample since its melting temperature (Tm = 961°C) is close to the sintering temperature.

Table 1. Elements composition (EDS analysis)

	Pellet		Multifilament	
Element	Mass %	Atom %	Mass %	Atom %
O	7.84	30.20	10.89	36.93
Ca	7.06	10.85	10.26	13.89
Cu	39.54	38.34	40.25	34.38
Sr	17.54	12.34	9.35	5.78
Ag			5.91	2.98
Pb	6.50	1.94	5.87	1.53
Bi	21.52	6.34	17.48	4.53

Table 1 shows the elemental composition of Bi, Pb, Sr, Ca, Cu, O and Ag in pellet and multifilament samples. There is no big difference between pellet and multifilament. In general Ca content in both pellet and multifilament forms is higher than that of Pb, whereas the Cu content is higher than those of Bi and Sr. In multifilament sample, Ag also was found in the region of BPSCCO.

3.3. Resistivity Analysis

Resistivity was measured using thefour-point probe method. In order to expose it to the electrical current, the sample was connected to the cryogenic system through four terminals and glued using Ag paste. Cryogenic cooling system was utilized to reach low temperature environment.

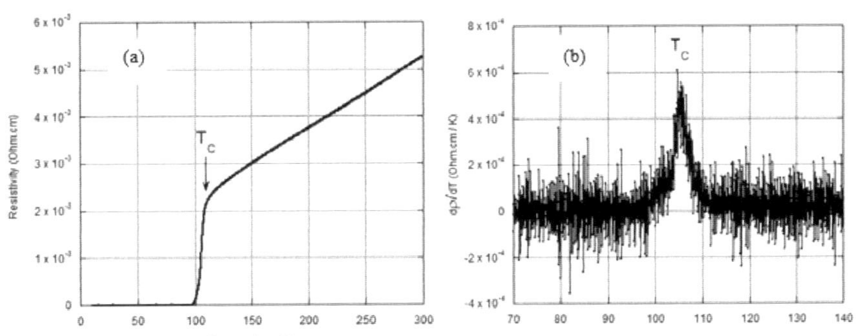

Figure 10. Change in resistivity vs temperature (a)resistivity vs temperature, and (b) gradien of resistivity vs temperature (around T_C)

Figure 10 shows the temperature dependence of resistivity of pellet sample. The resistivity starts to drop around 108 K. And it reached zero around 98 K. Hence the sample has been confirmed to have a superconductivity behavior. Figure 10(b) shows the gradient of resistivity vs temperature around the critical temperature. There is a clear peak at the temperature of 105 K.

The resistivity of conducting material can be analyzed by Matthiessen's rule [12,13,14]:

$$\rho(c,T) = \rho_0(c) + \rho_i(T) \quad (1)$$

Where $\rho_0(c)$ is the residual resistivity due to defect scattering and is essentially temperature independent, and $\rho_i(T)$ is the temperature-dependent part of the resistivity (or intrinsic resistivity).

The temperature-dependent part of the resistivity can be analyzed using Bloch-Grüneisen formula [15,16].

$$\rho_i(T) = \frac{C}{M\cdot\theta}\cdot\left(\frac{T}{\theta}\right)^5 \int_0^{\frac{\theta}{T}} \frac{z^5 e^z}{(e^z-1)^2} dz \quad (2)$$

Where T is temperature, θ is characteristic temperature of metal known as the electric resistivity Debye temperature, M is the atomic weight, and C is a constant of the metal.
For silver resistivity, Matthiessen's rule becomes [12,17]:

$$\rho(c,T) = 3.3 \times 10^{-7}(Ohm.cm) + \rho_i(T) \tag{3}$$

If, $x = \frac{\theta}{T}$, the intrinsic resistivity can be expressed as Grüneisen function, F(x). [18]

$$F(x) = \frac{4}{x^5}\int_0^x \frac{z^5 e^z}{(e^z-1)^2}dz \tag{4}$$

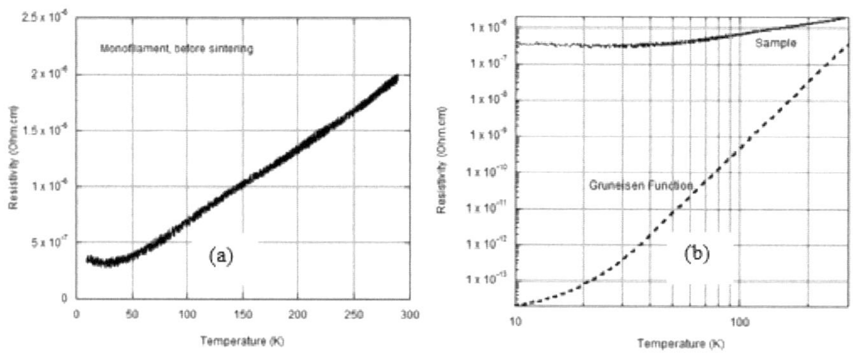

Figure 11. Resistivity of temperature dependence of monofilament, plotting (a) without and (b) with Grüneisen function

Figure 11 shows the temperature dependence of resistivity of monofilament sample. There is no resistivity drop observed. It shows only conductor behavior along with temperature

Figure 11b shows the resistivity analysis along with the plot of Grüneisen function for silver. Sample has larger resistivity compared to Grüneisen function. This result indicates that the silver has been contaminated and has high residual resistivity.

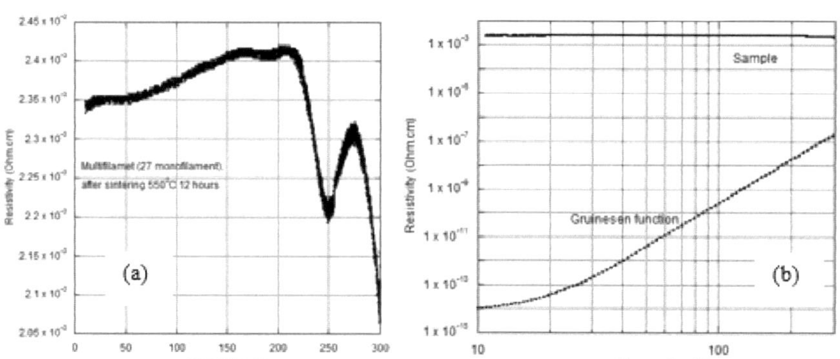

Figure 12. Resistivity of temperature dependence of multifilament, plotting (a) without and (b) with Grüneisen function

Figure 12 shows the temperature dependence of resistivity of multifilament sample. There is no resistivity drop observed.

Figure 12b shows the resistivity analysis along with the plot of Grüneisen function for silver. Sample has larger resistivity compared to Grüneisen function. This result indicates that the silver has been contaminated and has high residual resistivity. The residual resistivity in this sample is much higher than that in the monofilament one.

To analyze the conductivity behavior of both samples, the residual resistivity ratio (RRR) was calculated using the following equation,

$$RRR = \frac{\rho_{(297K)}}{\rho_{(4.2K)}}\ldots \quad (5)$$

RRR for monofilament sample is 6.21, and for multifilament sample is 0.87. It indicates that the multifilament sample has higher impurity than monofilament sample does. This suggests that in the process of making multifilament, the conductivity properties of silver was decreased, which is expected to occur due to high sintering temperature.

4. Conclusion

The pellet sample of BPSCCO was made by a solid state reaction, and used as precursor to make wire superconductor. The superconductor pellet was sintered at 865°C for 90 hours. The sample has phases of 2223 and 2212, with the critical temperature of 110 K. However, when it was used as precursor for making superconducting wire, the material lost its superconductivity. This situation is believed to occur due to the change of its crystallinity into amorphous state, when the sample was ground and sintered. Silver also was found to infiltrate material. Based on the results of resistivity measurements on monofilament and multifilament samples, it appears that the resistivity value of the samples are above the plot of Grüneisen function of the silver, which indicates that the samples have high residual resistivity. This result also indicates that the silver has been contaminated.

References

[1]. Bednorz J G, Muller K 1986 Possible High Tc Superconductivity in the Ba - La - Cu - O System *Zeitschrift für Phys B Condens Matter* **64** 189-193.
[2]. Yamaguchi T, Ueno E, Kato T, Hayashi K 2015 Quench Protection of DI-BSCCO Coil. *Phys Procedia* **65**:225-228.
[3]. Sato K 2015 *Bismuth-Based Oxide (BSCCO) High-Temperature Superconducting Wires for Power Grid Applications* Elsevier Ltd.
[4]. Vase P, Flükiger R, Leghissa M, Glowacki B 2000 Current status of high-T_c wire. *Supercond Sci Technol.* **13** R71-R84.
[5]. Sebaoun A I, Satre P, Mansori M, Honore A L 1997 PHASE DIAGRAM IN THE Bi-Pb--Sr-Ca-Cu-O SYSTEM Buffer effect of non-superconducting phases on oxygen exchange of the superconducting ones. *J Therm Anal.* **48** 971-980.
[6]. Staneva A D, Blagoev B S, Shivachev B L 2014 Effect of heat treatment on the phase transformation and magnetic properties of BPSCCO/LPMO composites. *J Alloys Compd.* **582** 223-228.
[7]. Patel R H, Nabialek A, Niewczas M 2005 Characterization of superconducting properties of BSCCO powder prepared by attrition milling *Supercond Sci Technol.* **18** 317-324.
[8]. Ibrahim M 2012 Preparation and Characterization of Hot-Pressed and Sintered BiPbSrCaCuO Superconductor *New J Glas Ceram.* **2** **2** 75-84.
[9]. Shen B, Li J, Geng J 2017 Investigation and comparison of AC losses on stabilizer-free and copper stabilizer HTS tapes. *Phys C Supercond its Appl.* **541** 40-44.
[10]. Guo Y C, Tanaka Y, Kuroda T, Dou S X, Yang Z Q 1999 Addition of nanometer SiC in the silver-sheathed Bi2223 superconducting tapes. *Phys C Supercond its Appl.* **311** 65-74.
[11]. Yildirim G, Varilci A, Terzioglu C 2013 Anisotropic nature and scaling of thermally activated dissipation mechanism in Bi-2223 superconducting thin film *J Alloys Compd.* **554** 327-334.
[12]. Kats Y, Klein L 2002 Can fractional power-law conductivity explain the deviations from Matthiessen's rule in SrRuO3 *Phys B Condens Matter.* **312-313** 793-794.

IOP Conf. Series: Journal of Physics: Conf. Series **1091** (2018) 012004 doi:10.1088/1742-6596/1091/1/012004

[13]. Timm C 2012 *Theory of Superconductivity* 108.
[14]. Kamimura H, Ushio H, Matsuno S, Hamada T 2013 *Theory of Copper Oxide Superconductors* **53**.
[15]. Mohammed N H 2012 Optimizing the Preparation Conditions of Bi-2223 Superconducting Phase Using PbO and PbO2. *Mater. Sci Appl.* **3** **4** 224-233.
[16]. Ansari I A 2010 Numerical solution of Bloch–Gruneisen function to determine the contribution of electron–phonon interaction in polycrystalline MgB2 superconductor *Phys C Supercond.* **470** **11-12** 508-510.
[17]. Akhoondali H, Yazdanpanah Goharrizi A, Javad Sharifi M 2015 Suppression of localized transmission peaks and deviation from Matthiessen's rule in disordered armchair graphene nano-ribbons due to acoustic phonons *Phys E Low-dimensional Syst Nanostructures* **74**:168-174.
[18]. Werheit H, Manghnani M H, Kuhlmann U, Hushur A, Shalamberidze S 2017 Mode Grüneisen parameters of boron carbide *Solid State Sci.* **72** 80-93.

IOP Conf. Series: Journal of Physics: Conf. Series **1091** (2018) 012005 doi:10.1088/1742-6596/1091/1/012005

Moderate thermal oxidation of iron oxide found in nature as raw materials for ferrite magnets

A Septiani, N Idayanti and N Sudrajat

Research Center for Electronics and Telecommunications – Indonesian Institute of Sciences, Jl. SangkuriangKomp-LIPI Gd.20, Bandung 40135, Indonesia

Abstract. Iron oxide found in nature with high iron oxide content has potential to be utilized as raw material for magnetic application. In this study, an attempt has been made to increase hematite content through thermal oxidation or roasting in moderate temperature with additional oxygen gas to accommodate oxidation. Iron oxide powders were milled in 6 hours then washed and dried. Roasting processwere performed at four varying temperatures of 480°C, 520°C, 560°C, and 600°C with oxygen influx 3L min-1, preceded by isothermal holding at 385°C. TGA-DSC (thermogravimetric analysis – differential scanning calorimetry) and XRD (X-Ray Diffraction) characterization has been performed to understand iron oxide behavior in response to thermal oxidation at moderate temperature. It is observed that two main reactions are clearly distinguished. In DTG curve, first peak indicate reduction followed by oxidation at the second peak. XRD results has shown that the highest content of α Fe_2O_3 was achieved by roasting at 600 °C for 1 hour which produced hematite portion of 31 wt. %, yet it is still insufficient to be utilized as raw material for ferrite based magnets fabrication. To prevent reduction reaction of Fe_3O_4 which will retard the oxidation process, it is suggested to avoid isothermal run or lower heating rate at temperature below 406°C

1. Introduction

Ferrite magnets belong to ceramic magnets with chemical formula $MO.Fe_2O_3$. Ferrite magnets are highly utilized in electronics applications such as transformer core, digital data storage, signal transfer, circulator and DC motor[1-3]. Different M has different spin magnetic moment and likely alter crystal structure. Therefore, by adjusting composition and phases, it is possible to tailor their magnetic properties [2,4]. Co addition in ferrite magnets produces hard magnetswith high coercivity, while Mn and Zn addition produces soft magnet with chemical formula $Mn_{(1-x)}Zn_xFe_2O_4$ [5,6].

Hematite or αFe_2O_3 has hexagonal or rhombohedral unit cell that is isostructural with corundum (αAl_2O_3). At room temperature it is weakly ferromagnetic, and at 260 K, it undergoes a phase transition to an antiferromagnetic state [7].Magnetite and wustite are other forms of iron oxide, magnetite has Fe bivalent and trivalent atom and has spinel crystal structure.Magnetite is ferrimagnetic while wustite is paramagnetic at room temperature [8].

Standard ceramic method which involve sintering process iswidely used in ferrite magnets fabrication because it suitable for mass production. Fabrication of ferrite magnets use hematite or αFe_2O_3 as main raw materials. On previous research [9], barium ferrite magnets was fabricated through ceramic sintering method using iron oxide found in nature, resulted in 350 Gauss of intrinsic remanent induction. Further effortsto increase hematite content are required to enhance its magnetic properties.One way to achieve this is through thermal oxidation that allow Fe with lower oxidation number to be oxidized by oxygen gas. The purpose of this research is tostudy the response of iron

IOP Conf. Series: Journal of Physics: Conf. Series **1091** (2018) 012005 doi:10.1088/1742-6596/1091/1/012005

oxide found in nature to moderate thermal oxidation and to measure hematite content achieved by this method.

2. Materials and Methods

Increasing hematite content from magnetite-dominant raw material was carried out by employing thermal oxidation method through roasting process. Powder of iron oxide found in nature was washed with water then dried, separated from dirt through magnetic separation method. Sample of 52.52% Fe content is examined by EDX characterization with remaining composition of 29.76% of O and minute content of Al, Si, Ti, and V. TGA-DSC characterization was performed to measure the mass change during heating. Iron oxide powders was milled using ball mill for 6 hour then proceeded with roasting. Heating started at room temperature then raised 10°C/minute, held at 385°C for 30 minutes, continued with heating at 10°C/minute, then maintained at four temperature variation for one hour, which were 480°C, 520°C, 560°C, and 600°C. Oxygen gas influx was added to accommodate oxidation in closed chamber with flow rate of 3L/minutes. XRD characterization was employed to determine phase composition ratio of initial powder and oxidation product and Rietveld refinement for quantification analysis.

3. Results and Discussion

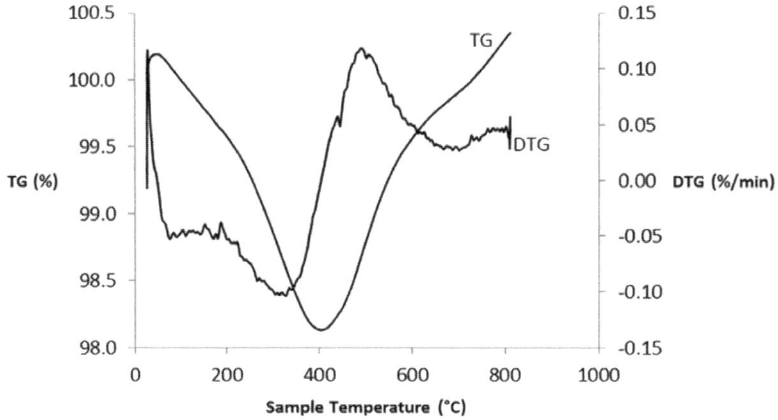

Figure 1. TG and DTG curve at heating rate of 10°C/minute

Figure 1 shows TG (mass change) and DTG (rate of mass change) versus temperature at heating rate of 10°C /min. It can be seen that TG curve has nearly symmetrical curve, where mass-losing reaction occursand peaked at 406°C then the powders start to gain mass. These two processes of losing and gaining mass indicate there are two reactions take place. DTG curve refers to rate of change of mass per minute, where peaks show higher mass change perminute than flat ones. It demonstrates that peaks of DTG are related to the kinetics of reactions. From the DTG curve, the first reaction is peaked at around 330° and second reaction is peaked at 507°C. After 507°C, mass change continuesto progress with lower rate. However, the mass changes versus temperature curve are heavily influenced by the heating rate used during measurement, as study done by Lysenko et al.[10].From the Arrhenius equation, we can infer that the rate of reaction is influenced by Arrhenius parameter, activation energy, and temperature. To generate these data, it is suggested to use model-free method from TG curve with different heating rate as recommended by Vyazovkin and Wight [11].

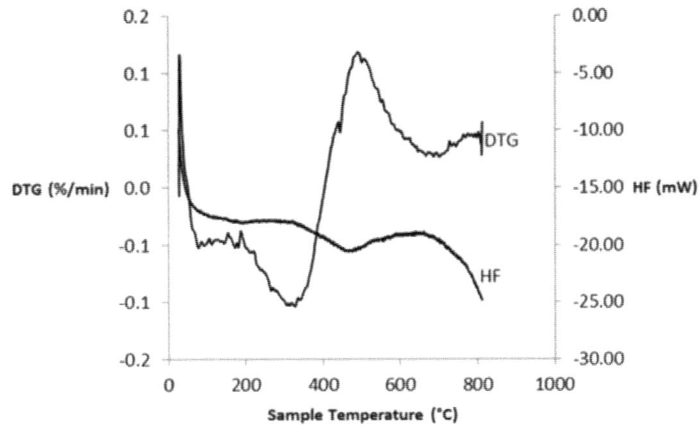

Figure 2. TG and Heatflow of at heating rate of 10°C/minute

Figure 2 depicts DTG and HF (heatflow) versus temperature at heating rate of 10°C /min from room temperature to 800°C. We can see that the endothermic peak for first reaction is rather flat and exothermic peak at 494°C corresponds to DTG peak at 507°C.The exothermic peak is followed by less exothermic curve, and after 648°C, the reaction is gradually increasing exothermically with slight change in rate of mass change.

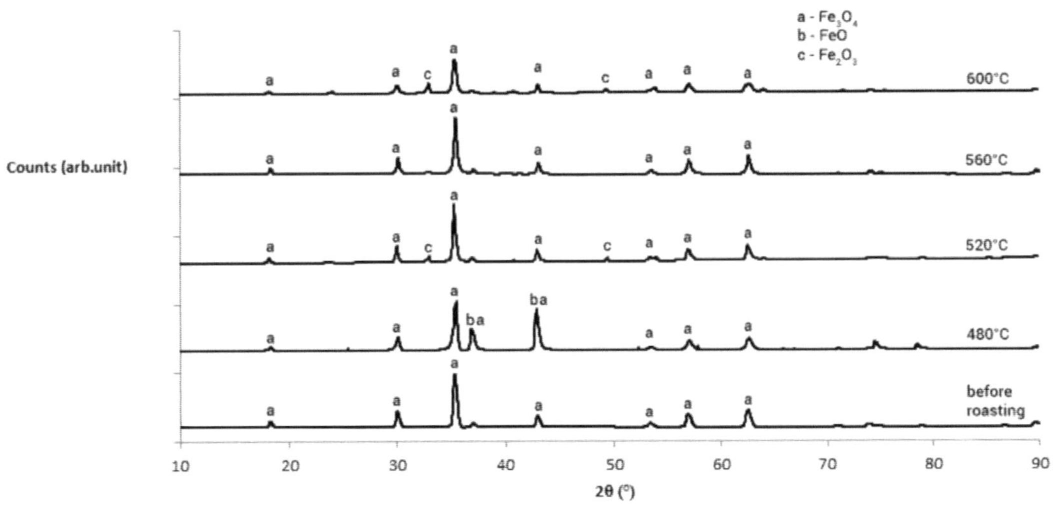

Figure 3. XRD patterns of different isothermal temperature of roasting process for 1 hour.

Figure 3 shows the XRD reflection intensity of initial powders and roasted powders with different isothermal temperatures. The initial powders constitute of Fe_3O_4, albeit the elemental composition of iron oxide found in nature reveals that there are other metals present such as Al, Si, Ti, and V. These metals have a low concentration thus their reflection intensity is much smaller than that of $Fe_3O_4$and therefore it is entirely burriedinthe background noise.Isothermal run at 480°C shows peaks of FeO emerging. It demonstrates that a reduction reactionhas occurred, in which the$Fe_3O_4 \rightarrow 3FeO + \frac{1}{2} O_2$reaction haspossibly taken place. This finding seems to be in agreement with the appearance of the endothermic peak in the HF curve, however, the temperature of the isothermal run is not in good agreement with DTG curve. According to DTG curve, at 480°C, mass-gaining reaction which indicate oxidation process should occur, yet XRD curve shows that reduction process has converted some

IOP Conf. Series: Journal of Physics: Conf. Series **1091** (2018) 012005 doi:10.1088/1742-6596/1091/1/012005

$Fe_3O_4$to FeO.Results of quantitative analysisemploying the Rietveld analysis code as shown inTable 1 indicate that 37.6 wt. % of Fe_3O_4has been reduced to FeO, and parts of it were also oxidized to Fe_2O_3.

Table 1. Iron Oxide content from XRD characterization

Isothermal Temperature	Phases (wt %)		
	FeO	Fe_3O_4	Fe_2O_3
Initial Powders	-	100	0
480°C	37.6	61.6	0.8
520°C	-	88	12
560°C	-	91	9
600°C	-	69	31

Isothermal run and non-isothermal run may have different kinetic parameters since they have different heating rates. It is also influenced by the pre-isothermal run at 385°C that according to DTG has reduced some Fe_3O_4 into FeO, thus alters kinetics parameter and differs from DTG curve. Furthermore, to prevent reduction reaction of Fe_3O_4which will retard the oxidation process, it is suggested to avoid isothermal run lower than 406°C. Low heating rate also provide sufficient time to reduce larger amount of Fe_3O_4 into FeO.

Hematite formed during isothermal 560°C shows a slight decrease compared to roasting at 520°C, then shows a significant increase in 600°C isothermal. The isothermal run at 600°C produces the largest ratio of hematite at 31 wt. % compared to other isothermal temperatures. However, the ratio of hematite formed during roasting process is still insufficient to be utilized as raw materials for magnets fabrication

4. Conclusion

During heating of iron oxide found in nature, it is observed that two main reactions are clearly distinguished. In DTG curve, first peak indicates reduction process followed by oxidation at the second peak. Pre-isothermal run at 385°C causes discrepancy between XRD and TGA-DSC results. XRD results have shown that the highest content of αFe_2O_3 was achieved by roasting at 600 °C for 1 hour which produced hematite portion of 31 wt. %, yet it is still insufficient to be utilized as raw material for ferrite based magnets fabrication. To prevent reduction reaction of Fe3O4 which will retard the oxidation process, it is suggested to avoid isothermal run or lower heating rate at temperature below 406°C.

Acknowledgements

This research was supported by TEMATIK DIPA LIPI. We would like to thankMr. R. Ibrahim Purawiardi for his assistance in the analysis of the XRD raw data.

References

[1] Macrelli E, Romani A, Wang N, Roy S, Hayes M, Paganelli R P and Tartagni M 2014 *Proc. Eurosensors* 87 (Italy: Proc. Eng./Elsevier) 1157
[2] Pullar R C 2012 *Prog. Mater. Sci.* **57** 1191-334
[3] Katayama K, Chinda Y, Shimizu O, Goto Y, Suzuki M and Noguchi H 2015 *J. Appl. Phys.* **117** 17E305
[4] Callister W D 2013*Material Science and Engineering: An Introduction, 9th Edition* (New York: John Wiley & Sons Inc.) w30
[5] Yi F 2014 *Ceram. Int.* **40** 7837-40
[6] Mitrovic N S, Djukic S R, Randjic S, Ristanovic Z and Danninger H 2012 *Sci. Sinter.* **44** 355-64
[7] Martin-Hernandez F and Hirt A M 2013*Geochem. Geophy. Geosy.* **14** 4444-57
[8] Lin Y F, Chen J Y, Xu C Y and Chung T W 2014 *Chem. Eng. J.* **250** 409-15
[9] Septiani A, Idayanti N and Sudrajat N 2014 *Proc. Sem. Nasional XXIII "Kimia dalam Industri dan Lingkungan* 53 1-4

[10] Lysenko E N, Surzhikov A P, Zhuravkov S P, Vlasov V A, Pustovalov A V and Yavorovsky N A 2014 *Journal of Thermal Analysis and Calorimetry* **115** 1447–52
[11] Vyazovkin S and Wight C A 1999 *Thermochimica Acta* **340–341** 53-68

IOP Conf. Series: Journal of Physics: Conf. Series **1091** (2018) 012006 doi:10.1088/1742-6596/1091/1/012006

Microwave Absorber Sheet of the Composite Silicon Rubber – Iron Oxide

Didin S. Winatapura, Deswita, and Wisnu Ari Adi

Centre for Science and Technology of Advanced Materials – National Nuclear Energy Agency, KawasanPuspiptekSerpong, Tangerang Selatan, Banten, Indonesia

Corresponding author : didinsw@batan.go.id

Abstract. Microwave absorber sheet is a composite of silicon rubber – iron oxide. The composite was prepared by silicon rubber, toluene, and magnetic powder of iron oxide. The raw material was blended in the beaker glass for 60 min and then pressed at temperature of 70 °C for 15 min. The unsure analysis showed that the sample contained carbon, oxygen, sulfur, zinc, lanthanum, barium, manganese, titanium and iron. The refinement results of x-ray diffraction pattern show that the sample is semicrystalline with the crystallinity of 46%. The sample consist of the matrix is amorphous phase and the filler is crystalline phase. The filler consist of two phases, namely iron oxide and iron metal phases. The functional groups analysis show that the sulfur addition modified the polymer by forming crosslink (bridges) between individual polymer chains and bonding between magnetic filler and rubber matrix. We concluded that this study has been successfully made a composite of silicon rubber – iron oxide for microwave absorber sheet application.

1. Introduction

Information technology has become an important element in everyday life. The role of information technology in human activity at the present timewas so greatandmodern devices, suchas mobile phones and computers are very complex and most of the electronic components are operating in the high frequencies range. However these electronic components are prone to leaks that resulted in the presence of interference frequency electromagnetic waves that can interfere with the performance of the equipment system [1]. Usually the electronic components mentioned aboveare mountedon a polymeric material that can absorb electromagnetic waves, also known as microwave absorber sheet. Microwave absorber sheet consists of a matrix made of polymer and filler material containing an electromagnetic wave absorbent material based upon magnetic materials. It means that the material is a composite that can absorb electromagnetic waves at a certain frequency range.

Fabricating absorbing materials provides an effective way to overcome these problems due to the dielectric loss or magnetic loss. Recently many authors have carried out studies on the absorbents and agents in order to improve absorbing properties of the composites [2–5]. It is demonstrated that a good absorbing material should satisfy two important conditions: 1) the intrinsic impedance of the absorbing materials is made to be equal to that of the free space; 2) the incident EM wave must enter and be rapidly attenuated through the material layer. Therefore, adjusting the complex permittivity and complex permeability of composites plays a key role in achieving excellent absorbing properties [6]. Presently, some research is concentrating on an attempt to modify certain magnetic materials such that the magnetic material can serve as an electromagnetic wave absorbent material [7–11]. One is the use of ferrite-based magnetic material, known as M-type hexagonal ferrite. This type of magnetic material is very promising because this material has some properties required as electromagnetic wave absorbent material. However, the synthesis of magnetic material of this kind is not easy to do and in order to meet the needs of mass production, there arise some constraints which result in a relatively high production cost. Beside that the development of electronic components is growing rapidly with a relatively cheaper price. Therefore, this study was conducted in order to find an alternative use of the product market of iron-based magnetic material and iron oxide. Iron particles (IPs) and iron oxide magnetic particle instant confirmation are used in absorbing sheet as an effective absorbent, they possess a good EM microwave absorption properties due to the large values of saturation magnetization and large Snokes limit in the gigahertz (GHz) frequencyrange.

The purpose of this work is to analyze the influence of silicone rubber composites filled with spherical iron/iron oxide composition on the microwave EM absorption properties. The main focus is on the effects ofweight content and shape of the iron and iron oxideunder silicone rubber on the reflection loss of the absorbing composites.

2. Materials and Methods

The absorber sheet in this study consists of silicon rubber, iron powder, and iron oxide. The matrix of absorber sheet were used was Methyl vinylsilicone rubber with toluene used as vulcanized assistants. The raw commercial of silicon rubber were supplied by best product, Germany, and raw commercial of iron oxide were supplied by pigment of local resources, Indonesia, which were prepared by a mechanical millingprocess using high energy millingmethod. The average particle diameter of the iron oxide was 100 – 300 nm and that of iron was 1 – 2 μm. The area of absorber sheet was about 25 mm2 and the thickness was about 1.5 mm. The 50% volume fraction of the iron and iron oxidewere added into the silicone rubber, andthen different amountof iron (the weight ratiowas 10wt%, 25wt% and 50 wt% relative to the iron oxide weight) was mixed in each composite. The three synthesized samples, each with a with different composition are shown in Table 1.

Table 1. The composition of absorber sheet

No.	Sample	Content (wt%)		
		Silicon rubber	Pigment (iron oxide)	Iron
1.	Sheet-1	50	40	5
2.	Sheet-2	50	37.5	12.5
3.	Sheet-3	50	25	25

The silicone rubber and absorbents were mixed according to two procedures. Firstly the iron oxidewas added to thesilicone rubber to guarantee a better dispersion. Secondly the ironwas added to the previous compounds. The preparation result of absorber sheet is shown in figure 1.

Figure 1. Sample of absorber sheet

The surface morphology- and elemental compositionexamination of the sample was carried out by using the JEOL scanning electron microscope (SEM) and energy dispersive spectroscopy (EDS) method, respectively. The phase qualitative and quantitative analysis were carried out using the Pan

Analytical diffractometer equipped (XRD) with a tube provided with a copper anode. The wavelength of X-ray radiation (CuKα) was 1.5406 Å. The phase analysis was performed by using GSAS program (Rietveld method) (Toby, 2001). The analysis of functional groups was carried out employing the Fourier transformation infra red (FTIR) method. The magnetic properties were evaluated using an Oxford instrument vibrating sample magnetometer (VSM) apparatus. Finally, the reflection loss of microwave was carried out using the vector network analyzer (VNA) with a frequency range of 300 kHz - 20 GHz.

3. Results and Discussion

Figure 2 shows the SEM image of absorber sheet which prepared by the mixture of silicon rubber elastomer thermoplastics, pigment (iron oxide), and iron powder.

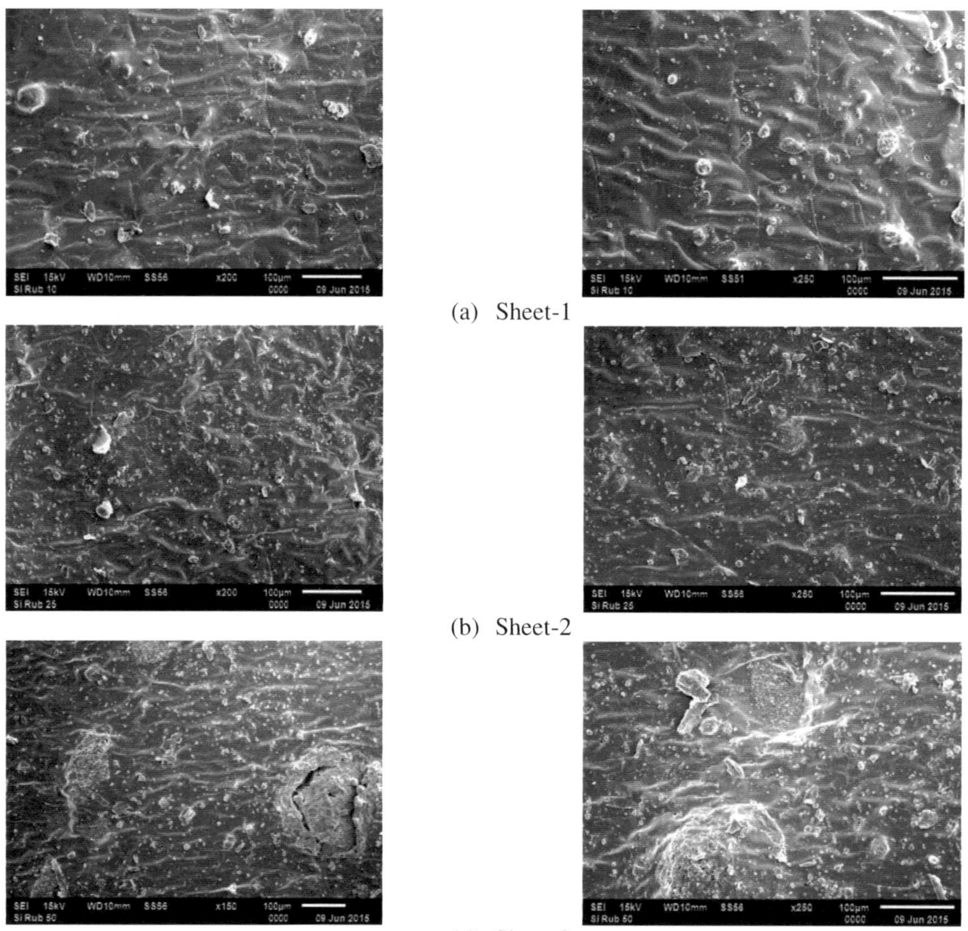

(a) Sheet-1

(b) Sheet-2

(c) Sheet-3

Figure 2. The SEM image of absorber sheet

In figure 2 it appears that the distribution of the filler in the form of iron oxide pigments in the silicone rubber matrix looks very homogeneous,in the form of polymer sheets,which is excellent. However, after being added to the filler metal, the composite looks like scattered granules, scattered over the surface of the sheet. The more filler added more iron granules are scattered over the surface of the sheet while the homogenization level decreases. This is presumably due to the higher density level composite, which makes the homogenization of the filler into the polymer matrix more difficult.

Figure 3 is showed the results of the elementary analysis by using energy dispersive spectroscopy on the filler and matrix of the composite material.

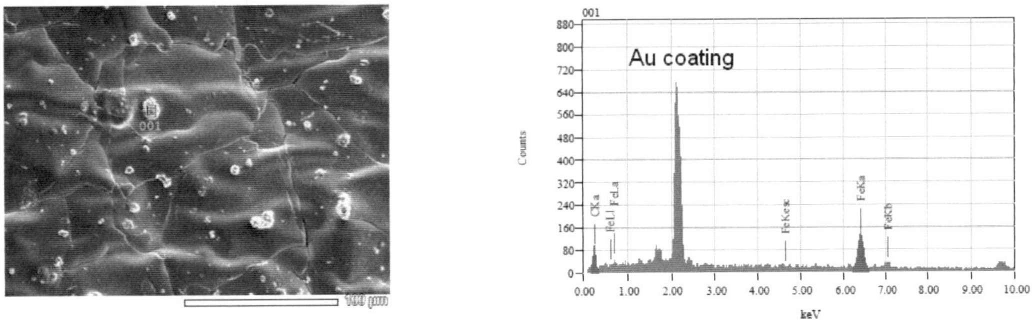

(a) Filler

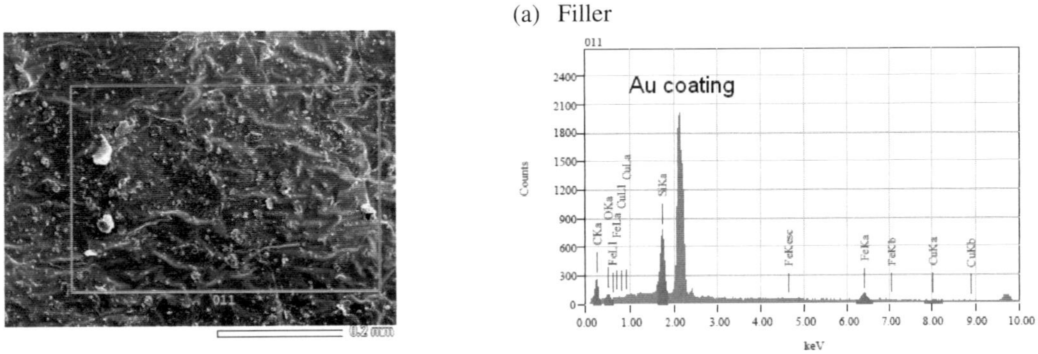

(b) Matrix

Figure 3. The elementary analysis of composite material by using energy dispersive spectroscopy

The energy dispersive spectrum as in figure1 shows that the dominant elements of the filler are carbon (C), and iron (Fe) while the dominant elements of the matrix are carbon (C), oxygen (O), silicon (Si), iron (Fe), and copper (Cu). The detailed content of the sample are showed in Table 2.

Table 2. The result of elementer analysis of the composite

No.	Unsure	Normalize of Content (wt%)	
		Filler	Matrix
1.	Copper (Cu)	-	1.32 ± 0.94
2.	Iron (Fe)	79.68 ± 1.60	22.36 ± 1.72
3.	Silicon (Si)	-	23.85 ± 1.59
4.	Oxygen (O)	-	7.49 ± 2.83
5.	Carbon (C)	20.32 ± 0.65	44.98 ± 1.44

Microstructure examination of the sample shows that the filler of magnetic particles are embedded in the silicon rubber matrix and are distributed on the sample's surface. Based on the results of the elemental analysis it was found that iron powder as filler contained in the composite have materialized. So that, it is needed a further confirmation about phases formed in the sample by using XRD as shown in figure 4.

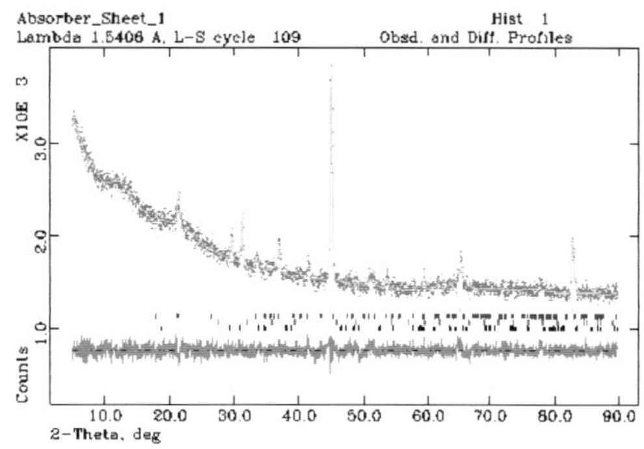

(a) Sheet-1

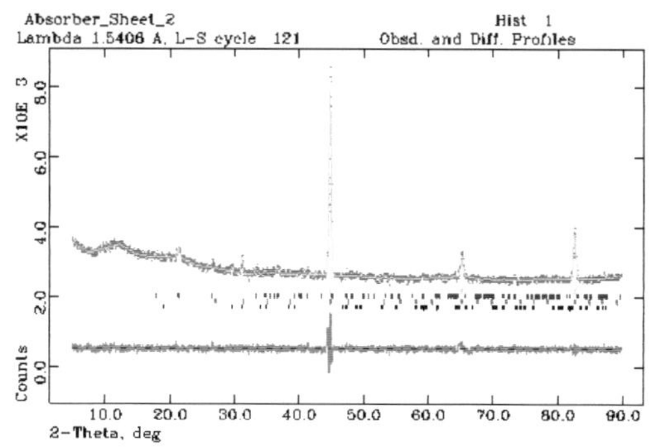

(b) Sheet-2

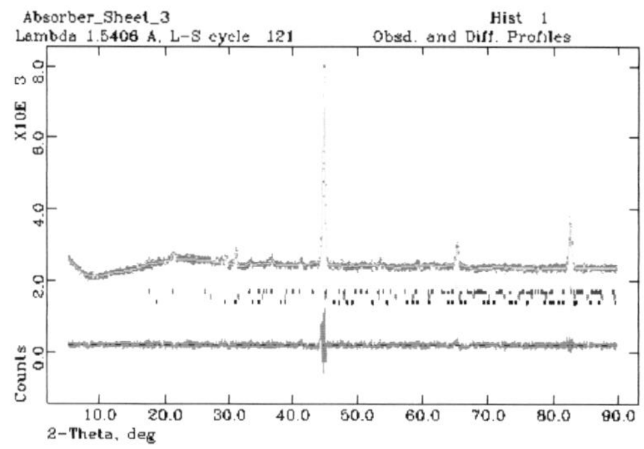

(c) Sheet-3

Figure 4. The refinement result of XRD pattern on the samples.

The refinement result of XRD pattern on the samplehas produced very good quality of fitting with R factor being very small. And goodness of fit value χ^2 (chi-squared) is found to be in agreement with the Rietveld standard [14]. The sample is semicrystalline with the crystallinity of 50%. And the sample consist of the matrix was amorphose phase and the filler was crystalline phase. The filler consist of four phases, namely $CuFeS_2$ (chalcopyrite), FeS_2 (pyrite), FeO_2 (goethite) and Fe(iron) phases. The chalcopyrite phase had tetragonal structure, space group of I-42d (COD-1010940), pyrite phase had cubic structure, space group of Pa-3 (COD-9006170), goethite phase had orthorombic structure, space group of Pbnm (COD-9016406), and iron phase had cubic structure, space group of Im-3m (COD-9013415). This result is evidenced by the calculation results of the mass fraction as shown in Table 3.

Table 3. Mass fraction of the samples

Sample	Phase	Mass fraction (%)	Rwp	χ2 (chi-squared)
Sheet-1	$CuFeS_2$	5.89(5)	2.77	1.135
	FeS_2	2.16(1)		
	FeO_2	72.69(1)		
	Fe	19.24(2)		
Sheet-2	$CuFeS_2$	3.97(3)	2.38	1.160
	FeS_2	1.08(8)		
	FeO_2	65.86(1)		
	Fe	32.09(9)		
Sheet-3	$CuFeS_2$	1.98(4)	2.70	1.182
	FeS_2	1.16(9)		
	FeO_2	43.32(1)		
	Fe	53.53(1)		

However, the XRD result could not explain the bonding mechanism including the formation of crosslink (bridges) between individual polymer chains in the composite, so that additional confirmation is needed by using FTIR equipment as shown in figure 5.

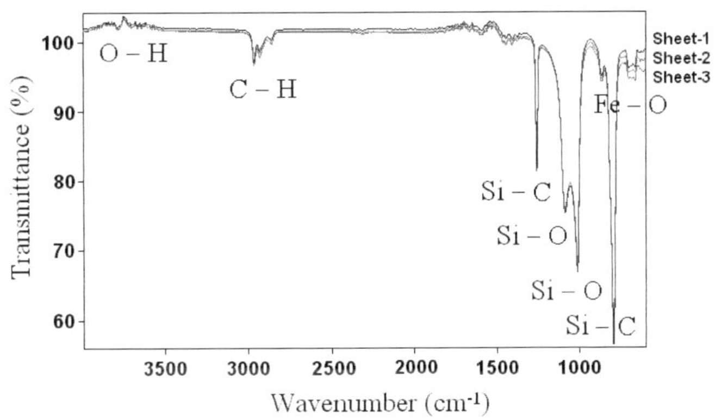

Figure 5. The FTIR transmittance spectrum of the sample.

The vibrational spectrum of a molecule is considered to be a unique physical property of the molecule and is characteristic of a molecule. A quantitative analysis of structure and confirmation of polymer chains requires the assignment of absorption bands and qualitative measurement of their intensities. The assignments of the fundamental frequencies have been made on the basis of magnitude and relative intensities of the observed bands.The vibrational band assignments of the rubber materials have been made in correlation with molecular composition and with their indices of stereo regularity. Figure 5 represented the FTIR spectrum between 4000 to 600 cm^{-1} of the composite sample. Transmittance peaks of the FTIR spectra on the composite sample indicate the presence of vibration of O–H, C–H, Si-C, Si-O, and Fe–O bond [12]. Natural rubber consists of suitable polymers of the organic compound isoprene with minor impurities of other organic compounds and water. According to the functional groups analysis by using FTIR showed that transmittance peaks of the silicon rubber appeared at wave numbers around 3000 cm^{-1}, 1250 cm^{-1}, and 1050 – 750 cm^{-1} that indicated the presence of the C–H, Si–C, and Si–O bond vibrations, respectively. However the transmittance peaksof the iron oxideare also observed at wave numbers around 3700 cm^{-1} and 600 cm^{-1} that indicate the presence of the H–O and Fe–O bond vibrations, respectively. It is suspected that there was bonding between iron oxide as filler and silicon rubber as matrix.

Characterization of the magnetic properties of the absorber sheet was performed using a vibrating sample magnetometer (VSM) that produced hysteresis curves of magnetic particles.

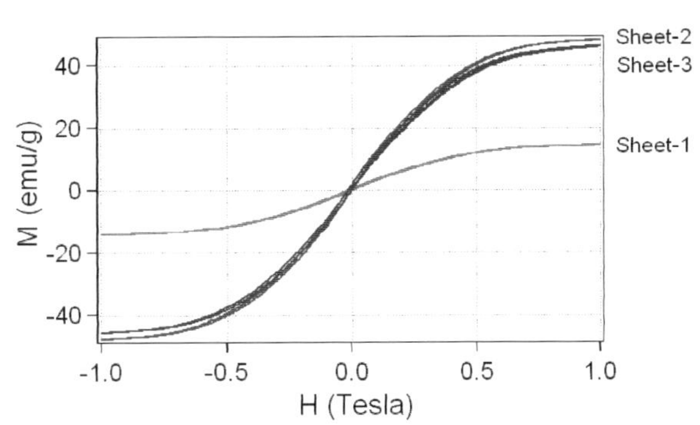

Figure 6. The hysteresis curves of absorber sheet

The hysteresis curves of synthesized absorber sheet are shown in figure 6 in which respective profiles for sheet-1, sheet-2, and sheet-3 are also compared. The hysteresis loop as shown in figure 6 is characterized with intrinsic saturation Ms, remanence field Mr and coercivity Hc. The intrinsic saturation, Ms is the state when the material is not able to absorb more magnetic field, so increasing magnetization force will not change magnetic flux density significantly. Meanwhile, the remanence field, Mr is the residual magnetization in a medium once the external magnetic field is removed and coercivity (also called coercive force) is a force required to demagnetize the material so the residual induction becomes zero after magnetizing up to saturation.

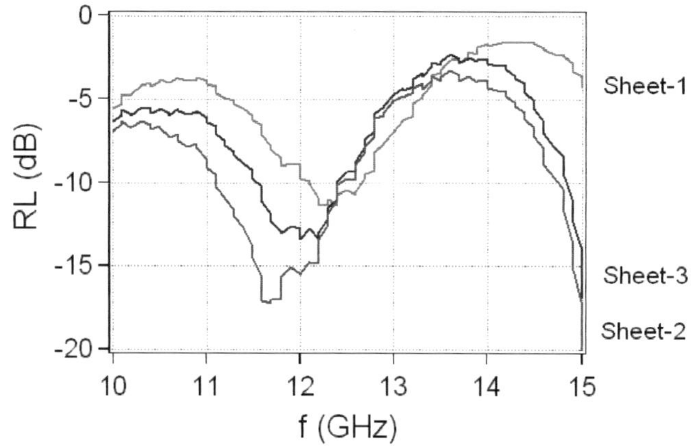

Figure 7. The reflection loss (RL) curve of absorber sheet

The microwave absorbing properties of synthesized absorber sheet is shown in figure 7 in which respective profiles for sheet-1, sheet-2, and sheet-3 are also compared. Figure 4 shows the relation between the reflection loss (RL) of absorber sheet and the microwave frequency X-band in the range of 10-15 GHz when the thickness of the sample is 1.5 mm. There are at least a absorption peaks observed which have a high RL within the frequency range for all samples. Base on the phase composition show that the absorber sheet has a low RL which then increases. This case shows that however the best absorber sheet is occurred at nearly the same thickness and at practically identical frequencies, there are differences. This can be explained by the effect of electromagnetic properties on the attenuation characteristic in each sample. This indicates that absorber sheet has certain microwave absorption properties in the frequency range of 10 – 15 GHz, absorption peak values of -15 dB at 12 GHz.

4. Conclusion

The synthesis and characterization of absorber sheet have been carried out using chemical process. Microstructure of the sample showed that the filler of magnetic particles embedded in the silicon rubber matrix and distributed homogeneously on the sample surface. The unsure analysis showed that the sample contained carbon (C), oxygen (O), silicon (Si), iron (Fe), and copper (Cu). The refinement results of x-ray diffraction pattern showed that the sample is semicrystalline with the crystallinity of 50%. And the sample consist of the matrix was amorphose phase and the filler was crystalline phase. The filler consist of two phases, namely $CuFeS_2$ (chalcopyrite), FeS_2 (pyrite), FeO_2 (goethite) and Fe(iron) phases. The functional groups analysis showed that the sulfur addition modified the polymer by forming crosslink (bridges) between individual polymer chains and bonding between magnetic filler and rubber matrix. This study has been successfully synthesized silicon rubber – iron oxide and has been understood changes in the microstructure, crystal structure and bonding mechanism of the composite.

Acknowledgements
This work is supported by the program for research and development of smart magnetic material (DIPA 2015), Center for Science and Technology of Advanced Materials, National Nuclear Energy Agency. The authors are thankful to Drs. Gunawan, M.Sc. and Edy Giri R.P., Ph.D., for their permission to publish this paper. Many thank to Dra. Mujamilah, M.Sc. for their kind help to characterize using VSM.

References

[1]. Eswaraiah V, Sankaranarayanan V and Ramaprabhu S 2011 Functionalized graphene-PVDF foam composites for EMI shielding Macromol. Mater. Eng. 296 10 894–898.
[2]. Liang J J, Huang Y, Zhang F, Li N, Ma Y F, Li F F and Chen Y S 2014 High microwave absorption performances for single-walled carbon nanotube-epoxy composites with ultra-low loadings Chinese Phys. B 23 8 89501.
[3]. Wan D, Bie S, Zhou J, Xu H, Xu Y and Jiang J 2015 A Thin and Broadband Microwave Absorber Based on Magnetic Sheets and Resistive FSS Prog. Electromagn. Res. C 56 93–100.
[4]. Savi P, Miscuglio M, Giorcelli M and Tagliaferro A 2013 Analysis of Microwave Absorbing Properties of Epoxy MWCNT Composites Prog. Electromagn. Res. Lett. 44 63–69.
[5]. Folgueras L C, Alves M A and Rezende M C 2014 Evaluation of a nanostructured microwave absorbent coating applied to a glass fiber/polyphenylene sulfide laminated composite Mater. Res.17 1 197–202.
[6]. Duggal S and Aul G D 2014 Review on Effect of Electric Permittivity and Magnetic Permeability over Microwave Absorbing Materials at Low Frequencies Int. J. Eng. Adv. Technol. 5 3 12–19.
[7]. Manaf A and Adi W A 2013 Microstructure and Phase Analysis of $La_{0.8}Ba_{0.2}Ti_xMn_{(1-x)}O_3$ System for Microwave Absorber Material (x = 0 – 0.7) Adv. Mater. Res. 789 2013.
[8]. Pullar R C 2012 Hexagonal ferrites: A review of the synthesis, properties and applications of hexaferrite ceramics Prog. Mater. Sci. 57 1191–1334.
[9]. Herme C A, Bercoff P G and Jacobo S E 2012 Nd–Co substituted strontium hexaferrite powders with enhanced coercivity Mater. Res. Bull. 47 3881–3887.
[10]. Zhang Z, Liu X, Wang X and Wu Y 2012 Effect of Nd–Co substitution on magnetic and microwave absorption properties of $SrFe_{12}O_{19}$ hexaferrites J Journal Mol. Struct. 525 5 2012.
[11]. Mohit K, Gupta V R and Rout S K 2014 Microwave Dielectric Properties of $Ni_{0.2}Cu_xZn_{0.8-x}Fe_2O_4$ for Application in Antenna Prog. Electromagn. Res. B 57 157–175.
[12]. Hamzah R, Bakar M A, Khairuddean M, Mohammed I A and Adnan R 2012 A structural study of epoxidized natural rubber (ENR-50) and its cyclic dithiocarbonate derivative using NMR spectroscopy techniques Molecules 17 9 10974–10993.

IOP Conf. Series: Journal of Physics: Conf. Series **1091** (2018) 012007 doi:10.1088/1742-6596/1091/1/012007

Magnetic Structure of Magnetite Phase of Iron Sand Retrieved from Banten, Indonesia

E Sukirman[1], Y Sarwanto[1], A Insani[1], Th Rina M[1] and A Purwanto[2]

[1]Centre for Science and Technology of Advanced Materials, Kawasan Puspiptek Serpong, Tangerang Selatan 15314, Indonesia
[2]Surya College of Education, Scientia Boulevard, Blok U/7, GadingSerpong, Tangerang 15810, Indonesia

Abstract. Among the many types of natural resources that are abundant in Indonesia is iron sand. One of the products of the iron sand processing is magnetite (Fe_3O_4). The stages of separation and purification process that has been done before were extracting magnetite phase and discarding the other phases. Magnetic structure of magnetite phase of iron sand retrieved from Banten, Indonesia has been invesigated. The aims of this research is to analyze the structure of the new magnetite phase, $Fe_xM_yO_4$. The magnetic properties of the sample was evaluated by means of vibrating sample magnetometer (VSM). Sample characterizations was carried out by neutron activation analysis (NAA) method. The analysis result from NAA shows that there are still 17 impurities remaining, namely: M=Ti, Al, Ce, Co, Cr, Eu, La, Mg, Mn, Na, Sc, Sm, Th, V, Yb and Zn, and there are about 94% of iron sites are filled by Fe elements. The structural analysis was carried out by the help of the high resolution neutron powder diffractometer (HRPD) using a wavelength of λ = 1.8223 Å. Neutron diffraction data was analyzed by means of FullProf software. The analysis result shows that the sample hasthe chemical formula: (Fe0.97M0.03)3O4 with a cubic crystal system, space group: Fd-3m and lattice parameters: a = b = c = 8.36834 (3) Å, $\alpha = \beta = \gamma = 90°$. The sample is magnetic at room temperature because the compound has a net magnetic moment, μ = 2.9(1) μB. The magnetic moments of the sample is lower than the magnetic moments of pure Fe3O4 (μ = 3.8 μB), because of about 3% of Fe sites are occupied by impurity elements. Magnetic properties of magnetite phase depend on the purity of the magnetite phase.

1. Introduction

Indonesia has great iron mineral resources, comprising primary iron ore (17 %), iron sand (8 %) and lateritic iron ore (75 %), indicated by Rochani et al.[1]. Ironsand is a type of sand with heavy concentrations of iron. It is typically dark grey or blackish in colour.It comprises mainly magnetic mineral such as magnetite (Fe_2O_3), hematite (α-Fe_2O_3), and maghemite (γ-Fe_2O_3) and also contains small amounts of titanium, silica, manganese, calcium and vanadium, described by Fleur et al.[2].Magnetite is a ferrimagnetic mineral. It possesses the same properties as a ferromagnetic mineral except instead of parallel alignment of the dipoles, it has an anti-parallel alignment of unequal spins, pointed by Chicazumi[3]. The Curie temperature of magnetite is 858 K (585°C).The Curie temperature is the temperature at which a magnetic material becomes paramagnetic; hence losing its magnetic properties. Once exposed to a magnetic field, it remains magnetized even when the field is removed, described by Schroeder[4]. Of all the magnetic minerals, it displays the strongest magnetism, explained by Nagata[5]. Therefore magnetite canbe used for various industrial applications. Some of these applications of magnetite are as a base material of making toner used in photocopier and laser printers. Maghemite is the main material of making the magnetic tape. Third mineral magnetite, hematite and maghemite can also be used in the manufacturing industry of permanent magnets, indicated by Toto[6]. Magnetite and maghemite nanoparticles are also widely used as ferrofluids, such as in rotary shaft sealing, dynamic loudspeakers, and computer hard drives, and also have medical use in, for example, magnetic resonant imaging and targeted drug transportation. Indicate references by Ildiko et al.[7] or Erzsebet et al.[8] or German et al.[9] in the text.

Preliminary study of the purification process, qualitative and quantitative phase analysis of iron sand that was retrieved from Banten, Indonesia has been performed by Wisnu et al. (2014). Based on the X-ray diffraction pattern, it has been obtained that the iron sand consists of there phases, namely $(Fe_{0.85}M_{0.15})_3O_4$ (magnetite), $(Fe_{0.36}M_{0.64})_2O_3$ (ilmenite), and α-Fe_2O_3 (hematite) phase, where M = Ti, Na, Sc, Cr, Co, Zn, Ga, As, La, Sm, Eu, Ho, Yb, Hf, W, Pt, U, Sb, Th, Ce, Mg, Mn, V, Al, and Ta,there are25elements totally. Furthermore, the second stage of separation and purification process were carried out, i.e., extracting magnetite phase and discarding the other two phases. Of this separation and purification process is expected to get a magnetite phase with a new composition, i.e.,$(Fe_xM_y)_3O_4$, where $0.85 < x < 1.0$ and $0.0 < y < 0.15$, indicated by Sari et al. (2015). The aims of this research is to analyze the magnetic structure of the $(Fe_xM_y)_3O_4$ phase using the neutron diffraction technique by means of High Resolution Neutron Powder Diffractometer (HRPD). The neutron diffraction has been indispensible for the magnetic structure determination. The quality of magnetite can be seen from its magnetic properties and magnetic properties depend on magnetic structure of the material

2. Materials and Methods

The elemental composition analysis of $(Fe_xM_y)_3O_4$ sample was performed using Neutron Activation Analysis (NAA). The sample was weighed in LDPE (low density polyethylene) vial using a microbalance with a weight of 20-50 mg. The standard reference material of coal fly ash is used as internal quality control and as flux monitor used Al-0.1% Au from the Institute for Reference Materials and Measurement (IRMM). Samples, SRM and flux monitor together put into irradiation capsules was made from polyethylene or aluminium.The target was irradiated in the irradiation facility of Multi Porpuse Reactor G.A. Siwabessy in Serpong at thermal neutron flux of about 1013 n.cm^{-2}.det^{-1}. Counting of irradiated samples have been done by a high resolution HPGe(ε =15 %, FWHM=1.89 keV on 1.33 MeV) detector from Cannbera couple to multi channal analyzer. Data have been analyzed by GENIE 2000 and k0-IAEA sofware. The Scheme of irradiation, decay and counting time for analysis of samples are listed in Table 1, pointed by Wisnu et al.[10].

Table 1. Scheme of irradiation, decay and counting time for analysis of iron sand sample.

Facilities	T_i	T_d	T_c	Radionuclides
Rabbit System	10 s	3-5 m	150 s	^{28}Al, ^{57}Mg, ^{56}Mn, ^{51}Ti, ^{52}V
	15 m	1-2 d	1800 s	^{24}Na, ^{76}As, ^{140}La, ^{153}Sm
	4 h	10 d	5000 s	^{51}Cr, ^{60}Co, ^{141}Ce, ^{152}Eu, ^{59}Fe, ^{46}Sc, ^{65}Zn, ^{182}Ta, ^{181}Hf, ^{169}Yb

The magnetic properties of materials were characterized using VSM (Vibrating Sample Magnetometer) Oxford Type 1.2H in PSTBM, BATAN Serpong. Measurement of magnetic properties of materials was made in an applied magnetic field to a maximum value of 1Tesla. VSM measurement was carried out at room temperature. The neutron powder diffraction experiment was undertaken with High Resolution Powder Diffraction (HRPD) at the Neutron Scattering Laboratory of BATAN, Serpong, Indonesia. The sample was loaded in cylindrical vanadium holder. The data were collected at λ = 1.8213(7) Å, angular range 2.5 – 162° in interval of approximately 0.05° at room temperature. The data reduction, including detectors efficiency correction was performed after the data measurements. The neutron diffraction data was refined using FullProf software, produced by Rodriguez-Carvajal[12].

3. Results and Discussion

The NAA analysis results of $(Fe_xM_y)_3O_4$ sample is shown in Table 2.The table shows that previously there are 25 of impurities inthe magnetite compounds, but after the second stage of the purification process, there are still 17 impurities remaining, namely: M=Ti, Al, Ce, Co, Cr, Eu, La, Mg, Mn, Na, Sc, Sm, Th, V, Yb and Zn, and there are about 94% of iron sites are filled by Fe elements. However of the 17 elements, only 4 elements that have a concentration greater than or equal to 1%, i.e.,Ti, Al, Mg and Mn.

Table 2. The result of elemental analysis of $(Fe_xM_y)_3O_4$ sample

No.	Element	Concentration (μg/g)	Concentration(%)
1.	Fe	530677 ± 2773	0.936
2.	Ti	4249 ± 112	0.007
3.	Al	14838 ± 359	0.026
5.	Ce	19.44 ± 0.53	0.000
6.	Co	144.78 ± 1.48	0.000
7.	Cr	652.08 ± 7.53	0.001
8.	Eu	0.39 ± 0.03	0.000
9.	La	8.14 ± 0.37	0.000
10.	Mg	7559 ± 760	0.013
11.	Mn	4534 ± 40	0.008
12.	Na	536.13 ± 10.18	0.001
13.	Sc	13.11 ± 0.06	0.000
14.	Sm	2.61 ± 0.07	0.000
15.	Th	0.70 ± 0.07	0.000
16.	V	3154 ± 83	0.005
17.	Yb	0.95 ± 0.05	0.000
18.	Zn	561.65 ± 21.08	0.001

The magnetization curve, M (emu/g) of $(Fe_xM_y)_3O_4$ sample as a function of the applied magnetic field, H (Tesla) which was measured at room temperature is shown in Figure 1. The coercivity (Hc), the remanent magnetization (Mr) and saturation magnetization (Ms) of the sample is determined from Figure 1 and their values are listed in Table 3. As shown in the table, the saturated magnetization of impure magnetite sample is slightly lower than the one of pure magnetite as described by Ghandoor et al.[13].The existence of impurities in the magnetic crystalline can cause the spin scanting disorder and any disorder could result in a decrease of saturated magnetization.This result is in agreement with the results of previous research indicated by Doan et al.[14] or Henrik et al.[15]. The Figure 1 does not actually shows a typical pattern or the soft ferromagnetic cor super-paramagnetic materials because it is still observed a magnetic hysteresis phenomenom where the coercive force and remanence are not zero. This is may be due to the existence of impurities in the magnetic crystalline.

Table 3. Magnetic properties of magnetite.

	M_s (emu/g)		M_r (emu/g)		H_{c1} (Tesla)		H_{c2} (Tesla)	
Sample	Presentwork	Ghandoor (2012)	Present work	Ghandoor (2012)	Present work	Ghandoor (2012)	Present work	Ghandoor (2012)
Magnetite	46.20	46.70	17.00	1.65	-0.002	0	-0.021	0

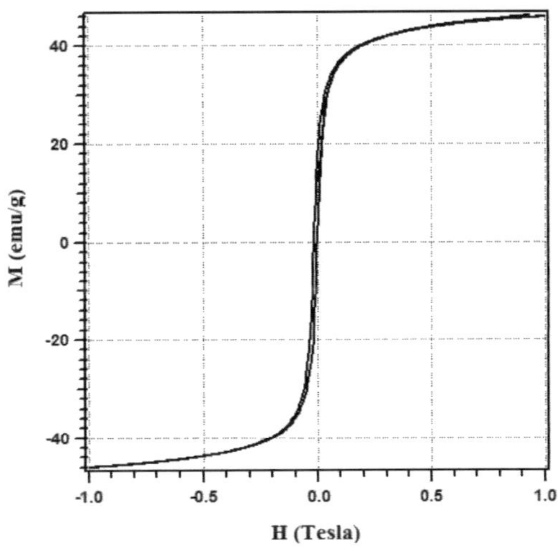

Figure 1. The magnetizationcurves, M (emu/g) vs applied magnetic field, H (T) at room temperature for the $(Fe_xM_y)_3O_4$ sample.

When the sample is placed in the monochromatic neutron beam, both magnetic and the nuclear Bragg peaks will appear because based on the VSM data tthe sample is ferromagnetic. The neutron diffraction data was analysed by the Rietveld method using the FullProf software, produced by Rodriguez-Carvajal[12]. Refinement was carried out by assuming that the sample has a magnetite phase with space group Fd-3m (Vol. I, No.227). The initial value of the atomic coordinate fractions of magnetite phases are illustrated in Table 4, indicated in Crystallography Open Data Base (2015). A fourth impurity elements, i.e.,Ti, Al, Mg and Mn were each added to Fe (A) and Fe (B) sites. The number of sites Fe (B) is twice the site Fe (A).

Tabel4. The initial value of atomic coordinate fraction (xj, yj, zj) and occupationfactor(gj)of the magnetite phaseindicated in Crystallography Open DataBase (2015).

	Occupation factor	Atomic coordinate fraction		
Atom	g_j	x_j	y_j	z_j
Fe(A)	0.0416	0.125	0.125	0.125
Fe(B)	0.0833	0.5	0.5	0.5
O	0.1667	0.255	0.255	0.255

The calculations were carried out step by step as follows: Iteration-1, refinement was carried out on scale factor(s) only. Iteration-2 was refinement on parameter in iteration-1, the background parameter (b0) and the zero point shift of detector (z0). Iteration-3 was refinement on parameters in iteration-2 and thebackgroundparameter (b1). Iteration-4 was refinement on parameters in iteration-3 and the lattice parameters (a,b,c). Iteration-5 was refinement on parameters in iteration-4 and the shape parameter. Iteration-6 was refinement on parameters in iteration-5 and the FWHM parameter (U, V, W).

Tabel5. The atomic coordinate fraction (xj, yj, zj) and occupation factor (gj)after refinement.

	Occupation factor	Atomic coordinate fraction		
Atom	g_j	x_j	y_j	z_j
Fe(A)	0.97(2)	0.125	0.125	0.125
Fe(B)	1.95(4)	0.5	0.5	0.5
O	4.00(0)	0.25487(1)	0.25487(1)	0.25487(1)

The Rietveld refinement results of atomic coordinate fraction (xj, yj, zj) and occupation factor (gj) are shown in Table 5. Of this Table5, it can be concluded that the Fe sites on the compound $(Fe_xM_y)_3O_4$ are occupied by 97.3% of Fe elements and the rest are occupied by impurity elements of M, therefore the chemical formula is $(Fe_{0.97}M_{0.03})_3O_4$. So, after the second stage of separation and purification process, the amount of impurity in the sample is reduced. The profile fitting of the calculated and observed diffraction pattern are shown in Figure2(a) with χ^2= 2.31.While Fig.2(b) is the Rietveld refinement results based on neutron diffraction pattern of $(Fe_xM_y)_3O_4$ composition, with assuming that x = 1.0, y = 0.0, and χ^2 = 2.53. It appears that the profile fitting result shown in Fig.2(a) is better than the profile fitting results presented in Figure 2(b). It means that the sample is more appropriate to be treated as a material that is not a pure magnetite.

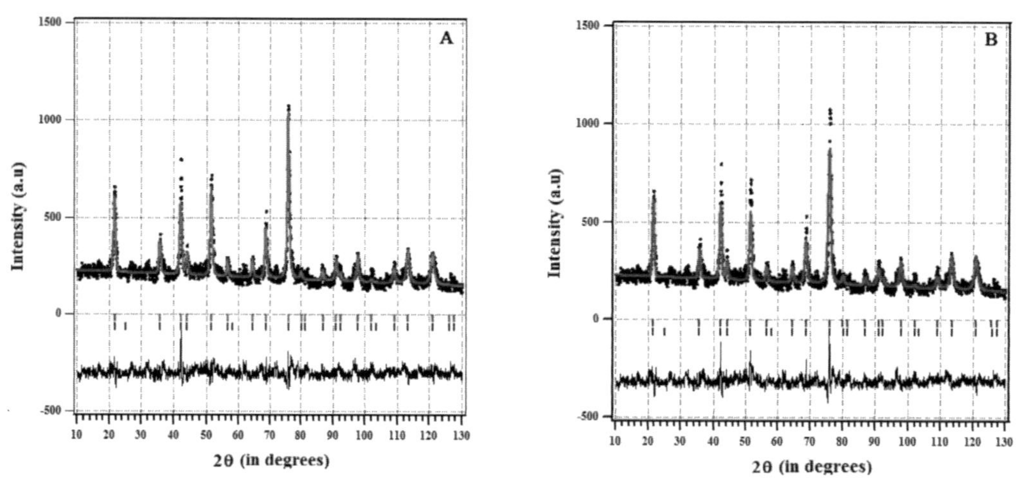

(a) (b)

Figure 2. TheRietveld refinement results based on neutrondiffractionpatternofsample (a) with composition $(Fe_xM_y)_3O_4$, where x = 0.97, y = 0.03, and χ^2 = 2.31, and (b) is the one with assuming that x = 1.0, y = 0.0, and χ^2 = 2.53.

The observed profile is indicated bythe dots and the calculated profile is described by solid line; vertical short lines indicate the position of Bragg peaks. Under the vertical short lines is the deviation profile between the observed and calculated intensities. It is shown in Figure 2 that almost all of the calculated diffraction profile coincides with the observed diffraction pattern. The magnetic structure of (Fe0.97M0.03)3O4 sample in one unit cell of two and three dimension is indicated in Figure 3 (a) and (b), respectively. It seems that the compound is a ferromagnetic material.

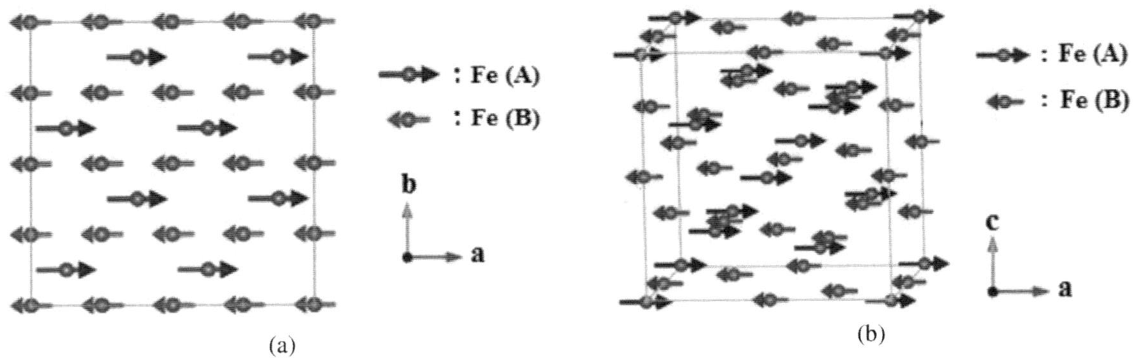

Figure 3. The magnetic structure of $(Fe_{0.97}M_{0.03})_3O_4$ sample in one unit cell of (a) two and (b) three dimension.

4. Conclusions

Magnetic structure of magnetite phase of iron sand, $(Fe_xM_y)_3O_4$ where M = Ti, Al, Mg and Mn, retrieved from Banten, Indonesia has been analyzed by the high resolution neutron powder diffractometer (HRPD) using a wavelength of λ= 1.8223 Å. Neutron diffraction data was analyzed by means of FullProf software. The analysis result shows that the sample has the chemical formula: $(Fe_{0.97}M_{0.03})_3O_4$ with cubic crystal system, space group: Fd-3m and lattice parameters: a = b = c = 8.36834 (3) Å, $\alpha = \beta = \gamma = 90^o$. The sample is magnetic at room temperature because the compound has a net magnetic moment, μ = 2.9(1) μB. The magnetic moments of the sample is lower than the magnetic moments of pure Fe_3O_4 (μ = 3.8 μB), because some of Fe sites are occupied by impurity elements. After the second stage of separation and purification process, the number and type of impurity M on the sample becomes less, and Fe content increases. Magnetic properties of magnetite phase depend on the purity of the magnetite phase.

Acknowledgement

We would like to thank to Mr. Herry Mugirahardjo and Dr. Aziz Khan Yahya for their help and good cooperation so as to enable R&D activity is running smoothly. Also gratefull to Head of Administration and staff who have helped from the administrative side of the procurement of materials and tools.

References

[1] Rochani S, Pramusanto, Sariman and Anugrah R I 2008 The Current Status of Iron Minerals in Indonesia *Journal of Indonesian Mining* **11 11** 1 – 17.

[2] Fleur Templeton, search December 22 2014 Iron and steel-The steel industry, Te Ara - the Encyclopedia of New Zealand.http://www.TeAra.govt.nz/en/graph/5892/chemical-composition-of-ironsands.

[3] Chikazumi S 1998 Physics of Ferromagnetism Oxford Science Publishing London 118.

[4] Schroeder D 2000 An Introduction to Thermal Physics. Addison Wesley Longman Boston 339.

[5] Nagata T 1953 Rock-Magnetism Maruzen Co. Ltd Tokyo 29.

[6] Rusianto T, Wildan W, Abraha K and Kusmono 2012 The Potential of Iron Sand from the Coast South of Bantul Yogyakarta as Raw Ceramic Magnet Materials *Jurnal Teknologi* **5 1** 62-69.

[7] Ildikó Y T, ErzsébetIllés, MártaSzekeres, Tombácz E 2015 Preparation and Characterization of Chondroitin-Sulfate-A-CoatedMagnetite Nanoparticles for Biomedical Applications *Journal of Magnetism and Magnetic Materials* **380** 168–174.

[8] ErzsébetIllés, MártaSzekeres, Kupcsik E, Ildikó Y T, Farkas K, Jedlovszky-Hajdúa A, Tombácz E 2014 PEGylation of Surfacted Magnetite Core–Shell Nanoparticlesfor Biomedical Application *Colloids and Surfaces A: Physicochemical and Engineering Aspects* **460** 429–

440.

[9] German S V, Navolokin N A, Kuznetsova N R, Zuev V V, Inozemtseva A A, Anis'kov A A, Volkova E K, Bucharskaya A B, Maslyakova G N, Fakhrullin R F, Terentyuk G S, Vodovozova E L, Gorin D A 2015 Liposomes Loaded with Hydrophilic Magnetite Nanoparticles:Preparation and Application as Contrast Agents for Magneticresonance Imaging *Colloids and Surfaces B: Biointerfaces* **135** 109-115.

[10] Adi W A and Rina M Th 2014 Composition Analysis of Titanomagnetite and Ilmenite from Iron Sand using Neutron Activation Analysis as Preliminary Study for Producing Iron Oxide and Titanium Dioxide *Proceedings International Conference on Materials Science and Technology 2014 (ICMST 2014)* National Nuclear Energy Agency of Indonesia Puspiptek Serpong Indonesia.

[11] Sari H D and Adi W A 2015 Synthesis and Characterization of High Purity Fe_3O_4 and α-Fe_2O_3 from Local Iron Sand *to be presented at The 9th Seminar on Magnetic Materials.Palembang* Indonesia paper #PP-44.

[12] Rodriguez-Carvajal J 2001 An introduction to the Program FullProf 2000 (Version July 2001).Laboratoire Leon Brillouin (CEA-CNRS) Bât. 563 CEASaclay 91191 Gif sur Yvette Cedex.

[13] Ghandoor H El, Zidan H M, Mostafa M H, Khalil and Ismail M I M 2012 Synthesis and Some Physical Properties of Magnetite (Fe_3O_4) Nanoparticles *International Journal of Electrochemical Science* **7** 5734-5745.

[14] Dung D T K, Hai T H, Phuc L H, Long B D, Vinh L K and Truc P N 2009 Preparation and Characterization of Magnetic Nanoparticles with Chitosan Coating *Journal of Physics: Conference Series* **187** 012036.

[15] Jacobsen H, Lefmann K, Brok E, Frandsen C, Mørup S 2012 Temperature Dependence of the Magnetization of Canted Spin Structures *Journal of Magnetism and Magnetic Materials* **324** **19** 3218–3222.

[16] Crystallography Open Data Base search January 20 2015 (http://www.Crystallogra-phy.net/ search. Html).

IOP Conf. Series: Journal of Physics: Conf. Series **1091** (2018) 012008 doi:10.1088/1742-6596/1091/1/012008

Time milling influence on the size of the Lemabang iron sand powder synthesized by using high energy milling method

S Firza, S Nita, S A Fitri, J Akmal

Physics Department – Universitas Sriwijaya, St.Palembang-Prabumulih 32, Indralaya 30662, Indonesia

Corresponding author: firzaseptian@issp.u-tokyo.ac.jp

Abstract. Has been synthesized iron sand from the Lemabang-Sumsel Area by using High Energy Milling Method. Before synthesized, extraction of Lemabang iron sand in advance by using permanent magnets and Methanol-Soap Bathed Method in order to separated iron sand and impurities. After it, iron sand milled by using High Energy Milling with variations in milling time 2 hours, 4 hours, and 6 hours. Milling time optimization done in order to see the effect of time milling on powder size and surface morphology. Then Iron sand already in milling characterized by using XRD to see crystal structure and crystal size, and SEM-EDS used to see surface morphology and composer elements. XRD's result show that, the longer of the milling time resulting the shorter of the powder size. Whereas the results of the SEM-EDS's photo, its seem that the grain morphology of the iron sand powder after milling is smoother and more homogeneous be compared before milling

1. Introduction

Iron Sand is one of the abundant natural resources found in Indonesia. There are many uses or applications for this material, starting from magnetic based materials to high quality steel materials. In the application, iron sand has been used in a variety of length scale, ranging from the millimeter-down to the nanometer length scale. The main problem in the synthesis and processing of nanoparticle iron sand, is the lengthy time it takes to finish the process. Nanoparticle synthesis variously has been done but it was not efficient in the conventional scale. By using High Energy Milling especially Shaker Mill PPF-UG, iron sand nanoparticle could be synthesized with a higher efficiency as far the duration of processing time is concerned. The rotational speed of this apparatus is up to 800 rpm. The synthesis of iron sand nanoparticle then takes up only a few hours.

Iron sand is a mineral containing various oxide iron compounds such as magnetite, ilmenit, hematite, and also other minerals (but in lesser quantities), such as silica and titania with varying concentration depending upon the location Zulfalina et al.[1]. Experiment about magnetization still exist at synthesize. Its cause more benefit and it have nice prospect by various sector. In Indonesia several research groups are involved in the expeimental work of iron sand synthesis. Presently iron sand synthesis technology has become more rapidly developed Zaehir et al.[2].

High Energy Milling (HEM) is a method of synthesis using an apparatus which gives priority to transfer high mechanical energy to the material. HEM using collision of balls in a vial to crush the material until the smallest size material is obtained. This type of apparatus does not need high temperature smelting. This process produces smooth nanoparticle powder by maximizing operation, by Wank et al.[3].Theoretically, nanocrystalline metals didn't deform at high temperaturesby Darling et al.[4].

Several ways done to achieve nano-scale, such as a nanoparticle self-stabilization mechanism in molten metal by Chen et al.[5], layer-by-layer assembly of two-dimensional buildingblocks under vacuum by Kang et al.[6],a molecular-level liquid–liquidmixing/doping technique by Liu et al.[7].Complicated way to synthesize nano-scale material.

Innovative tools and technology made their effort go furtherby Pain et al.[8].Several conventional synthesis have reached certain limits in further improving the propertiesof metals by Nie et al.[9].

Shaker Mill PPF-UG is a new innovation in milling system which was developed by the HEM-E3D system milling company (previous product). This apparatus has a rotational speed in the range of 700 rpm to 800 rpm, by Sukarto et al.[10]. It has been the authors'experience, that the milling's duration time parameter is also important in this case. High energy milling is a very suitable method to minimize particles size in a sample and then observe the physical changes occuring in the milled sample.

By virtue of XRD reflection intensity data, the particle size could be obtained by calculating the FWHM (Full Width Half Maximum). Particle sized be found by using the *Scherrer Formula*, by Hadiati et al.[11],

$$D = \frac{0.9\,\lambda}{B\cos\theta} \tag{1}$$

With D is the particle size, B is the *Full Width Half Maximum*, $\lambda$is the X-Ray wave length, and $2\theta$is the Bragg diffraction angle. This method is an accurate technique to calculatethe particle size in nanoparticles but it is not suitable to calculate particle size in bulk material. Comparator parameter has been calculated by observing the SEM data

2. Experimental Method

The sand material has been obtained from a location in Lemabang-South Sumatera, Indonesia. This sample was rinsed with water and and then dried. First extraction of iron sand has been carried out repeatedly forthirty-five times using a permanent magnet to separate it from the main impurity. And then the sample was milled by mortar in order to separate it from its sticky impurity, followed by the second extraction by a permanent magnet to extricate the iron sand. By assuming that the process has been going smoothly as planned, the authors have expected to be able to obtain least 48 grams of ironsand sample. The sample is then rinsed with 25 ml water and 1 mg detergent, and then stirred until it turns foamy. Twenty-five ml of technical methanol is then added to the sample and the stirring continues until all foam has been cleansed from the sample. The iron sand is then collected using a permanent magnet device. After dehydrating the collected ironsand sample, it isthen divided into four equal parts; the first part of the sample is destined for comparator parameter (0 hours milling), thesecond sample for 2 hours milling, the third sample for 4 hours milling, and the fourth sample for 6 hours milling. The next step is characterization of the first sample or the comparator parameter by using SEM-EDS. Followed by characterization of the second, the third, and the fourth sample by using XRD method. Finally analyzing the collected experimental SEM-EDS data in order to get particle size, bulk material by observed data, nanomaterial by observed XRD data. XRD tabulation has done by using Match Application.

3. Results and Discussion

3.1. Particle Size Calculation of 0 hours milling sample

IOP Conf. Series: Journal of Physics: Conf. Series **1091** (2018) 012008 doi:10.1088/1742-6596/1091/1/012008

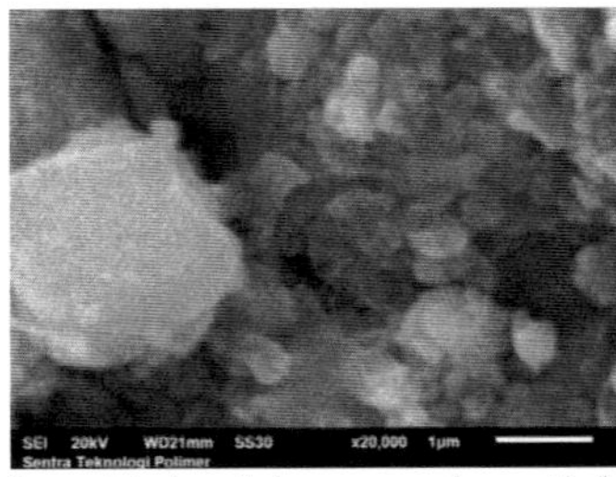

Figure 1. SEM-EDS surface area for 0 hours milling

By comparator line, be obtained iron sand particle size for about 0.5 μm. Minority particle more than 1 μm but majority particle has 0.5 μm.

3.2. Calculation Particle Size of 2 hours milling sample

Before execution of particle size calculation by way of the Scherrer Formula, XRD data must first be refined. Refining has been done by using the Match! Application. And then the corresponding FWHM value was computed for each of the XRD reflection peak. The XRD refinement reflection intensity (two- hours milled sample) are shown in Figure 2 below.

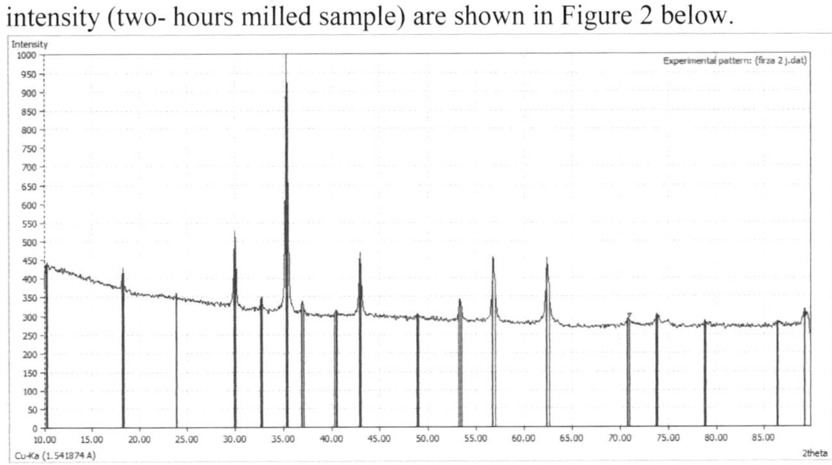

Figure 2. XRD of 2 hours milling after refine

The experimental XRD results was used to calculate the particle size by using Scherrer Formula. Particle Size Calculation results for the two hours milling sample are shown in Table 1.

Table 1. Particle Size Calculation of 2 hours milling

2θ	B_{FWHM} (Degree)	B_{FWHM} (Radian)	D (particle size), nm
10.3100	0.0900	0.001570796	88.70164659
18.2100	0.1800	0.003141593	44.73509856
23.8700	0.2300	0.004014257	35.33274379
29.9800	0.1200	0.002094395	68.5912472
32.7000	0.2000	0.003490659	41.42971027
35.3300	0.2300	0.004014257	36.27964325
36.9300	0.2300	0.004014257	36.44524975
40.4300	0.2600	0.004537856	32.58753274
42.9500	0.2400	0.00418879	35.60002363
48.9300	0.1200	0.002094395	72.7929078
53.3000	0.3300	0.005759587	26.95737754

56.8000	0.3200	0.005585054	28.24586286
62.3800	0.3500	0.006108652	26.5551639
70.8400	0.3100	0.005410521	31.47272919
73.7900	0.1200	0.002094395	82.84867431
78.7700	0.1400	0.002443461	73.47893311
86.4900	0.0800	0.001396263	136.4381314
89.3900	0.1600	0.002792527	69.90519402
Particle Size Average			53.79988166

The average value of particle size after two hours of milling is found to be 53.79988166 nm, and the round-off value is 53.8 nm. It is amazing to get nanoparticles after milling two hours only from the bulk material.

3.3.Calculation Particle Size of 4 hours milling sample.

Before execution of particle size calculation by way of the Scherrer Formula, XRD data must first be refined. Refining has been done by using the Match!Application. And then the corresponding FWHM value was computed for each of the XRD reflection peak. The XRD refinementreflection intensity (4 hours of milling sample) are shown in Figure 3 below.

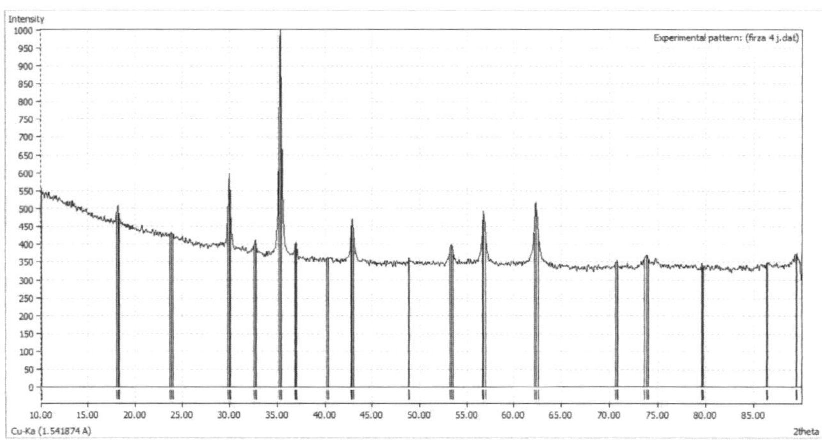

Figure 3. XRD of 4 hours milling after refine

By this result, the angular position of the diffraction peaks was then obtained. This data was then used to calculate the particle size by using the Scherrer Formula, and the results for the sample milled for four consecutive hours.are shown in Table 2.

Table 2. Particle Size Calculation of 4 hours milling

2θ	B_{FWHM} (Degree)	B_{FWHM} (Radian)	D (particle size), nm
10.1300	0.0600	0.001047198	133.0337818
18.1700	0.2600	0.004537856	30.96872227
23.8300	0.3300	0.005759587	24.62403641
29.9600	0.2400	0.00418879	34.29402145
32.6600	0.1900	0.003316126	43.60575856
35.3000	0.2800	0.004886922	29.7986521

36.9300	0.1900	0.003316126	44.1179339
40.3000	0.1600	0.002792527	52.93266274
42.9200	0.3200	0.005585054	26.69726899
48.8900	0.1200	0.002094395	72.78135303
53.2900	0.3700	0.006457718	24.04201363
56.7700	0.3800	0.006632251	23.78262406
62.3400	0.4300	0.007504916	21.61010299
70.8200	0.2800	0.004886922	34.84048325
73.9000	0.4400	0.007679449	22.61139731
79.6500	0.1700	0.00296706	60.89779093
86.4200	0.0800	0.001396263	136.3598121
89.4700	0.0800	0.001396263	139.9070613
Particle Size Average			53. 16141538

The average particle size after four hours of milling time is 53.16141538 nm., rounded-off to the nearest decimal point to 53.2 nm. The average particle size of 4 hours milled sample is smaller than the value of the average particle size in the two-hours milled sample, which is about 0.6 nm smaller.

3.4. Calculation Particle Size of 6 hours milling sample

Before execution of particle size calculation by way of the Scherrer Formula, XRD data must first be refined. Refining has been done by using the Match! Application. And then the corresponding FWHM value was computed for each of the XRD reflection peak. The XRD refinement reflection intensity (six hours of milling sample) are shown in Figure 4 below.

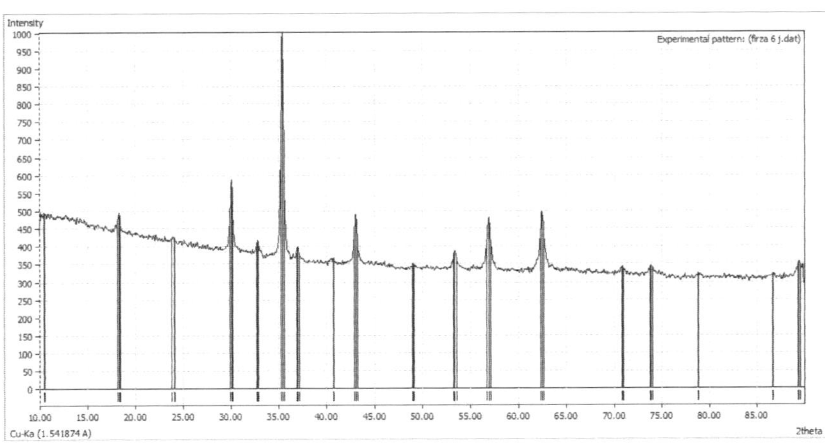

Figure 4. XRD of 6 hours milling after refine

By this result, has been obtained data of diffraction angle. This data was using to calculate particle size by using Scherrer Formula. Here table for 6 hours milling.

Table 3. Particle Size Calculation of 6 hours milling

2θ	B_{FWHM} (Degree)	B_{FWHM} (Radian)	D (particle size), nm
10.5	0.0700	0.00122173	114.0621927
18.2800	0.2300	0.004014257	35.01351147
23.9700	0.3700	0.006457718	21.96765793

30.0600	0.2800	0.004886922	29.40175213
32.8000	0.2300	0.004014257	36.03507415
35.3900	0.3200	0.005585054	26.08034604
36.9800	0.2700	0.004712389	31.05048045
40.7500	0.1000	0.001745329	84.8151294
43.0300	0.3400	0.005934119	25.13633825
49.0200	0.1400	0.002443461	62.4162445
53.3500	0.3400	0.005934119	26.17024663
56.9000	0.3800	0.006632251	23.79722752
62.4400	0.4000	0.006981317	23.24313916
70.8800	0.1000	0.001745329	97.58969325
73.8400	0.3000	0.005235988	33.15033121
78.8200	0.1000	0.001745329	102.9073792
86.7300	0.1000	0.001745329	109.3661815
89.4600	0.2500	0.004363323	44.76638915
Particle Size Average			51. 49829526

Particle size average after 6 hours milling is 51.49829526 nm. The round-off value is 51.5 nm. Particle size of 6 hours milling smaller than 4 hours milling, it is for about 1.7 nm smaller.

4. Conclusion

A longer milling time would result in the smaller particle size and by a significant disparity, as shown at table 4.

Table 4. Particle Size of Iron Sand

No.	Milling Time	Particle Size
1.	0 Jam	0.5 μm
2.	2 Jam	53.8 nm
3.	4 Jam	53.2 nm
4.	6 Jam	51.5 nm

After a milling time of only two-hours, the authors are able to obtain iron sand nanoparticle, furthermoreit is observed that there is a big alteration in size. However by consecutively increasing the milling-time by two-hours for each sample,no big alteration in size has been observed. Only changes ranging from 0.6 nm up to 1.7 nm has been found in this experiment. Therefore based upon this data the authors make the assessment that after 2 hours of milling no further minimization with respect to the particle size would take place, but only a better and more homogenous distribution of particles in the sample.

References

[1] Zulfalina and Manaf A 2004 *Identification of Mineral Compound and Titanium Dioxide Extraction From Mineral Sand* **5 2** (Jakarta : Indonesian Journal of Material Science 5) 46-50

[2] Zaehir L M, Yulianto A and Sulhadi 2013 *Low Density Polyethylene (LDPE) Application In Making Ferrite Magnetic Composite* **2** (Semarang :J. Sains Dasar) 72-78

[3] Wank A and Wielage B 2003 *High Energy Ball Milling – A Promising Route For Production Of Tailored Thermal Spray Consumables* (Chemnitz : Conference On Modern Wear and

Corrosion Resistant Coating Obtained by Thermal Spraying)

[4] Darling K A, Rajagopalan M, Komarasamy M, Bhatia M A, Hornbuckle B C, Mishra R S, Solanki K N 2016 *Extreme creep resistance in a microstructurally stable nanocrystalline alloy* **537** (Texas : Macmillan Publishers Limited, part of Springer Nature) 378 – 381

[5] Chen L Y, Xu J Q, Choi H, Pozuelo M, Ma X, Bhowmix S, Yang J M, Mathaudhu S, Li X C 2015 *Processing and properties of magnesium containing a dense uniform dispersion of nanoparticles* **528** (London : Nature) 539 – 543

[6] Kang K, Lee K H, Han Y, Gao H, Xie S, Muller D A, *Park J 2017 Layer-by-layer assembly of two-dimensional materials into wafer-scale heterostructures* **550** (New York : Macmillan Publishers Limited, part of Springer Nature) 229 – 233

[7] Liu J, Zhang G J, Jiang F, Ding X D, Sun Y J, Sun J, Ma E 2013 *Nanostructured high-strength molybdenum alloys with unprecedented tensile ductility* **12** (Maryland : Nature Materials) 344 – 350

[8] Pain S 2017 *Power Through The Ages* **551** (Macmillan Publishers Limited, part of Springer Nature) 134 – 137

[9] Nie J F 2012 *Precipitation and Hardening in Magnesium Alloys* **43A** (Clayton : Metallurgical And Materials Transactions A) 3891 - 3939

[10] Sukarto A W 2014 *Shaker Mill PPF-UG* (Jakarta : Ball Milling Local High Energy) Online (http://blog.sivitas.lipi.go.id/blog.cgi?isiblog&1136659685&&&1036006479&&1395314337&agus046&1318847657)

[11] Hadiati S, Ramelan A H, Variani V I, Hikam M, Soegijono B, Saputri D F, Iriyani Y 2013 *Annealing Temperature Variation And Holding Time at Thin Film Emergence* $BaZr_{0.15}Ti_{0.85}O_3$ *By Sol Gel Method* (Original : Kajian variasi suhu annealing dan holding time pada Penumbuhan lapisan tipis $BaZr_{0,15}Ti_{0,85}O_3$ dengan metode sol gel) **1** (Surakarta : MIPA Journal 1, 20-27.

Influence of Ti addition on the microwave absorbing behavior of $Ba_{0.6}Sr_{0.4}Fe_{12-x}Ti_xO_{19}$ system at X-band frequencies

Y.E. Gunanto[1], **L. Cahyadi**[2] **and W. A. Adi**[3]

[1]Department of Biology Education, Faculty of Education, University of Pelita Harapan, Karawaci, Tangerang 15811, Indonesia
[2]Faculty of Science and Technology, University of Pelita Harapan, Karawaci, Tangerang 15811, Indonesia
[3]Centre for Science and Technology of Advanced Materials – National Nuclear Energy Agency, Kawasan Puspiptek Serpong, Tangerang Selatan, Banten, Indonesia

Corresponding author : yohanes.gunanto@uph.edu

Abstract. The synthesis and characterization of $Ba_{0.6}Sr_{0.4}Fe_{12-x}Ti_xO_{19}$ system (x = 0, 1, 2, 3) have successfully performed by mechanical milling through solid state method. Stoichiometric quantities of $BaCO_3$, $SrCO_3$, Fe_2O_3, and TiO_2 were mixed. The mixture were first milled for 48 h and sintered at 1000 °C for 5 hrs. The refinement of x-ray diffraction shows that a single phase material with a hexagonal structure and lattice parameters appears to increase with increasing Ti doping. The mechanical milling resulted in powders of all samples with mean size of 90 nm. Results of hysteresis curve shows that the samples are ferromagnetic and have low coercivity after Ti doping. Results of VNA evaluation show that the sample after Ti doping has absorption peaks with high reflection loss values at frequency ~ 11 GHz. We concluded that this study has been successfully known effect of Ti addition on the microwave absorbing behavior of $Ba_{0.6}Sr_{0.4}Fe_{12-x}Ti_xO_{19}$system at X-band frequencies.

1. Introduction

Electromagnetic wave absorber material is a material that can weaken the electromagnetic wave energy. Externally, this absorber can reduce the reflection or transmission of a particular object and can be used internally to reduce oscillations caused by the cavity resonance. One of the applications in various electronic devices which are working in the high frequency-rangeis as signal booster or amplifier, but often have a problem caused by the emission of high frequency noise such as electromagnetic wave interference (EMI). To eliminate the EMI requires a magnetic material which is able to resonate at a specific frequency that is expected to absorb undesirable electromagnetic radiation.

The prerequisites are essential since electromagnetic wave absorber material is a material of high permeability and permittivity. Presently, absorber materials are being developed through the modification of ferrite-based magnetic materials because they have a relatively high permeability [1-4]. Through the structural modification, the material is expected to be utilized as a superior material for electromagnetic wave absorber applications [5-7]. The structural modificationis implemented primarily to enhance the permittivity of the material. The magnetic properties of this material are influenced by the presence of Ti atoms. Due to the presence of coupling Ti^{4+} is greatly contributing to the magnetic properties of this material. In the present study, the authors have carried out a modification of the hexagonal barium strontium hexaferrite system by titanium substitution of the iron atoms. Therefore the objectives of this study is to understand the changes in the parameters' values of the crystal structure of this material and the influence of Ti addition on the microwave absorbing behavior.

2. Materials and Methods

The synthesis of $Ba_{0.6}Sr_{0.4}Fe_{12-x}Ti_xO_{19}$ system (x = 0, 1, 2, 3) material was performed by using the solid reaction method. This magnetic material is prepared by oxide materials, namely $BaCO_3$, $SrCO_3$, Fe_2O_3, and TiO_2. The four starting materials are mixed by using a Spex 8000 high-energy milling (HEM) apparatus for 10 hours at room temperature. The finely mixed powder was compacted at 5000 psi into pellets and sintered in the THERMOLYNE electric chamber furnace at 1000°C in the air at atmospheric pressure for 10 hours. The qualitative and quantitative phase analysis were carried out using the PW1710 Philips diffractometer equipment (XRD). The Rietveld analysis was performed by employing GSAS special application code [8,9]. The surface morphology and elemental identification of the sample were analyzed by using the JEOL scanning electron microscope (SEM) and energy dispersive spectroscopy (EDS), respectively.

3. Results and Discussion

The identification results of XRD pattern on the samples show that the $Ba_{0.6}Sr_{0.4}Fe_{12-x}Ti_xO_{19}$ system (x = 0, 1, 2, 3) has a diffraction intensity pattern that is similar to one another as shown in figure 1.

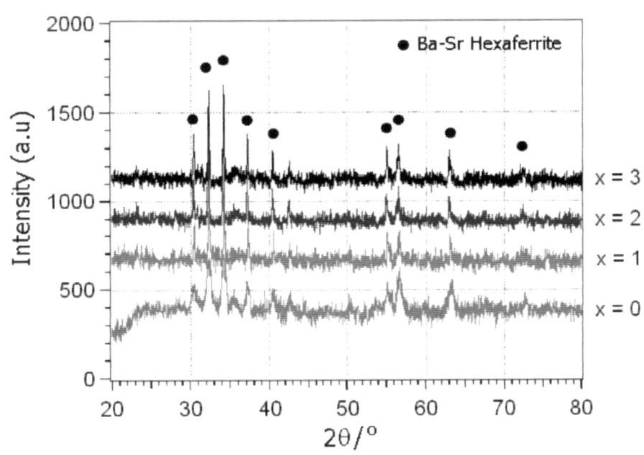

Figure 1. Phase identification of x-ray diffraction pattern of the $Ba_{0.6}Sr_{0.4}Fe_{12-x}Ti_xO_{19}$ (x = 0, 1, 2, 3) samples

The diffraction pattern of the product samples shown in Fig. 1 is that of the samesingle phase for all samples. Therefore, it was necessary for the analysis of phase content in the samples using the GSAS software as shown in figure 2 for the Ti content of x = 0, 1, 2, 3. The parameters of crystal structure referred to COD-1008841for the phases of $BaO.6Fe_2O_3$.

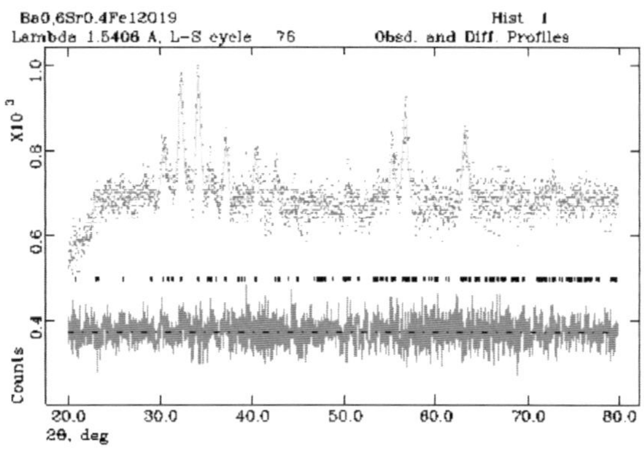

(a) $Ba_{0.6}Sr_{0.4}Fe_{12-x}Ti_xO_{19}$ (x = 0)

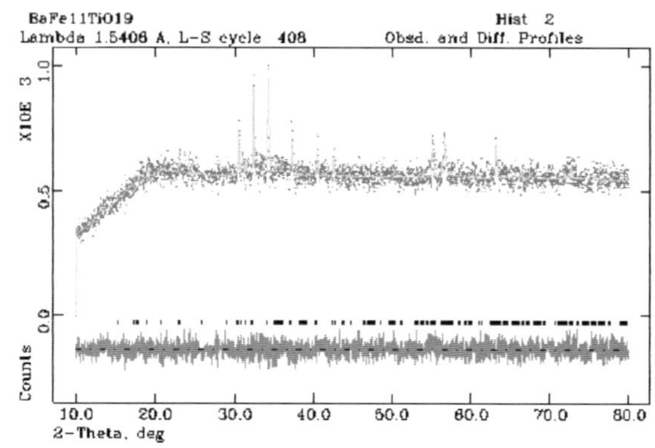

(b) $Ba_{0.6}Sr_{0.4}Fe_{12-x}Ti_xO_{19}$ (x = 1)

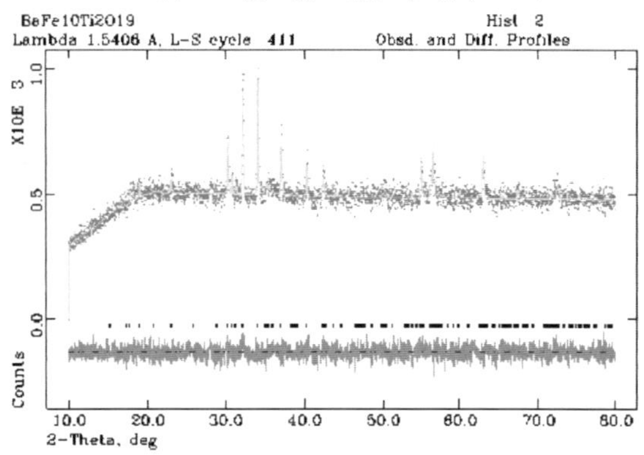

(c) $Ba_{0.6}Sr_{0.4}Fe_{12-x}Ti_xO_{19}$ (x = 2)

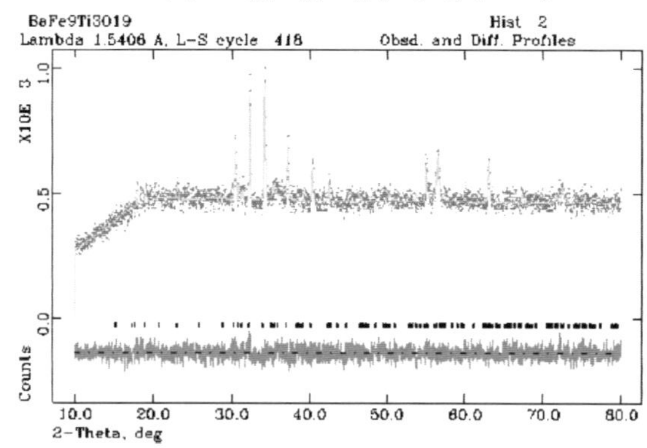

(d) $Ba_{0.6}Sr_{0.4}Fe_{12-x}Ti_xO_{19}$ (x = 3)

Figure 2. The refinement of XRD pattern of the samples (x = 0, 1, 2, 3).

The refinement results have produced very good quality of fitting with the R-factor and the χ^2 (chi-squared) values being very small [9]. The refinement results of the $Ba_{0.6}Sr_{0.4}Fe_{12-x}Ti_xO_{19}$ (x = 0, 1, 2, 3) crystalline structural parameters are shown in table 1.

Table 1. Structure parameter, factor R, *goodness of fit* (χ^2).

x = 0			x = 1		
Space group: P 63/m m c (194), system: hexagonal $a = b = 5.902(8)$ Å, $c = 23.23(4)$ Å, α= β = 90°, γ = 120°, V = 700.9(2) Å^3 and ρ= 5.110 gr.cm^{-3}			Space group : P 63/m m c (194), system : hexagonal $a = b = 5.86(1)$ Å, $c = 23.34(5)$ Å, α = β = 90°, γ= 120°, V = 695.1(1) Å^3 and ρ= 5.124 gr.cm^{-3}		
R factor	*w*Rp = 6.96	$\chi^2 = 1.187$	Factor R	*w*Rp = 6.68	$\chi^2 = 1.201$
x = 2			x = 3		
Space group: P 63/m m c (194), system: hexagonal $a = b = 5.902(8)$ Å, $c = 23.23(4)$ Å, α = β = 90°, γ= 120°, V = 700.9(2) Å^3 and ρ= 5.110 gr.cm^{-3}			Space group : P 63/m m c (194), system : hexagonal $a = b = 5.86(1)$ Å, $c = 23.34(5)$ Å, α = β = 90°, γ= 120°, V = 695.1(1) Å^3 and ρ= 5.124 gr.cm^{-3}		
Factor R	*w*Rp = 6.57	$\chi^2 = 1.240$	Factor R	*w*Rp = 6.42	$\chi^2 = 1.180$

The refinement results provide the experimental evidence that titanium substitution into iron atoms was able to reach the value of x = 3 and that with the doping concentration of uptox ≤ 0.3,the product formed is still of a single phase.

In figure 3 are shown the microstructures of the $Ba_{0.6}Sr_{0.4}Fe_{12-x}Ti_xO_{19}$ (x = 0, 1, 2, 3) sample. The surface morphology results of observations by scanning electron microscope on the samples show that the $Ba_{0.6}Sr_{0.4}Fe_{12-x}Ti_xO_{19}$ system (x = 0, 1, 2, 3) samples has been obtained and that the substitution increases boththe density and the inter grain connectivity in the microstructure of sintered samples.

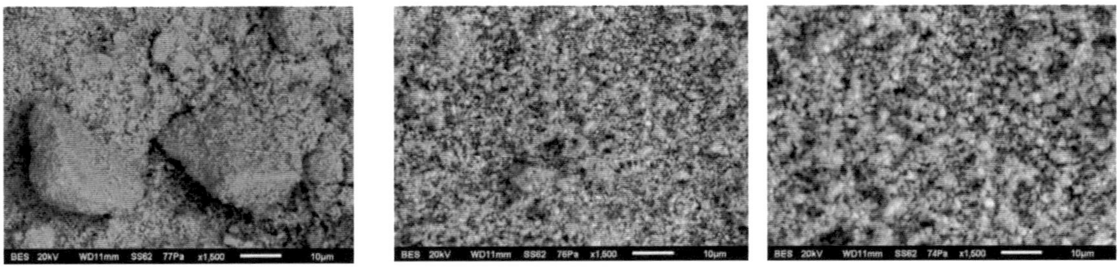

(a) $x = 1$ (b) $x = 2$ (c) $x = 3$

Figure 3. Surface morphology photo of the $Ba_{0.6}Sr_{0.4}Fe_{12-x}Ti_xO_{19}$ system ($x = 0, 1, 2, 3$)

The microstructure analyses shows that the particles' shape was polygonal with the varied particle sizes distributed homogeneously on the surface of the samples.

Characterization of the magnetic properties of the $Ba_{0.6}Sr_{0.4}Fe_{12-x}Ti_xO_{19}$ ($x = 0, 1, 2, 3$) magnetic particles has beenperformed using a vibrating sample magnetometer (VSM) that produced hysteresis curves of magnetic particles as shown in figure 4.

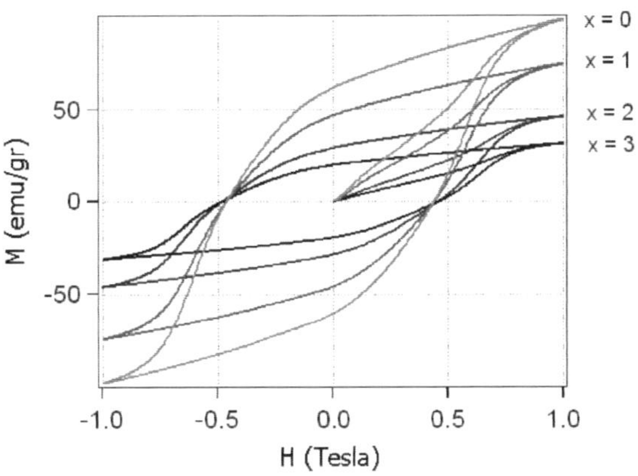

Figure 4. The hysteresis curves of $Ba_{0.6}Sr_{0.4}Fe_{12-x}Ti_xO_{19}$ system ($x = 0, 1, 2, 3$)

The hysteresis curves of synthesized $Ba_{0.6}Sr_{0.4}Fe_{12-x}Ti_xO_{19}$ system are shown in Figure 4 in which respective profiles for x of 0, 1, 2, and 3 are also compared. The hysteresis loop as shown in Figure 4 is characterized with intrinsic saturation Ms, remanence field Mr and coercivity Hc. The intrinsic saturation, Ms is the state when the material is not able to absorb more magnetic field, so increasing magnetization force will not change the magnetic flux density significantly. Meanwhile, the remanence field Mr, is the residual magnetization in a medium once the external magnetic field is removed and coercivity (also called coercive force) is a force required to demagnetize the material so the residual induction becomes zero after magnetizing up to the saturation point. In reference to Figure 6 it seems that Hc is constant however the remanence magnetization decreases along with increasing doping concentration of Ti [10]. Thus, the doping concentration valueof $x = 1$ is the optimal composition for which the magnetic properties was still able to maintain their strength, becauseafter x exceeds the value of 0.1, the remanence magnetization would decrease significantly. A doping concentration of $x = 1$ is required for futher confirmation about the presence of Ti in the sample. Therefore an elemental content analysis of the sample is needed to find out whetherthe substitution of Ti has been succesful in the composition.

Figure 5 shows the elemental analysis result of $Ba_{0.6}Sr_{0.4}Fe_{12-x}Ti_xO_{19}$ (x = 1) samples using the energy dispersive spectroscopy (EDS) method.

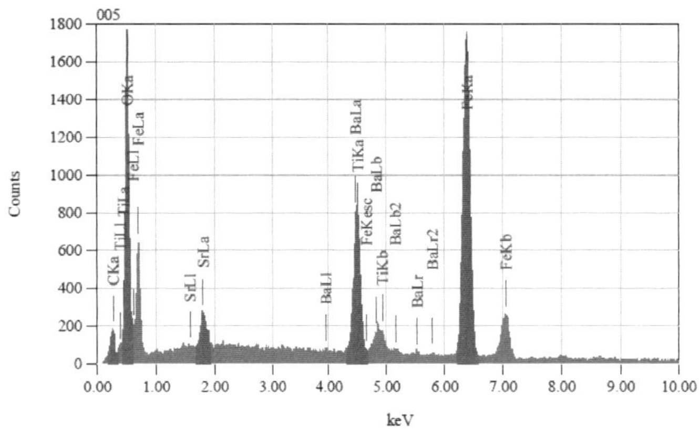

Figure 5. The element analysis result of the $Ba_{0.6}Sr_{0.4}Fe_{12-x}Ti_xO_{19}$ system (x = 1).

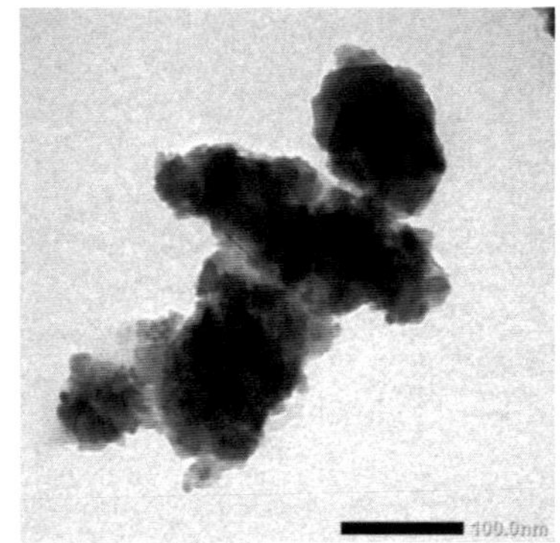

Figure 6. TEM image of the $Ba_{0.6}Sr_{0.4}Fe_{12-x}Ti_xO_{19}$ system (x = 1).

The elements content of the $Ba_{0.6}Sr_{0.4}Fe_{12-x}Ti_xO_{19}$ system (x = 1) is shown in table 2.

Table 2. The results of element analysis by using energy dispersive spectroscopy

No.	$Ba_{0.6}Sr_{0.4}Fe_{12-x}Ti_xO_{19}$ (x = 1)	
	Unsure	Mass Fraction (wt%)
1.	Barium (Ba)	13.23 ± 0.43
2.	Strontium (Sr)	2.82 ± 0.32
3.	Iron (Fe)	44.69 ± 0.39
4.	Titanium (Ti)	6.47 ± 0.18
5.	Oxygen (O)	28.48 ± 0.15
6.	Carbon (C)	4.31 ± 0.10

Based on the Table 2 it could be shownthat the sample has a Ti content of around 6.47 wt.%. It means that substitution of Ti has been succesfully achieved in the composition. Additional supporting data is provided by micrograph results from the transmission electron microscope (TEM) observation as shown in figure 6.

In Figure 6 it seemsthat the particleswith adoping concentration of x = 1 are nanoparticles with a good homogeneity, uniformity, and having an aggregate particle shape. The range of the particle size is around 50-100 nm.

The microwave absorbing properties of synthesized $Ba_{0.6}Sr_{0.4}Fe_{12-x}Ti_xO_{19}$ system (x = 0, 1, 2, 3) are shown in figure 7 in which respective profiles for x of 0, 1, 2, and 3 are also compared.

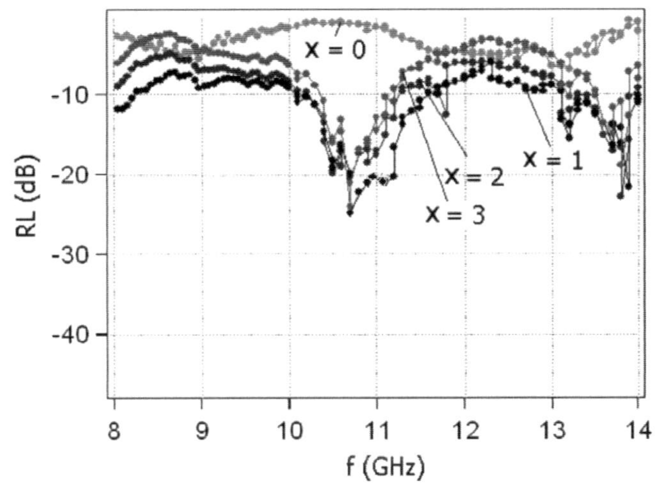

Figure 7. The reflection loss (RL) curve of $Ba_{0.6}Sr_{0.4}Fe_{12-x}Ti_xO_{19}$ system (x = 0, 1, 2, 3)

Figure 7 shows the relation between the reflection loss (RL) of $Ba_{0.6}Sr_{0.4}Fe_{12-x}Ti_xO_{19}$ system (x = 0, 1, 2, 3) particlesand the microwave frequency X-band in the range of 8-14 GHz when the thickness of the sample is 1.5 mm. There are at least two absorption peaks observed which have a high RL within the frequency range for all samples. Based on the phase composition shown in figure 7, it is evident that the $Ba_{0.6}Sr_{0.4}Fe_{12-x}Ti_xO_{19}$ system (x = 0, 1, 2, 3) has a low RL value, which then increases for the Tidoping concentration x = 1 and decreases again for x = 2 and 3. However, this case shows that the best absorber turns out to be the Ti substituted Ba-Srhexaferrite sample with the Ti doping concentration of x = 0, 1, 2, and 3 and which occurred at nearly the same thickness and at practically identical frequencies, there are differences. This can be explained by the effect of electromagnetic properties on the attenuation characteristic in each sample. It was found that substitution with Ti effectively increased the attenuation constant in the entire range of frequencies used here, compared with non-substitution Ti doping concentration x = 0, and the optimum value of microwave absorption was found on the phase composition of Ti doping concentration x = 1. This indicates that $Ba_{0.6}Sr_{0.4}Fe_{12-x}Ti_xO_{19}$ (x = 1) has certain microwave absorption properties in the frequency range of 8 – 14 GHz, absorption peak values of -25 dB at 10.7 GHz and -20.0 dB at 13.8 GHz.

4. Conclussion

Synthesis of $Ba_{0.6}Sr_{0.4}Fe_{12-x}Ti_xO_{19}$ system (x = 0, 1, 2, 3) was successfully performed by solid state reaction through a mechanical milling method. It was found that the best phase composition which was a high performance absorber was $Ba_{0.6}Sr_{0.4}Fe_{11}TiO_{19}$. The refinement results of X-ray diffraction patterns showed that the $Ba_{0.6}Sr_{0.4}Fe_{11}TiO_{19}$ was a single phase, which has a hexagonal structure (P 63/m m c), a = 5.9587(3) Å, b = 5.9587(3) Å and c = 23.406(1) Å, α=β= 90° and γ= 120°, V = 719.7(1) Å^3 and ρ = 5.568 gr.cm^{-3}. The mechanical milling resulted in $Ba_{0.6}Sr_{0.4}Fe_{11}TiO_{19}$ powders with a mean particle size of~850 nm. The TEM micrograph image revealsthat the particles' morphology is an aggregate of fine grains. The magnetic properties of the particle $Ba_{0.6}Sr_{0.4}Fe_{11}TiO_{19}$ shows a low coercivity and a high remanent magnetization. The $Ba_{0.6}Sr_{0.4}Fe_{11}TiO_{19}$ has microwave absorption properties in the frequency range of 8 – 14 GHz, with absorbing peak values of -8 dB at 8.5 GHz and -10.0 dB at 12.5 GHz.

Acknowledgments
This work was supported by the Dikti Competitive Grants 2015 No. DIPA : SP DIPA-023.04.2.189705/2015, The Development of Magnetic Materials Based Composite Systems $(Ba_{(1-y)}Sr_y)O.6Fe_{2(1-x)}(Mn,Ti)_xO_3$ for Interference Absorption of Electromagnetic Wave son Cell Phones., Y.E. Gunanto.

References

[1] Kumar A, Agarwala V and Singh D 2013 Effect of Particle Size of $BaFe_{12}O_{19}$ on The Microwave Absorption Charateristics in X-Band *Progres In Electromagnetics Research M* **29** 223-236.

[2] Tudorache F, Popa P D, Brinza F and Tascu S 2012 Structural Investigations and Magnetic Properties of $BaFe_{12}O_{19}$ Crystals *Acta Physica Polonica A* **121** 95-97.

[3] Kanapitsas A, Tsonos C, Psarras G C and Kripotou S 2016 Barium ferrite/epoxy resin nanocomposite system : Fabrication, dielectric, magnetic and hydration studies *eXPRESS Polymer Letters* **10 3** 227-236.

[4] Gunanto Y E, Cahyadi L and Adi W A 2016 Composition and Phase Analysis of Nanocrystalline $Ba_xSr_{1-x}Fe_{12}O_{19}$ (x = 1.0; 0.6; and 0.4) by Using General Structure Analysis System *AIP Conference Proceedings* **1725** 020023.

[5] Gunanto Y E, Jobiliong E and Adi W A 2016 Microwave Absorbing Properties of $Ba_{0.6}Sr_{0.4}Fe_{12-z}Mn_zO_{19}$ (z = 0 – 3) Materials in X-Band Frequencies *J. Math. Fund. Sci.* **48 1** 55-65.

[6] Qin X, Cheng Y, Zhou K, Huang S and Hui X 2013 Microwave Absorbing Properties of W-Type Hexaferrite $Ba(MnZn)_xCo_{2(1-x)}Fe_{16}O_{27}$ *J. Materials Science and Chemical Engineering* **1** 8-13.

[7] Adi W A and Manaf A 2012 Structural and Absorption Characteristics of Mn-Ti Substituted Ba-Sr Hexaferrite Synthesized by Mechanical Alloying Route J. Basic. Appl. Sci. Res. 2 8 7826-7834.

[8] Toby B H 2001 EXPGUI a graphical user interface for GSAS *Journal of Applied Crystallography* **34** 210.

[9] Idris M S, Osman R AM 2013 Structure Refinement Strategy of Li-Based Complex Oxides Using GSAS-EXPGUI Software Package *Advanced Materials Research* **795** 479-482.

[10] Al-Saie A M, Al-Shater A, Arekat S, Jaffar A, Bououdina M 2013 Effect of annealing on the structure and magnetic properties of mechanically milled TiO_2-Fe_2O_3 mixture *Ceramics International* **39 4** 3803-3808.

Magnetically Separable and Reusable Fe_3O_4/ZnO/Nanographene Platelets Photocatalyst for the Removal of Dye

Ardiansyah Taufik[1,2], Iqriah K. Susanto[1,2], and Rosari Saleh[1,2,*]

[1]DepartemenFisika, Fakultas MIPA-Universitas Indonesia, 16424 Depok, Indonesia.
[2]Integrated Laboratory of Energy and Environment, Fakultas MIPA-UniversitasIndonesia, 16424 Depok, Indonesia.

[a]Corresponding's author: rosari.saleh@gmail.com ;rosari.saleh@ui.ac.id

Abstract. In this study, we report the sol-gel synthesis of nanocomposite Fe_3O_4/ZnO with nanographene platelets (NGP)and characterized by suitable techniques, including field emission scanning electron microscope (FESEM), X-ray diffraction (XRD), energy dispersive X-ray analysis (EDX), UV-visible diffuse reflectance spectroscopy and vibrating sample magnetometer. The as-synthesized Fe_3O_4/ZnO/NGP nanocomposites materials were used for photocatalyst under UV as well as visible light irradiation to remove methylene blue as a model of organic polutan. The XRD spectra reveal peaks that can be attributed to cubic spinel Fe_3O_4, hexagonal wurtziteZnO and nanographene platelets structure. The FESEM images show Fe_3O_4/ZnO uniformly decorating NGP sheets. EDX analysis signifies the formation ofFe_3O_4/ZnO/NGP. The photocatalytic activity was found to be ZnO loading dependent. Higher ZnO loading lead to higher rate of photocatalytic activity under UV light irradiation, while the opposite trend observed under visible light irradiation. The addition of NGP effectively enhanced the photocatalytic performance of Fe_3O_4/ZnO nanocomposites. Separation of Fe_3O_4/ZnO/NGP nanocomposites from the solution could be achieved by applying an external magnetic field, demonstrating the magnetic properties of the nanocomposites.

1. Introduction

Oxide semiconductor-based catalyst utilizing solar energy as energy source is a promising material for environmental purification of waste water and air [1-2]. To date, among various oxide semiconductors, most of investigations have focused on TiO_2 and ZnO. Between both metal oxides, it is proved that ZnO may exhibit better efficiency than TiO_2in photocatalytic degradation [3]. However, in practical application ZnOhas not yet fully been utilized due to their wide band gap, high recombination rate of electrons and holes and the low quantum yield of photocatalytic reactions in aqueous solutions [4]. To address this issue, several strategies and approaches such as nanocomposites construction and hetero-structure assembly have been studied [5].

The combining of graphene with ZnO nanoparticles to form nanocomposites has been regarded as an effective approach to promote the efficiency of photocatalysis [6]. It is believed that the two-dimensional layered structure of graphene can inhibit the recombination of photogenerated electron-hole pairs and can also significantly enhance the adsorption of organic pollutant [7-8]. In addition, incorporating Fe_3O_4nanoparticles with ZnOnanoparticles not only efficiently improve the separation of photogenerated electron and hole pairs but also enhances the absorption ability of ZnO nanoparticles in longer wavelength region [9]. Moreover, Fe_3O_4nanoparticles with separation function using via external magnetic field have been proved to effective in reused [10].Therefore, in the present study, nanocomposites comprising ZnO, $Fe_3O_4$and nanographene platelets have been synthesized using sol-gel techniques followed by hydrothermal. The effects of the molar ratios of ZnO to Fe_3O_4on the structural and magnetic properties, especially on the photocatalytic activity, are investigated systematically.

2. Experimental Method

All chemical reagents were of analytical grade and were used without further purification. Iron (II) sulfate heptahydrate ($FeSO_4$. $7H_2O$, 99%), zinc sulfate heptahydrate ($ZnSO_4$. $7H_2O$, 99%), sodium hydroxide (NaOH), ethanol and ethylene glycol (EG) were purchased from Merck. Nanographene platelets (N008-100) were purchased from Angstrom Materials. Fe_3O_4nanoparticles were synthesized using sol-gel method. In a typical procedure, 0.027 mole of FeSO4.7H2O was dissolved in 100 ml of deionized water to produce a solution that was then vigorously stirred at room temperature. The desired amount of CH_3COOH and 30 ml of ethylene glycol were added to the solution while stirring continuously until a pH value of 3 was obtained. Then, 0.054 mole of NaOH in 150 ml of deionized water was added dropwise to the mixture. The final solution was maintained at 80°C with stirring for 3h to form a gel. Afterward, the gel was dried at 80°C for 4 h before it was annealed at 125°C for 8h.

To synthesize Fe_3O_4/ ZnO nanocomposites, 0.0125 mole of $ZnSO_4.7H_2O$ was dissolved in 30 ml of distilled water with magnetic stirring. Then, 0.025 mole of NaOH in 65 ml of deionized water was added dropwise to the solution. This solution was designated as solution A, which was then stirred and heated at 80°C. The synthesized Fe_3O_4nanoparticles were dispersed in 30 ml of ethanol and were designated as solutions B. Subsequently, solutions B was added to solution A, and the mixtures were stirred continuously at 80°C for 2h. These mixtures were then centrifuged and washed several times with ethanol and distilled water to remove residual and unwanted impurities. Afterward, the final product was allowed to stand overnight at room temperature and was then heated at 100°C for 1h under vacuum. A series of Fe_3O_4/ZnO nanocomposites were prepared with molar ratios of 1:1:1 and 1:1:5. The Fe_3O_4/ZnO/NGP nanocomposites were prepared by a simple hydrothermal method based on Khalid's work [19]. Briefly, 200 mg of NGP was dissolved in a solution of water (80 mL) and ethanol (40 mL) through ultrasonic treatment for 2h, followed by the addition of 2g of Fe_3O_4/ZnO nanoparticles to the NGP solution and an additional 2h of stirring to achieve a homogeneous suspension. The suspension was then heated at 120°C for 3h to affect the deposition of Fe_3O_4/ZnO on the NGP sheets. The nanocomposite was isolated by centrifugation and dried at 70°C for 12h.

The samples were characterized by X-ray diffraction (XRD) measurements using a Rigaku Miniflex 600 and monochromatic Cu-Kα (λ = 1.54060 Å) radiation operated at 30 kV and 15 mA in the range of 10° to 80°.The instrumental broadening, including the instrumental symmetry, was calibrated using a Si powder standard. The X-ray diffraction patterns were analyzed by means of the MAUD program using Rietveld whole profile fitting to determine the crystal structure and lattice parameters. Elemental analyses and the morphology of the samples were studied with energy dispersive X-ray (EDX)spectroscopy and field emission scanning electron microscope (FESEM). The UV-Vis diffuse reflectance spectrum was measured using a Shimadzu spectrophotometer with an integrating sphere and a spectral reflectance standard over a wavelength range of 200-800 nm. Magnetic characteristics were measured by Oxford Type 1.2T vibrating sample magnetometer.

Methylene blue was selected as a model of organic pollutant. The photocatalytic activity was evaluated by degradation of organic pollutant in the presence of The Fe_3O_4/ZnO/NGP nanocomposites in aqueous solution under UV and visible light irradiation. The photocatalytic experiments were performed using the same method described in our previous work [11].The photocatalytic activities were then controlled, by adding different radical-scavengers into the solution of organic pollutant.

Ammonium oxalate, natrium sulfate, and tert-butyl alcohol were used as scavenger for photogenerated holes, photogenerated electrons and hydroxyl radical species, respectively. In the experiment the 10 mM of scavenger was added to the solution of organic pollutant and the same measurements as above are performed. To evaluate the photostability of the nanocomposites, a recycled photocatalytic activity measurement was performed. Typically, after the reaction of the first run under light irradiation, a permanent magnetic bar was used to separate the liquid and nanocomposites. The fresh organic pollutants, namely, methylene blue was mixed with the used nanocomposites to perform photocatalytic activity measurements for the second run. The experiment was repeated four times under identical conditions.

3. Results and Discussion

The magnetic properties of Fe_3O_4nanoparticles and Fe_3O_4/ZnO/NGP nanocomposites measured by VSM at room temperature are shown in Fig.1a. Saturation magnetization, remnant magnetization and coercivity values obtained from the hysteresis curves are given in Table.1. As can be seen from Fig.1a and Table. 1 all samples exhibited room temperature ferromagnetic. The decrease of saturation magnetization in order of Fe_3O_4, Fe_3O_4/ZnO =1:1 and Fe_3O_4/ZnO =1:5 is consistent with an increase of ZnO component in the nanocomposite samples, indicated that the strong ferromagnetic behavior of the nanocomposites depends on the content of the magnetic phase. It is reported that the magnetization of a ferromagnetic material was dependent on the strength of super exchange interaction between Fe^{3+} cation in octahedral sites [12]. ZnO being diamagnetic reduces the super exchange interaction among Fe^{3+} cation, therefore the saturation magnetization decreased. However, it was observed that the value of saturation magnetization increases as Fe_3O_4/ZnOis coupled with NGP.

Table 1. Magnetization value of Fe_3O_4 nanoparticles, Fe_3O_4/ZnO and Fe_3O_4/ZnO/NGP composites

Sample	Magnetic Saturation (emu/g)
Fe_3O_4	82
Fe_3O_4/ZnO	78
Fe_3O_4/5ZnO	29
Fe_3O_4/ZnO/NGP	76
Fe_3O_4/5ZnO/NGP	30

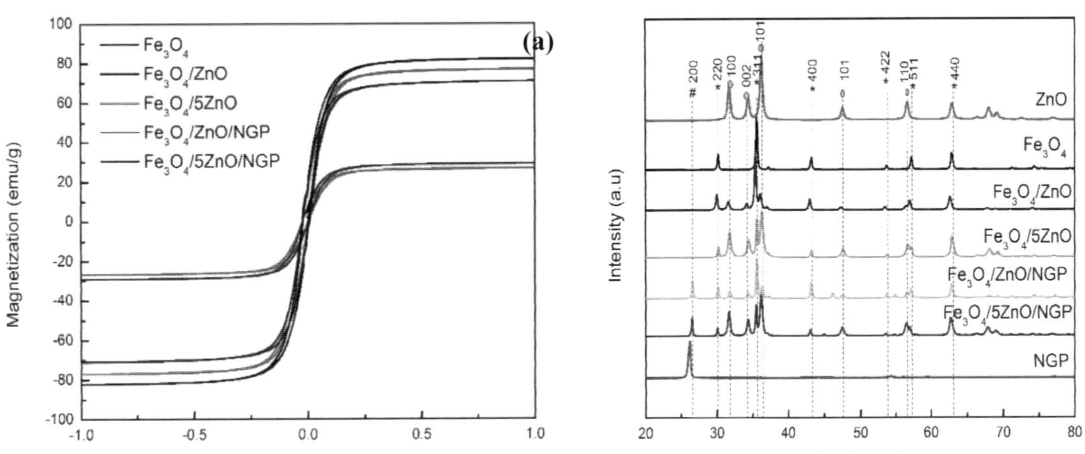

Figure 1. (a) VSM Spectra of Fe_3O_4 nanoparticles, Fe_3O_4/ZnO nanocomposites, and Fe_3O_4/ZnO/NGP composites with two different molar ratio, (b) XRD spectra of ZnO and Fe_3O_4 nanoparticles, Fe_3O_4/ZnO Nanocomposites, and Fe_3O_4/ZnO/NGP composites with two different molar ratio

Figure 1b illustrates the XRD patterns for Fe_3O_4/ZnO/NGP nanocomposite with a Fe_3O_4toZnOmolar ratio of 1:1 and 1:5. Also shown in the figure the XRD reflection intensityof pure Fe_3O_4, ZnO and NGP. The patterns of pure NGP, $Fe_3O_4$and ZnO nanoparticles were marked and all peaks can be indexed as NGP, the typical cubic spinel and hexagonal wurtzite, respectively. No other diffraction peaks have been detected in the pure $Fe_3O_4$and ZnO nanoparticles. In Fe_3O_4/ZnO/NGP nanocomposite, the characteristic 2Θ values of the XRD patterns of the Fe_3O_4/ZnO/NGP nanocomposites all have been identified as individual structure and phase of Fe_3O_4, ZnO and NGP with no other diffraction peaks being detected in the nanocomposite. Six characteristic Fe3O4peaks were observed at the 2Θ diffraction angles of 30.01°, 35.36°, 43.02°, 53.33°, 56.98° and 62.48° in the nanocomposite samples. These peaks correspond to the (220), (311), (400), (422), (511) and (440) planes in the cubic spinel structures. Besides those peaks, other peaks located at the 2Θ diffraction angles of 31.7°, 34.4°, 36.2°,47.5°, 56.5°,62.8°, 66.3°,67.9°and 69° were indexed to the (100), (002), (101),(102), (110), (103), (200), (112) and (201) diffraction plane indices of the ZnO structures with hexagonal wurtzite phase, while peaks located at the 2Θ diffraction angle of 26.4o has been attributed to the (200) diffraction planes in NGP structures. The diffraction analysis provides no evidence of the formation of mixed oxides in the sample. The values of crystallites'size calculated usingt he Scherrer formula are tabulated in Table 2.

Table 2. Grain Size, atomic percent and band gap energy of Fe_3O_4/ZnO and Fe_3O_4/ZnO/NGP composites

Sample	<D> (nm)		(at).%			
	ZnO	Fe_3O_4	Zn	Fe	O	C
ZnO	19	-	60	-	40	-
Fe_3O_4/ZnO	19	44	30	40	30	-
Fe_3O_4/5ZnO	17	38	55	14	32	-
Fe_3O_4/ZnO/NGP	16	35	24	23	39	14
Fe_3O_4/5ZnO/NGP	15	34	50	10	27	13

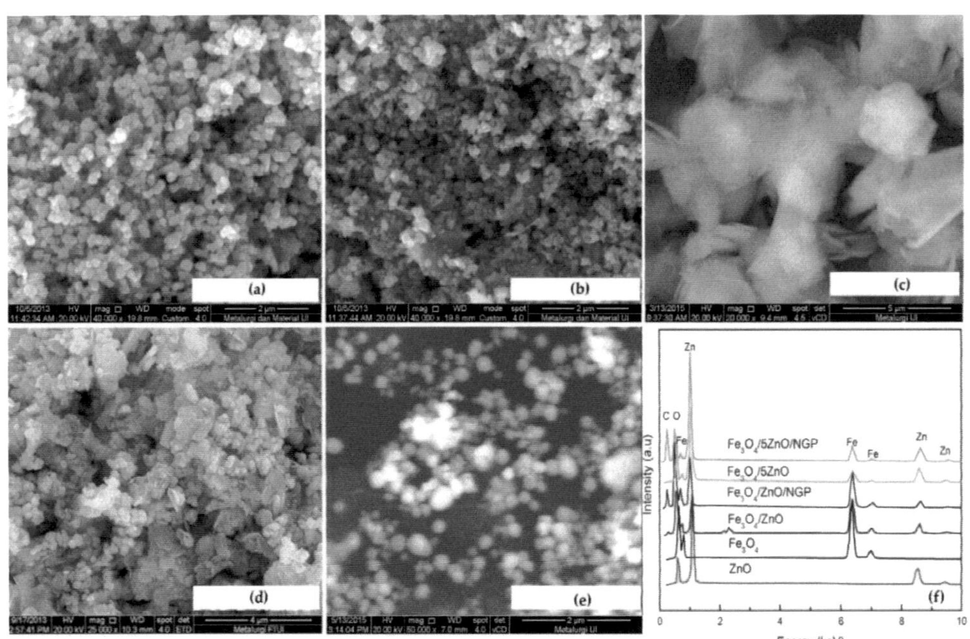

Figure 2. FESEM image of (a) Fe_3O_4 and (b) ZnO nanoparticles, (c) NGP materials, (d) Fe_3O_4/ZnO composites, (e) Fe_3O_4/ZnO/NGP composites, (f) EDX spectra of Fe_3O_4 and ZnO nanoparticles, Fe_3O_4/ZnO and Fe_3O_4

The surface morphology and elemental analysis of pure NGP, $Fe_3O_4$and ZnO nanoparticles as well as Fe_3O_4/ZnO and Fe_3O_4/ZnO/NGP nanocomposites has been studied by FESEM and EDX measurements and represented in figure 2. FESEM images revealed that pure $Fe_3O_4$and ZnO have sphere-like shapes. It is also observed that the sphere-like shape is retained in Fe_3O_4/ZnO nanocomposite. The EDX analysis shows that the sample is composed of Fe, Zn, O and C atoms. It is shown that as the ZnO loading increased the atomic ratio of Zn/Fe was also increased. It is found that the stoichiometry in the composite samples, in the final product is nearly the same as in starting. The compositions of these samples were displayed in Table 2.

The band gap energies can be calculated from the diffuse reflectance spectra by performing a Kubelka-Munk analysis [13]. The values of the band gaps of Fe_3O_4/ZnO and Fe_3O_4/5ZnOwere 2.91 eV and 3.31 respectively. These band gap were lower than ZnO nanoparticles band gap (3.37 eV). Itcan be due to the 4s orbital of Fe may be mixed with the 4s orbitalof Zn that results in formation of conduction band of ZnO at lowerenergy [14].

To evaluate the photocatalytic performance of Fe_3O_4/ZnO/NGP nanocomposite, the reduction of methylene blue in aqueous solution was investigated. figure 3 shows UV-vis spectra of methylene blue in the presence of Fe_3O_4/ZnO/NGP nanocomposite under UV (figure 3a) and visible light (figure 3b) irradiation. The intensity of the characteristic absorption peak aroun 635 nm gradually decreased with irradiation time and reaches maximum of 98 % and 76 % after 120 minutes, for UV and visible light irradiation, respectively. Moreover, the apparent rate constant of photocatalytic degradation of methylene blue in the presence of Fe_3O_4/ZnO/NGP was obtained from the slope of the plot between ln C0/Ct and irradiation time as shown in figure 4, by assuming that photocatalytic activity followed the first-order kinetic equation: ln (Ct/C0) = - kappx t, where Ct is the reaction concentration of methylene blue with irradiation time t and C0 is the initial concentration at t = 0. For comparison, the degradation of methylene blue in the presence of Fe_3O_4/ZnO was also plotted in the figure. The apparent rate constant, kapp is displayed in the Table.2. It is seen from figure 4 and table. 2, coupled NGP with Fe_3O_4/ZnO nanocomposites results greatly improvement in the photocatalytic performance.
The stability of Fe_3O_4/ZnO/NGP with respect to degradation of methylene blue as an organic pollutant model was examined in four consecutive runs. The results are shown in figure 5. After four cycles, photodegradation efficiencies of 85 % and 72 % were obtained for methylene blue under UV and visible light, respectively. The stability of Fe_3O_4/ZnO was also shown in the figure. Compared with Fe_3O_4/ZnO, it suggested that Fe_3O_4/ZnO/NGP appears to be more stable and can be reused without loss in photocatalytic activity.

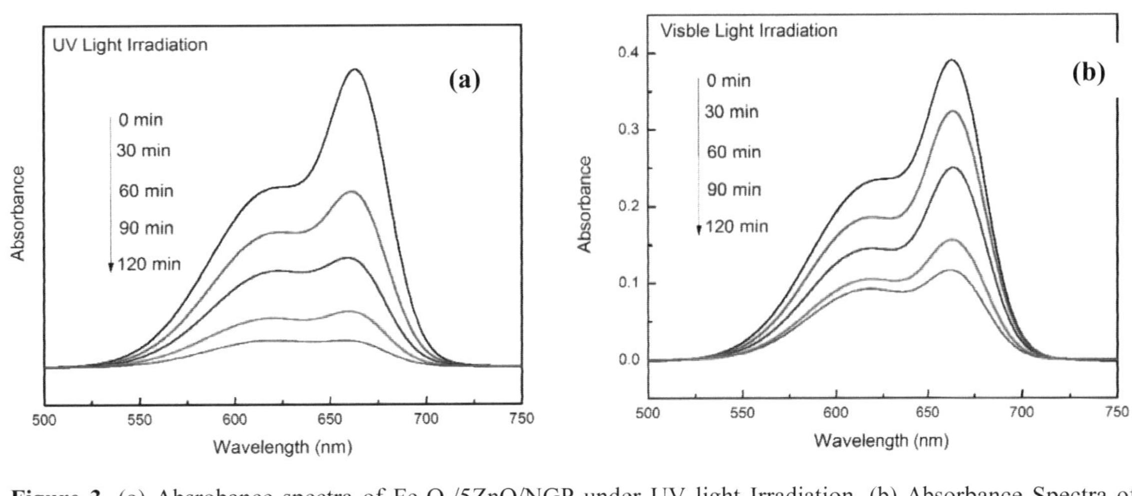

Figure 3. (a) Absrobance spectra of Fe_3O_4/5ZnO/NGP under UV light Irradiation, (b) Absorbance Spectra of Fe_3O_4/ZnO/NGP under Visible light irradiation

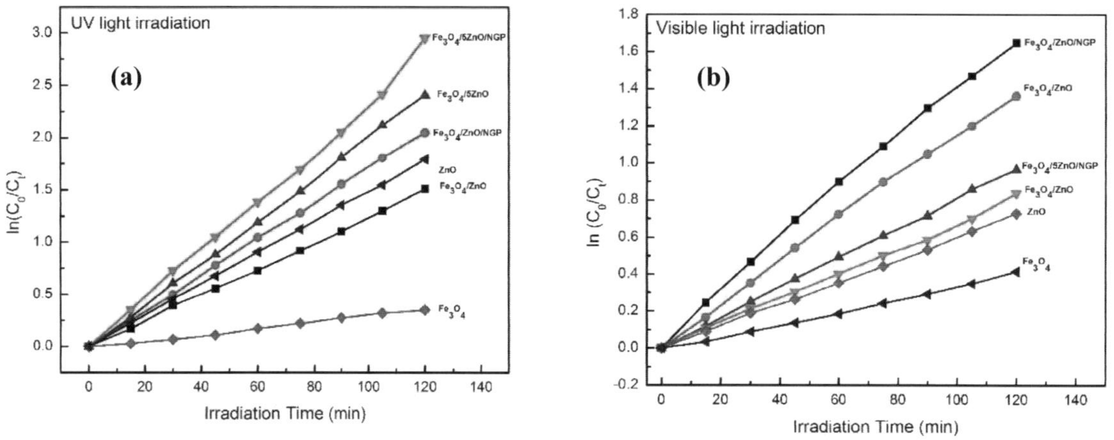

Figure 4. (a) Degradation rate of photocatalytic activity for all sample under (b) UV light irradiation and visible light irradiation.

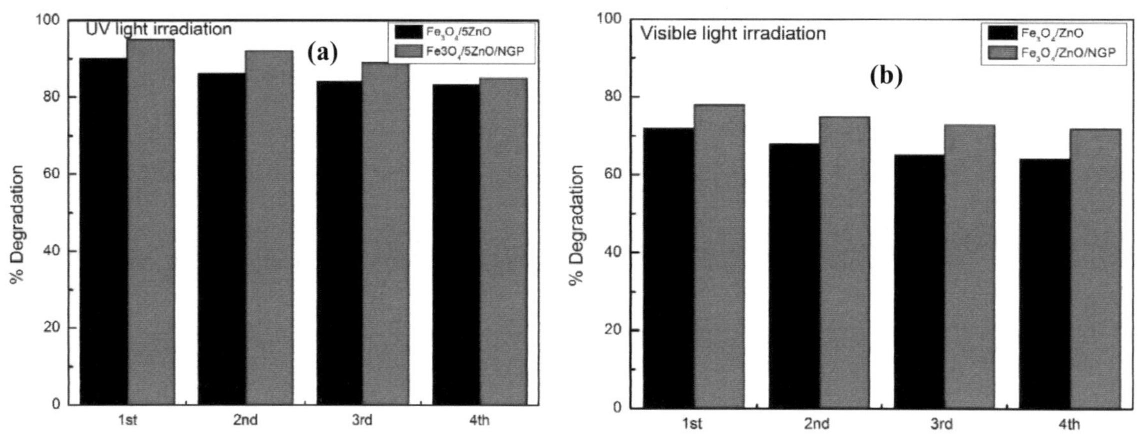

Figure 5. (a) Reusability of Fe_3O_4/5ZnO and Fe_3O_4/5ZnO/NGP under UV light irradiation, (b) Reusability of Fe_3O_4/ZnO and Fe_3O_4/ZnO/NGP under UV light irradiation

To detect the main active oxidative species responsible for the degradation of methylene blue, tert-butyl alcohol, ammonium oxalate and $Na_2S_2O_8$ scavenger for hydroxyl radicals, holes and electrons, respectively, was addedand displayed in figure 6.It can be seen that the addition of ammonium oxalate scavenger for hole inhibits the photocatalytic removal of methylene blue, suggested that the photocatalytic activities of methylene blue are driven by holes under UV and visible light irradiation.

According to the above results, the photocatalytic degradation mechanism is summarized in figure 5. The photogenerated electrons (e) and holes (h) were produced when Fe_3O_4/ZnO/NGP was illuminated. Narrow light-response range of UV can be improved by coupled ZnO nanoparticles with $Fe_3O_4$and adding NGP. From figure 5, it can be seen that the photo-excited electrons are transported to NGP instead of going back to the holes. NGP with delocalized π structure could efficiently transfer electrons, which can easily migrate for the inner region to the surface to take part in the surface reaction to form radical such as $\bullet O_2-$by reacting with O_2 adsorbed on the surface. Due to higher separation efficiency of electron and hole resulting in the increase of the number of holes participating in the photo oxidation process, and generating •OH radicals and degrade the target pollutant.

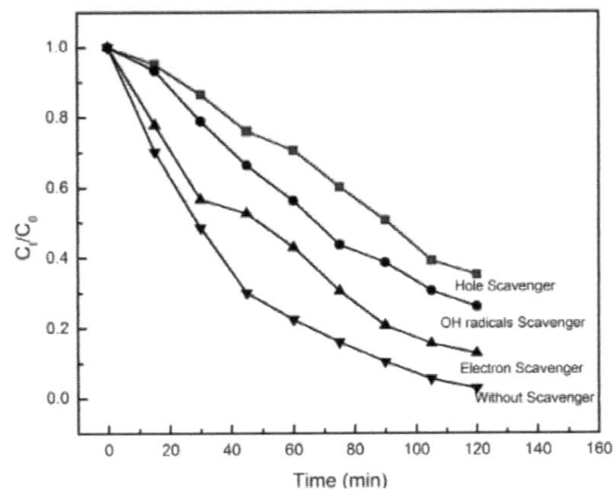

Figure 6. Effect of Different scavenger on photodegradation of methylene blue under UV light irradiation.

4. Conclusion

In summary, Fe_3O_4/ZnO/NGP photocatalysts were prepared by sol-gel followed by hydrothermal process. The obtained photocatalyst were characterized by XRD, EDX, FESEM, diffuse reflectance spectroscopy- and VSM method. The influence of NGP and ZnO loading were investigated by measuring the photocatalytic performance for degradation of methylene blue under UV and visible light irradiation. The Fe_3O_4/ZnO/NGP with a molar ratio of Fe_3O_4/ZnO is 1:1 exhibited enhanced photocatalytic activity for degradation of MB under visible light irradiation, while enhancement of photocatalytic activity under UV light irradiation was observed for photocatalyst with a molar ratio of 1:5. Compared with photocatalysts with the same molar ratio under identical condition, it can be concluded that incorporation of NGP resulted in an enhancement of photocatalytic efficiency. Moreover, the photocatalyst can be easily removed from the treated water and showed a better stability compared to the one synthesized without NGP.

References

[1] Qi K, Cheng B, Yu J, Ho W 2017 *Journal of Alloys and Compounds* **727** 792–820.
[2] Archana B, Manjunath K, Nagaraju G, Sekhar K B C, Kottam N 2017 *International Journal of Hydrogen Energy* **42** 5125–5131
[3] Kumaran N N, Muraleedharan K 2017 *Journal of Water Process Engineering* **17** 264–270
[4] Chen F, Tang Y, Liu C, Qian J, Wu Z, Chen Z 2017 *Ceramics International* **43** 14525–14528
[5] Samadi M, Zirak M, Naseri A, Khorashadizade E, Moshfegh A Z 2016 *Thin Solid Films* **605** 2–19.
[6] Cai R, Wu J, Sun L, Liu Y, Fang T, Zhu S, Li S, Wang Y, Guo L, Zhao C, Wei A 2016 *Materials and Design* **90** 839–844.
[7] Zhu L, Liu Z, Xia P, Li H, Xie Y 2018 *Ceramics International* **44** 849–856
[8] Zhao D, Gao X, Wu C, Xie R, Feng S, Chen C 2016 *Applied Surface Science* **384** 1–9
[9] Ghanbarnezhada S, Baghshahia S, Nematib A, Mahmoodi M 2017 *Materials Science in Semiconductor Processing* **72** 85–92
[10] Wang J, Yang J, Li X, Wang D, Wei B, Song H, Li X, Fu S 2016 *Physica E* **75** 66–71
[11] Taufik A, Kalim I, Saleh R 2015 *Materials Science Forum* **827** 37–42
[12] Xie T, Xu L, Liu C, Wang Y 2013 *Applied Surface Science* **273** 684-691
[13] Hapke B, Theory of Reflectance and Emittance Spectroscopy, University Press, Cambridge, 1993.
[14] Mirzaie R A, Kamrani F, Firooz A A, Khodadadi A A 2012 *Materials Chemistry and Physics* **133** 311

Decomposition of organic pollutant in waste water using magnetic catalyst nanocomposite

H Tju[1], A Taufik[1], M Fauzian[1]and R Saleh[1]

[1]Departemen Fisika, Fakultas MIPA-Universitas Indonesia, 16424 Depok, Indonesia.
[2]Integrated Laboratory of Energy and Environment, Fakultas MIPA-Universitas Indonesia, 16424 Depok, Indonesia.

[a]Corresponding's author: rosari.saleh@gmail.com; rosari.saleh@ui.ac.id

Abstract. In this study, magnetic recyclable catalyst Fe_3O_4/CuO/ZnO/nanographene platelets (Fe_3O4/CuO/ZnO/NGP) and Fe_3O_4/CuO/TiO_2/nanographene platelets (Fe_3O_4/CuO/TiO_2/NGP) composites were synthesized by simple hydrothermal method. Methylene blue was used as a model of textile dye to evaluate their catalytic activities. A range of analytical techniques including X-ray diffraction, energy-dispersive X-ray spectroscopy and vibrating sample magnetometer were employed to reveal the crystal structure, composition and property of the nanocomposites. The catalytic performance was evaluated by degradation of methylene blue in aqueous solution under UV light and ultrasonic irradiation simultaneously. X-ray diffraction results revealed that cubic spinel Fe_3O_4, monoclinic CuO, hexagonal wurtzite ZnO and graphene platelets exist in Fe_3O_4/CuO/ZnO/NGP, while in Fe_3O_4/CuO/TiO_2/NGP nanocomposites instead of hexagonal wurtzite ZnO, anatase TiO_2 is observed. These results confirmed that the nanocomposites were the desired materials. In addition, all samples exhibited ferromagnetic behavior at room temperature and could be rapidly separated from aqueous solution for repeated use under external magnetic field. From the degradation of methylene blue, it is found that the as-prepared nanocomposites exhibited excellent catalytic activity compared with nanocomposite synthesized without nanographene platelets. The nanocomposites still retain the 100% of the initial activity after it has been used four times repeatedly.

1. Introduction

Recently, great interest has been stimulated in the treatment of dye-contaminated household wastewater and industrial effluents through photocatalytic and sonocatalytic methods, since these methods can effectively and efficiently destruct the hazardous dyes completely [1-2]. The metal oxide semiconductors such as ZnO and TiO_2 are generally used as catalyst to remove organic pollutants due to their photosensitivity, strong oxidizing ability, non-toxic nature, favorable band gap energy and excellent chemical and mechanical stability [3-5]. In addition, ZnO and TiO_2 nanocatalyst are also environmental friendly and relatively inexpensive [6-7]. Unfortunately, the light response of ZnO and TiO_2 nanocatalyst are limited due to its large band-gap and high recombination of electron and hole is very fast [8-9]. Therefore, several methods have been employed to reduce recombination of charge carrier by coupling ZnO and TiO with other metal oxide to form heterogeneous semiconductor catalyst [10]. Apart from this strategy, recently carbon-based nanomaterials such as graphene has been investigated to be the effective material that can support coupled metal oxides to minimize charge carrier recombination [11]. It is reported that as a new kind of two-dimensional carbon materials constructed by a single layer of carbon atoms, graphene is an excellent supporting and electron-transport materials compared with other carbon materials in the process of photo and sonocatalytic [11-12].

In this study, we attempt to synthesize $Fe_3O_4/CuO/ZnO$ and $Fe_3O_4/CuO/TiO_2$ coupled with nanographene platelets (NGP) using simple sol-gel method [13-15] followed by hydrothermal method [16] in order to obtain synergetic effect towards the degradation of organic pollutant model, methylene blue. Such methods are inexpensive and and are among the versatile approaches that allow the synthesis of a great variety of crystalline metal oxides at low temperature. $Fe_3O_4/CuO/ZnO/NGP$ and $Fe_3O_4/CuO/TiO_2/NGP$ composites with different ZnO and TiO_2 loadings were used in catalytic reactions to investigate the role of graphene in the photocatalytic performance. The effect of ZnO and TiO_2 loading was also investigated.

2. Experiments

All chemical reagents were of analytical grade and were used without further purification. Iron (II) sulfate heptahydrate ($FeSO_4$. $7H_2O$, 99%), copper sulfate pentahydrate ($CuSO_4$. $5H_2O$, 99%), zinc sulfate heptahydrate ($ZnSO_4$. $7H_2O$, 99%) titanium dioxide (TiO_2, 99%), sodium hydroxide (NaOH), ethanol and ethylene glycol (EG) were purchased from Merck. Nanographene platelets (NGP-N006-P) were purchased from Angstron Materials. $Fe_3O_4/CuO/ZnO$ and $Fe_3O_4/CuO/TiO_2$ nanocomposites were synthesized using a sol-gel method described in our previous study [13-15]. A series of $Fe_3O_4/CuO/ZnO$ and $Fe_3O_4/CuO/TiO_2$ nanocomposites were prepared with molar ratios of 1:1:1, 1:3:1 and 1:1:5. The $Fe_3O_4/CuO/ZnO/NGP$ and $Fe_3O_4/CuO/TiO_2/NGP$ nanocomposites were prepared by a simple hydrothermal method based on Khalid's work [16]. Briefly, 200 mg of NGP was dissolved in a solution of water (80 mL) and ethanol (40 mL) through ultrasonic treatment for 2 h, followed by the addition of 2 g of $Fe_3O_4/CuO/ZnO$ to the NGP solution and an additional 2 h of stirring to achieve a homogeneous suspension. The suspension was then heated at 120 °C for 3 h to affect the deposition of $Fe_3O_4/CuO/ZnO$ on the graphene sheets. The resulting nanocomposite was isolated by centrifugation and dried at 70 °C for 12 h. The same procedure was also for $Fe_3O_4/CuO/$ TiO_2/NGP nanocomposites

The samples were characterized by X-ray diffraction (XRD) measurements using a Rigaku Minflex 600 and monochromatic Cu-Kα ($\lambda = 1.5$ Å) radiation operated at 30 kV and 15 mA in the range of 10° to 80°. The instrumental broadening, including the instrumental symmetry, was calibrated using a Si powder standard. Elemental analyses and the morphology of the samples were determined using energy dispersive X-ray (EDX) spectroscopy (LEO 420). Magnetic properties were investigated using vibrating sample magnetometry (Oxford Type 1.2T). UV-Vis diffuse reflectance spectra were measured using a Shimadzu spectrophotometer with an integrating sphere and a spectral reflectance standard over a wavelength range of 200 to 800 nm.

The experimental setup and conditions used for evaluation of sonocatalytic and photocatalytic activities were identicals. After dark absorption-desorption equilibrium, the solution in the presence of $Fe_3O_4/CuO/ZnO/NGP$ and $Fe_3O_4/CuO/$ TiO_2/NGP nanocomposites were irradiated using UV light and ultrasonic irradiation simultaneously using two 40 W UV lamps and ultrasonic bath operated at fixed frequency of 40 kHz and power of 150 W. In all experiments, 0.3 g/L $Fe_3O_4/CuO/ZnO/NGP$ and $Fe_3O_4/CuO/TiO_2/NGP$ nanocomposites were suspended in 100 mL methylene blue solution with a concentration of 20 mg/L. During the sonocatalytic as well as photocatalytic reaction, samples were collected for 2h at regular intervals. The solution was analyzed using a Dynamica UV-visible spectrophotometer with a quartz cuvette with an optical length of 10 mm.

3. Experiments

Figure 1 shows the XRD patterns of $Fe_3O_4/CuO/ZnO/NGP$ and $Fe_3O_4/CuO/TiO_2/NGP$ nanocomposites with different concentrations of ZnO and TiO_2. For comparison, XRD patterns of pure Fe_3O_4, CuO, ZnO and TiO_2 nanoparticles are also shown in the figure. As can be seen in figure1 all the reflection peaks can be readily indexed to a combination of reflection peaks that can be ascribed to cubic spinel Fe_3O_4, monoclinicCuO, hexagonal wurtzites ZnO and anatase TiO_2. The characteristic 2θ values in the XRD patterns of the Fe_3O_4/CuO/ZnO/NGP nanocomposites exhibited a series of characteristic diffraction peaks of Fe_3O_4, CuO, ZnO and NGP, while the XRD patterns of the Fe_3O_4/CuO/TiO_2/NGP nanocomposites revealed diffraction peaks of Fe_3O_4, CuO, TiO_2 and NGP.

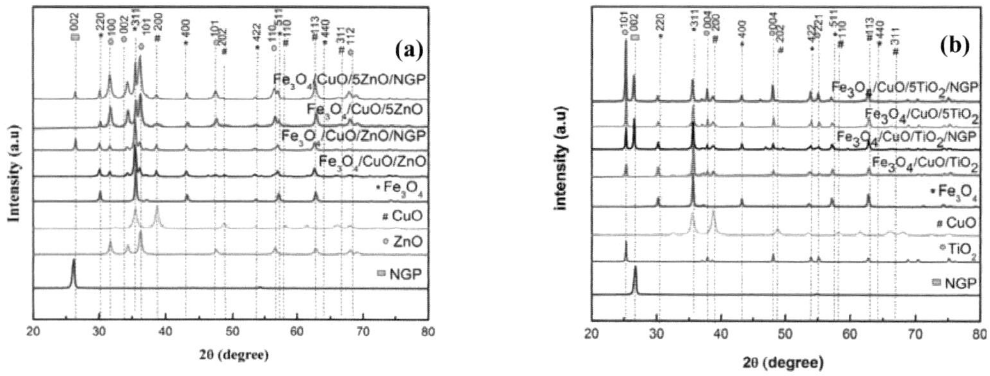

Figure 1. XRD Patterns of (a) Fe_3O_4/CuO/ZnO/NGP and (b) Fe_3O_4/CuO/TiO_2/NGP nanocomposites with different concentrations of ZnO and TiO_2.

Table 1. Lattice Parameter of Fe_3O_4, ZnO, CuO, and TiO_2 nanoparticles, Fe_3O_4/CuO/ZnO/NGP and Fe_3O_4/CuO/TiO_2/NGP nanocomposites

Samples	Lattice Parameter									Grain Size			
	Fe_3O_4	CuO			ZnO		TiO_2		NGP	Fe-$_3O_4$	CuO	ZnO	TiO_2
	a=b=c	a	b	c	a=b	c	a=b	c	a=b=c				
Fe_3O_4	8.364	-	-	-	-	-	-	-	3.610	45	-	-	-
CuO	-	4.691	3.425	5.138	-	-	-	-	-	-	15	-	-
ZnO	-	-	19	-	3.253	5.214	-	-	-	-	-	19	-
TiO_2	-	-	-	33	-	-	3.781	9.509		-	-	-	33
NGP	-	-	-	-	-	-	-	-	3.678				
Fe_3O_4/CuO/ZnO/NGP	8.374	4.7	3.432	5.136	3.242	5.214	-	-	3.614	37	15	20	-
Fe_3O_4/CuO/5ZnO/NGP	8.375	4.681	3.432	5.136	3.255	5.215	-	-	3.647	34	15	22	-
Fe_3O_4/CuO/TiO_2/NGP	8.369	4.848	3.394	5.019	-	-	3.784	9.513	3.614	34	15	-	52
Fe_3O_4/CuO/5TiO_2/NGP	8.379	4.832	3.391	5.083	-	-	3.786	9.518	3.656	30	15	-	78

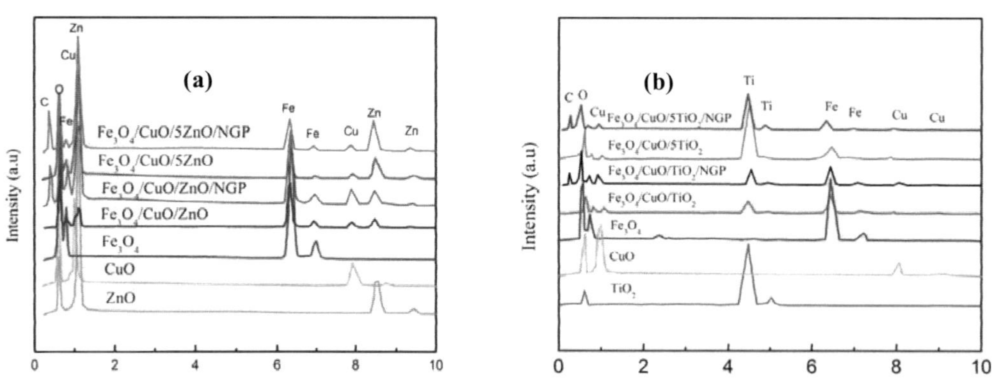

Figure 2. EDX Patterns of (a) Fe_3O_4/CuO/ZnO/NGP and (b) Fe_3O_4/CuO/TiO_2/NGP nanocomposites with different ZnO and TiO_2 loading.

The spectra confirmed that the synthesized nanocomposites were the desired materials, Fe_3O_4/CuO/ZnO/NGP and Fe_3O_4/CuO/TiO_2/NGP. The values of crystallite size calculated from Scherrer's formula and the values of the lattice parameters refined using the Rietveld refinement are tabulated in Table 1.

To confirm the presence of corresponding components in Fe_3O_4/CuO/ZnO/NGP and Fe_3O_4/CuO/TiO_2/NGP nanocomposites, EDX measurements were performed. Figure 2 shows the EDX spectra of Fe_3O_4/CuO/ZnO/NGP and Fe_3O_4/CuO/TiO_2/NGP nanocomposites with molar ratios of 1:1:1 and 1:1:5. The spectra of pure nanoparticles of Fe_3O_4, CuO, ZnO,TiO_2 and NGP are also shown in the figure. The EDX spectra indicate that Fe_3O_4/CuO/ZnO/NGP and Fe_3O_4/CuO/TiO_2/NGP nanocomposites were successfully formed.

Magnetic hysteresis curves of Fe_3O_4/CuO/ZnO/NGP and Fe_3O_4/CuO/TiO_2/NGP nanocomposites with different ZnO and TiO_2 loadings measured at room temperature are displayed in figure 3. All samples exhibited ferromagnetic behavior. For comparison the hysteresis curves of Fe_3O_4/CuO/ZnO, Fe_3O_4/CuO/TiO_2 and pure Fe_3O_4 nanoparticles were also plotted in the figure. It is exhibited that the value of saturation magnetization decrease as Fe_3O_4 is coupled with CuO and ZnO nanoparticles. However, it is seen that due to the incorporation of NGP, the Fe_3O_4/CuO/ZnO/NGP and Fe_3O_4/CuO/TiO_2/NGP nanocomposites show a relatively higher magnetization values compared with that of the Fe_3O_4/CuO/ZnO and Fe_3O_4/CuO/TiO_2 nanocomposites. According to the literature [17], this might be due to Fe_3O_4 nanoparticles attached to the surface of the NGP materials acting as defects and creating a stable NGP-based magnetic material [17].

Magnetic property is of great importance to the reusable property of catalyst. To study the magnetic separability of the material, a magnetic bar was placed near the sample holder supporting the Fe_3O_4/CuO/ZnO/NGP nanocomposites in water. As shown in the illustration of figure 3b, the Fe_3O_4/CuO/ZnO/NGP catalyst can be easily separated from the aqueous solution within a few minutes and supernatant was colorless. This experiment proved that Fe_3O_4/CuO/ZnO/NGP and Fe_3O_4/CuO/TiO_2/NGP possessed the required adsorption and magnetic properties and can be used also for magnetic adsorbent to remove organic pollutant in liquid phase.

The catalytic activities of Fe_3O_4/CuO/ZnO/NGP and Fe_3O_4/CuO/TiO_2/NGP nanocomposites were evaluated by degradation aqueous solution of MB under UV and ultrasonic irradiation simultaneously. The degradation of MB in aqueous solution in the presence of 20 mg of Fe_3O_4/CuO/ZnO/NGP and Fe_3O_4/CuO/TiO_2/NGP nanocomposites under both UV light and ultrasonic irradiation are presented in figure 4. Plotted is the time dependent UV-vis absorption spectra of MB in the presence of Fe_3O_4/CuO/ZnO/NGP (figure 4a) nanocomposites. It can be seen that after irradiation with both UV and ultrasonic for 120 min, the fraction of degradation for Fe_3O_4/CuO/ZnO/NGP nanocomposites is approximately 100%. The relative concentrations as a function of irradiation timeis shown in figure 4b and 4c. For detailed analysis the photocatalysis kinetics of the MB degradation, the pseudo-first model was applied. The rate constants were evaluated from the data and plotted in the figure 4d for Fe_3O_4/CuO/ZnO/NGP and Fe_3O_4/CuO/TiO_2/NGP nanocomposites. It can be seen from figure 4d, that an increase in ZnO and TiO_2 loadings results in a progressive increase in catalytic efficiency. Moreover, as shown in the figure, the incorporation of NGP was able to remove approximately 99% and 99% of MB within 90 and 75 min for Fe_3O_4/CuO/5ZnO/NGP and Fe_3O_4/CuO/5TiO_2/NGP nanocomposites, respectively. This result indicated that ternary nanocomposites prepared with incorporation of NGP showed excellent catalytic activity under both UV and ultrasonic irradiation, which may be due to improved charge transmission between each metal oxide and NGP thus prolonging the charge carrier life time and hindering charge carrier recombination [18].

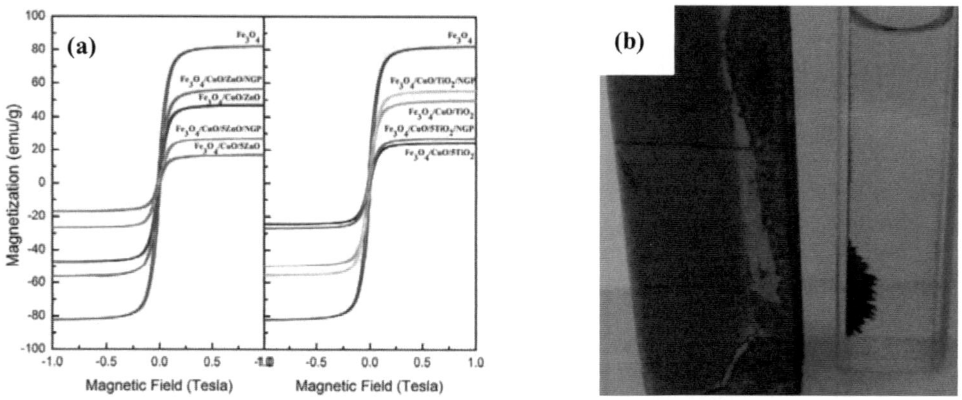

Figure 3. VSM Patterns of (a) Fe_3O_4/CuO/ZnO/NGP and Fe_3O_4/CuO/TiO_2/NGP nanocomposites with different concentrations of ZnO and TiO_2. (b) Magnetic separation process.

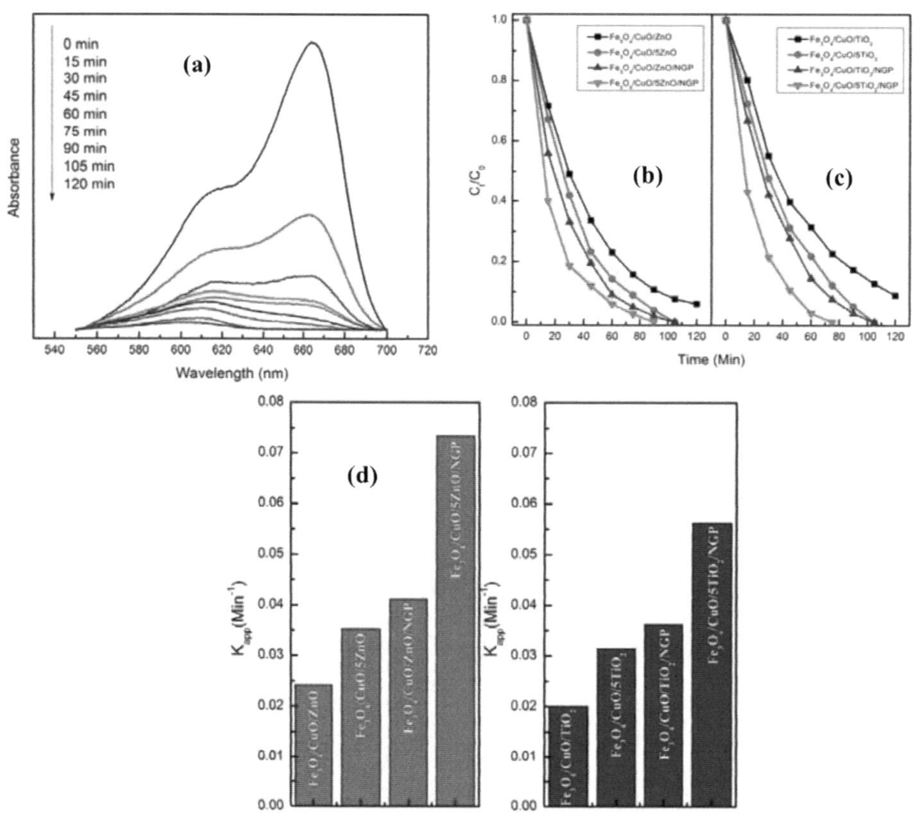

Fig. 4. Time dependent UV-vis absorption spectra of MB in the presence of (a) Fe_3O_4/CuO/ZnO/NGP and Fe_3O_4/CuO/TiO_2/NGP, (b) Photosonocatalytic activity of Fe_3O_4/CuO/ZnO/NGP and (c) Fe_3O_4/CuO/TiO_2/NGP nanocomposites, (d) the rate constants for Fe_3O_4/CuO/ZnO/NGP and Fe_3O_4/CuO/TiO_2/NGP nanocomposites.

In order to study the mechanism of MB removal in the photosonocatalytic activity, it is important to detect the main active oxidative species responsible in the catalytic activity. In this case, control experiments with added different type of scavengers such as tert-butyl alcohol, ammonium oxalate and $Na_2S_2O_8$ scavenger for hydroxyl radicals, holes and electrons, respectively, is deployed and the results are plotted in figure 5a and figure 5b. It can be seen that ammonium oxalate a large degree of inhibition of methylene blue, which confirmed the indispensable role of holes in the degradation process of MB.

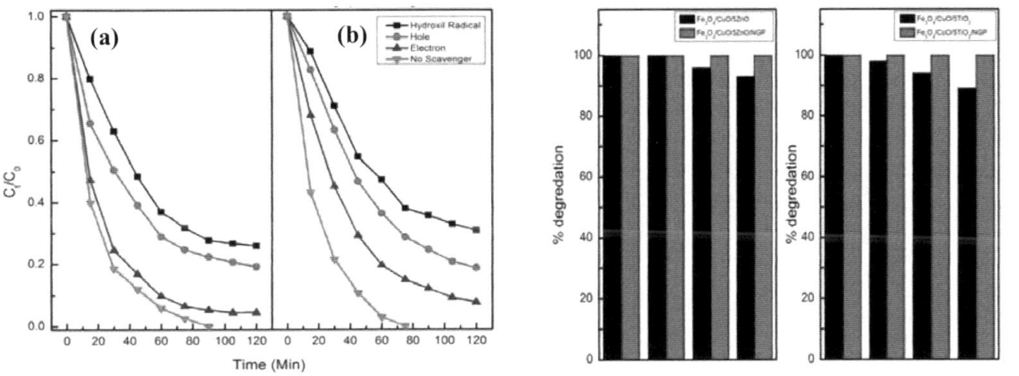

Figure 5. Effect of different scavenger on photosonocatalytic of (a) $Fe_3O_4/CuO/5ZnO/NGP$ and (b) $Fe_3O_4/CuO/5TiO_2/NGP$ nanocomposites and (c) reusability of prepared samples.

The mode of cycling used as well as maintaining high activity of catalyst is a critical issue in catalytic activity. Therefore it is necessary to investigate the reusability and the stability of the catalyst. $Fe_3O_4/CuO/ZnO/NGP$ and $Fe_3O_4/CuO/TiO_2/NGP$ nanocomposites in this study could be easily recovered from the solution using a permanent bar magnet as can be seen in figure 6a. To evaluate the stability of the catalyst under UV and ultrasonic irradiation simultaneously, the $Fe_3O_4/CuO/ZnO/NGP$ and $Fe_3O_4/CuO/TiO_2/NGP$ nanocomposites with a maximum loading of ZnO and TiO_2 were repeatedly used for four cycles. The results are shown in figure 6b. For comparison the stability of nanocomposite without the incorporation of NGP was also plotted. After four cycles, degradation efficiencies of 100% and 100% were exhibited for $Fe_3O_4/CuO/ZnO/NGP$ and $Fe_3O_4/CuO/ZnO$, respectively, while 100% and 100%, were obtained for $Fe_3O_4/CuO/TiO_2/NGP$ and $Fe_3O_4/CuO/TiO_2$ nanocomposites, respectively.

4. Conclusion

In this paper, $Fe_3O_4/CuO/ZnO/NGP$ and $Fe_3O_4/CuO/TiO_2/NGP$ nanocomposites were synthesized using a two step methods, sol-gel followed by hydrothermal. The catalytic properties of the synthesized samples were investigated under UV and ultrasonic irradiations simultaneously. The results showed that the incorporation of NGP in $Fe_3O_4/CuO/ZnO$ and $Fe_3O_4/CuO/TiO_2$ nanocomposite showed higher catalytic efficiency compared with the same catalyst synthesized without NGP. In addition, $Fe_3O_4/CuO/ZnO/NGP$ and $Fe_3O_4/CuO/TiO_2/NGP$ nanocomposites displayed an excellent stability during four reaction cycles, indicating that the catalysts have potential application in water purification.

References

[1] Kumar R., Kumar G., Akhtar M. S., Umar A. 2015 *J. of Alloys and Compounds* **629** 167-172
[2] Markovic D., Saponjic Z., Radoicic M., Radetic T., Vodnik V., Potkonjak B., Radetic M. 2015 *Ultrasonic Sonochemistry* **24** 221-229
[3] Chiang Y. J., Lin C. C. 2013 *Powder Technology* **246** 137-143
[4] Andreozzi R., Caprio V., Insola A., Moratta R. 1999 *Catalysis Today* **53** 51-59

[5] Cozzoli P.D., Fanizza E., Comparelli R., Curri M.L., Agostiano A., Laub D. 2004 *J. Phys. Chem.* B **108** 9623-9630
[6] Thompson T.L., Yates J.T. Jr. 2006 *Chem. Rev.* **106** 4428-4453
[7] Xu L.P., Hu Y.L., Pelligra C., Chen C.H., Jin L., Huang H., Sithambaram S., Aindow M., Joesten R., Suib S.L. 2009 *Chem. Mater.* **21** 2875-2885
[8] Xia J., Wang A., Liu X., Shu Z. 2011 *Applied Surface Science* **257** 9724-9732
[9] Ling Pang Y., Zuhairi Abdullah A. 2013 *Chemical Engineering* **214** 129-138
[10] Taufik A., Susanto I.K., Saleh R. 2015 *Materials Science Forum* **827** 37-42
[11] Susanto I.K., Taufik A., Saleh R. 2015 *Advance Materials Research* **1123** 227-232
[12] Jalaludin S. A, Arifin S. A, Saleh R. 2015 *Material Science Forum* **827** 13-18
[13] Li B., Wang Y. 2010 *Superlattices and Microstructures* **47** 615-623
[14] Fu D., Han G., Chang Y., Dong J. 2012 *Materials Chemistry and Physics* **132** 673-681
[15] Zhu L., Bok-Jo S., Ye S., Ullah K., Chun-Oh W. 2014 *Chinese Journal of Catalysis* **35** 1825-1832
[16] Khalid N.R., Ahmed E., Zhanglian Hong, Sana L., Ahmed M. 2013 *Current Applied Physics*, **13** 659-663
[17] Heidari E. K., Ataie A., Sohi M. H., Kim J. K., 2015 *J. of Magnetism and Magnetic Materials* **379** 95-101
[18] Tan L. L., Ong W. J., Chai S. P., Muhamed A. R. 2015 *Applied Catalysis B: Environmental* **166-167** 251-259

IOP Conf. Series: Journal of Physics: Conf. Series **1091** (2018) 012012 doi:10.1088/1742-6596/1091/1/012012

Baking process of mineral carrier of rare earth metals with sodium hydroxide using muffle furnace

M Setyadji[1] and H F Nurly[2]

[1]Center for Accelerator Science and Technology-BATANBabarsari Street Pos Office Box 6101 ykbb,Yogyakarta 55281, Indonesia
[2]Department Metallurgy Engineering, Sultan Ageng Tirtayasa University Jendral Soedirman Street Km03, Cilegon 42435, Indonesia

E-mail : mstyadji@batan.go.id

Abstract. Rare Earth Elements (REE) is one of the natural resources that are very strategic and high economic value. REE applications in industry are continues to grow such as LCD (liquid crystal display), LED (light emitting diode), magnet and hybrid batteries (NiMH batteries; Misch metal; Ce, La, Nd). This resulted in REE demand continues to increase up to 10%. Increased demand is not matched by the number of Rare Earth production in the world. This happens because the Chinese as the main actor in the producer REE world production (95%) restricting the export of Rare Earth up to 35% so that the world price of LTJ bounced up. Research & development processing, and purification REE continue to obtain high purity Rare Earth Metals. Baking process the mineral carrier of REE (raw material minerals) with solid NaOH using a muffle furnace has been carried out. The research aims to determine the effect of temperature and grain size of the monazite sand on the decomposition reaction monazite sand and reaction kinetics. The raw materials used are monazite sands containing rare earth oxides as much as 55.1% and 11.01% phosphate. At monazite sand baking using solid NaOH will occur monazite sand decomposition process and the release of the existing phosphate. The results showed that the optimal conditions baking process at a temperature of 400°C and 200 mesh particle size obtained 91.40% phosphate inseparable.

1. Introduction

Rare earth metals are a group of 15 lanthanide metals including yttrium and scandium. Rare earth metals have unique characteristic in particular spectroscopic and magnetic properties [1]. Rare earth metals have very important application in industry such as permanent magnetic materials, electronics, super-conductors, medical and nuclear technology. Rare earth metals have some unique properties of saturated 4f electronic structure that allow them to be used in luminescence-, magnetic- and electronic devices [1,2,3]. Geologically speaking, Rare Earth Minerals are not found on the earth as a free element, but as alloys in the form of complex compounds. Rare Earth Minerals are grouped into mineral carbonates, phosphates, oxides, silicates and fluorides [4-6]. One of the Rare Earth Minerals that has been extracted or mined in Indonesia in the form of phosphates are the monazite sand ((Ce, La, Nd, Pr, Sm) PO_4) and the xenotim sand ((Y, Dy, Gd, Sm) PO_4) [7]. Indonesia has an abundance of raw materials of local natural resources that have the potential to be processed into useful products for human need which has high economic value.

According to the results of the review in geology in 2013, the potential reserve of rare earth metal based monazite sand mineral commodity is relatively abundant on the islands of Bangka and Belitung, and is quantitatively estimated to be 1,564,707,280 tonnes (inferred resources) and 3,100,000 tonnes (probable reserves). With this vast reserve and also in reference to the conditions, rules and regulations specified in the Department of Energy and Mineral Resources Ministerial decree (ESDM Minister No. 7 of 2012) regarding the required specification that the final products of the metallurgical process must achieve high purity limits; and also in accordance with the specifications outlined in the Indonesian Mining Regulation (UU. MINERBA) No. 4 of 2009 on the ban on raw mineral exports, is expected to be the solution to encourage the domestic processing of rare earth minerals. In order to increase the added value of natural resources, the Government encourages the development of the domestic processing industry. The uses of rare earth metal in terms of their economic values are shown in fig. 1.

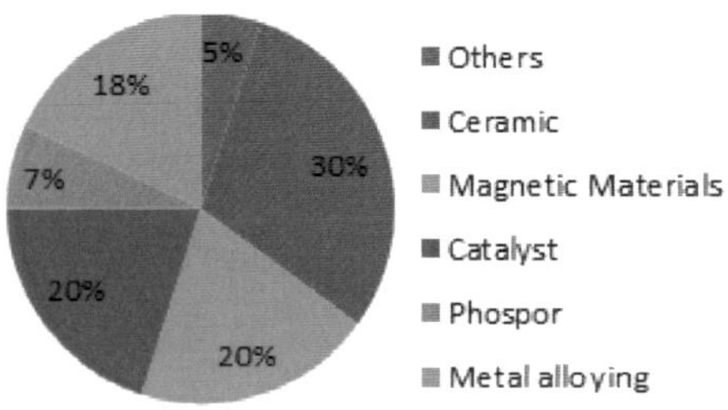

Figure 1. Applications of rare earth metals [8].

Phosphates contained in the rare earth metal minerals are difficult to separate, and this condition has caused some constraints in the purification process of rare earth metals [5,9]. According to the literature of Korea monazite sand baking, processing and refining of rare earth metals in the beginning of the separation of phosphate from minerals by baking process [10].

2. Materials and Methods

2.1. Materials

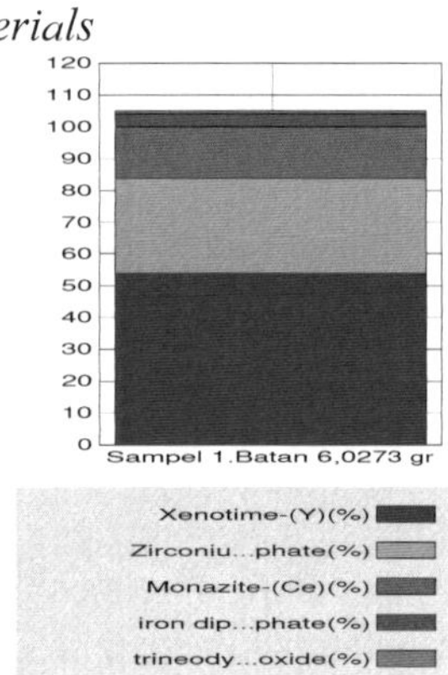

Figure 2. Qualitativeanalysisofrawmaterials mineral.

In this work, the raw sand material obtained as the side products or derivatives of Bangka-Belitung Island tin industrial processing is used as raw material. Minerals contained in the raw material are shown in figure 2. Results of qualitative analysis using X-Ray Diffraction method, indicates that the raw material is dominated by the xenotime mineral. The xenotime mineral fractional content is at least 55% in a 6.0273 gram sample, while the monazite mineral fractional content is 30%. In this research, there are 4 particle size variations. The first variation is -50 100 mesh which is a direct result of the classification of mineral raw materials. The second variation of the particle size of -100+120, the third of -120+170 and the fourth of -200 mesh.

IOP Conf. Series: Journal of Physics: Conf. Series **1091** (2018) 012012 doi:10.1088/1742-6596/1091/1/012012

2.2. Methods

The decomposition of phosphate in Korean monazite using NaOH have been carried out by other authors and it is a well proven method and has been used as a reference in comparative literature [8]. Under these same conditions, for the Indonesian raw mineral materials, the authors have examined the effects of temperature and have conducted classification of particle size of the mineral towards phosphate decomposition process using NaOH. The baking process of the raw sand materials was performed at different temperatures (350°C, 400°C, 450°C and 500°C) using a muffle furnace for 120 minutes. After the baking process for a certain period of time, the samples were taken out from the furnace. The mass of sample resulting from the baking process was then leached in hot distilled water (80°C) [10-14] using a magnetic stirrer of 300 rpm (constant agitation) for 60 minutes to dissolve the phosphate ions in the solution. The filtered slurry was dried at 1000°C in order to yield rare earth hydroxide. The experimental method is shown in figure 3.

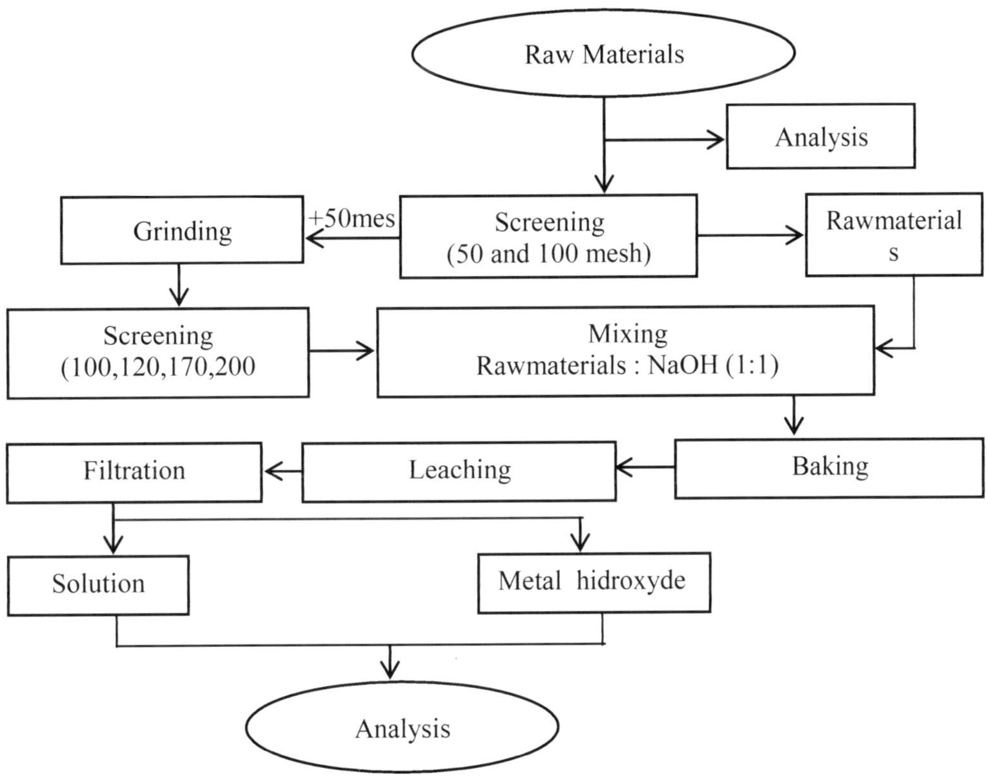

Figure 3. Experimental methods design.

3. Results and Discussion

The main purpose of the baking process is phospate separation from mineral by decomposition [10]. The effect of particle size and baking temperature on baking process using NaOH was studied. The reaction of mineral decomposition as follow:[1,15]

$$REPO_4 + 3NaOH \rightarrow RE(OH)_3 + Na_3PO_4$$

Where RE : Rare Earth

Decomposition process of phosphate from mineral is based on the result of baking Korean monazite. Table 1 show the composition of oxide in the raw material used in the experiment.

Table 1. Raw materials minerals oxide content.

Compound	(%)
P_2O_5	11.01
La_2O_3	6.51
CeO_2	16.11
Nd_2O_3	6.51
Y_2O_3	13.44
Pr_6O_{11}	1.18
Er_2O_3	1.12
Dy_2O_3	1.67
Yb_2O_3	1.43
Gd_2O_3	1.04
ThO_2	4.36
U_3O_8	0.32

3.1. Phospate Decomposition

The washing process of materials produced from baking was carried out using hot distilled water t 80°C. It is based on $NaPO_4$ solubility in H_2O. The concentration of phosphate produced from baking process (decomposition) was measured using high presure liquid chromatography (HPLC). Figure 4 (a) shows the percentage of phosphate produced from baking process with a 200 mesh particle size and a temperature variation. Figure 4 (a) shows the percentage of phosphate decomposition resulting from the baking process of raw material containing 11.01% of P_2O_5. Figure 4 shows that the optimum condition of baking process for the 200 mesh-particle size classification is to be found at the 400°C baking temperature. The values of the decomposing fraction of phosphate (percentage value) and the energy consumption are the basis for selecting the optimum temperature. At a temperature of 500°C decomposition of phosphate reached 92.96%, meanwhile, at a temperature of 400°C decomposition of phosphate reached of 91.40%. The difference of decomposition fraction between the temperature of 400°C and 500°C is only 1.56%, while the energy requirements to reach a temperature of 500°C greater than the temperature of 400°C. In Figure 4it is shown that at 450°C there is an anomaly, in which there is a decrease in the fraction of decomposed phosphate while theoretically the decomposition fraction should increase, as it is proportional to the increasing temperature. Figure 4 (b) shows effect of particle size was with respect to the generated decomposition fraction. The smaller particle size is related to lower fraction of phosphate decomposition. In figure 4 (b), it is shown that maximum decomposition fraction from the 200 mesh particle size is 91.40%. While the lowest phosphate decomposition fraction with the biggest particle size (-50+ 100mesh) is 74.10%. This occurs because one of the factors that support the reaction rate is the cross-sectional area. Therefore, the temperature of 400°C and the 200 mesh-particle size are set to be the optimum decomposition parameters for phosphates. The qualitative manner with which one could prove that the phosphate has completely decomposed is through the results of an x-ray diffraction analysis, as is shown in figure 5. In figure 5 it can be seen that the quantitative results of XRD seems to demonstrate that phosphate (P_2O_5) has been separated. Comparing the two figures, Figure 6 (a) and (b), there is the undetectable phosphate compound bonding with rare earth elements, and the compounds detected based on the results of the analysis are in fact rare earth oxides.

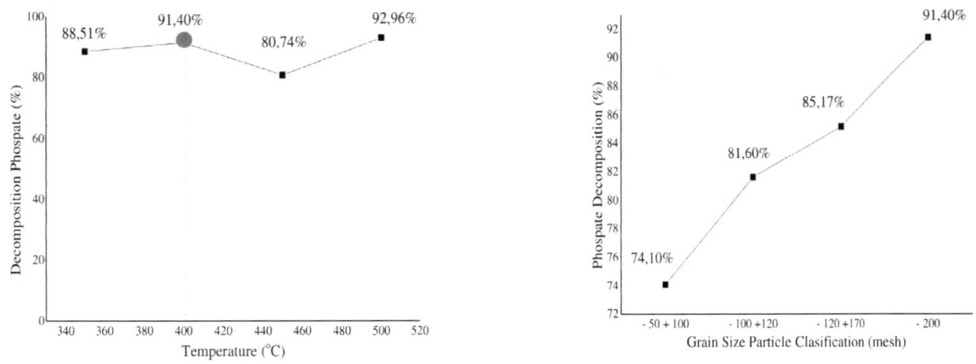

Figure 4.(a) Effect of temperature on the phosphate decomposes percent against particle size classification 200 mesh (b) Effect of particle size classification against decomposed phosphate at 400°C

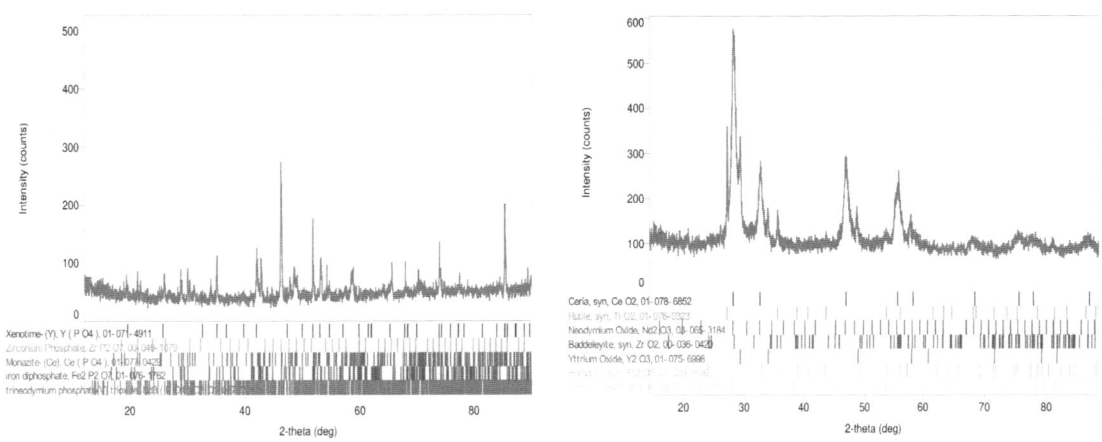

(a) (b)

Figure 5. X-Ray diffractionresults (a) raw materials mineral (b) metal hydroxyde (cake filtration from baking process leaching product)

3.2. Thermodynamic and Kinetic Analysis

Thermodynamic data or parameters such as Gibbs energy, enthalpy and entropy, in relation to the lanthanide (REM) formation reaction process and which is illustrated through the Ellingham diagram is not available at this time. Because Gibbs energy plays a predominant role in the determination of the metallurgical process reaction with the study of thermodynamic aspects, and therefore requiring a more in-depth study to reveal the thermodynamics aspects involved in the baking process of the monazite sand decomposition. Study reinforces the theory of thermodynamics aspects according the reference journals as well as several other clams journals. Based on the reference journal lanthanide hydroxide obtained by the reaction formation as follows:

$$\text{Lantanida (OH)}^{2+} + 2\text{OH-} \longrightarrow \text{Lantanida (OH)}_3 \text{ [11]}$$

On the literature gibbs acquired energy data is unavailable for the formation of all the elements contained in monazite sand or phospate minerals. Lanthanide elements are accumulated based on research data of the literature is only La, Ce, Pr, Nd, Sm, Eu, Gd, Tb, Dy, Ho, Er, Tm, Yb, Lu, and Y while the elements of U and Th energy data gibbs formation of hydroxides not available. Assuming the calculations based on data obtained from the literature and that the raw materials minerals are assumed to contain lanthanide series elements, the equation corresponding to the Gibbs energy in the lanthanide hydroxide formation implies that the reaction can take place spontaneously. Lanthanide element contained in the mineral monazite is the main element in the mineral, and there are 18 elements altogether in the sample, however in the available literature mentions only 15 elements. Based on the similarity of the nature of the lanthanide elements obtained through the calculation of dominant elements Ce in raw materials mineral (according to the composition analysis results) includes the following [11-13]:

$Ce(OH)^{2+}$ + OH-→ $Ce(OH)_3$ ΔG_o^{298} = -1296.91 kJ/mol

3NaOH(s)→ Na_3+ + OH^{3-} ΔG_o^{298}= 379,4 x 3 = +1138.2 kJ/mol

Na^{3+} + PO^{4-}→ Na_3PO_4(s) ΔG_o^{298}= -1819 kJ/mol

Ce + PO^{4-} + 3NaOH(s) → $Ce(OH)_3$ + Na_3PO_4(s) ΔG_o^{298} = -1977.71 kJ/mol

Based on the calculated Ce formation data ΔG_o^{298}, the formation of lanthanides with NaOH and $NaPO_4$ reactions with NaOH mineral, as the raw starting material, the results thermodynamic analysis shows that this is a spontaneous reaction ($\Delta G_o^{298}<0$). From these calculations are not done thoroughly formation of lanthanide compounds due to the similarity of the nature of the lanthanide spontaneous reactions can already be assumed that $\Delta G_o^{298}<0$ on the baking process raw material mineral decomposition using NaOH.

Kinetics metallurgy is one of the factors the length of time or the speed of reaction of a reaction mechanism. Bases the determination of the reaction rate calculation is based on the data percent decomposed. Mechanical rate control equation be an expressed as follows [14-17]:

$$1-\frac{2}{3}X-(1-X)^{\frac{2}{3}}=\frac{2.m.D.C}{\beta.r_o^2.p}xt \quad (1)$$

$$1-(1-X)^{1/3}=\frac{k.C}{r_o.\rho}.t \quad (2)$$

To predict the baking reaction mechanisms controlling the rate of decomposition, percent decomposition substituted at 1 equations and 2. The results obtained from the equation and the data obtained is then plotted a graph. Data from the equation results have been obtained and classified in table 2.

Table 2. Calculation Data Results Control Rate Reaction Based Phosphate Decomposition.

Description	Rate Control	
Clasiffication Size Particle (mesh)	Chemical Reaction $1-(1-X)^{\frac{1}{3}}$	Difussion Control $1-\frac{2}{3}X-(1-X)^{\frac{2}{3}}$
-50 +100	1.819	0.819
-100+120	1.873	0.873
-120+170	1.898	0.899
-200	1.942	0.942

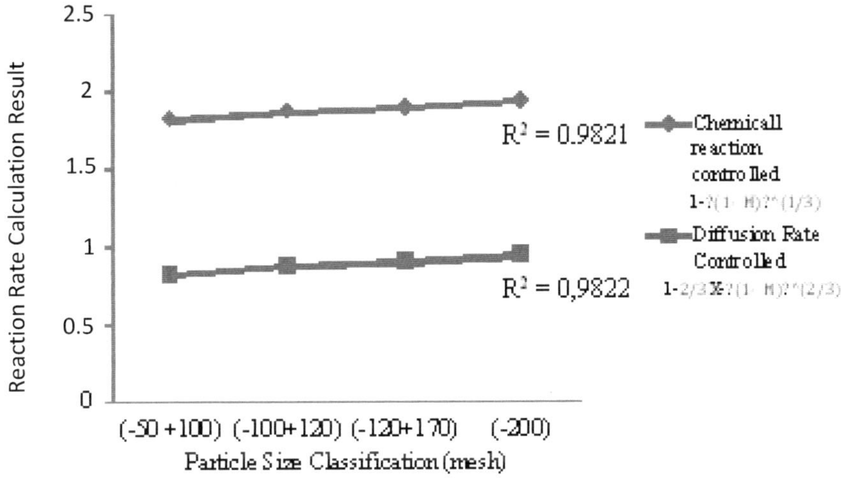

Figure 6. Rate reaction control based on data classification against phosphate decomposed on particle size classfication

Based on figure 6 can be seen in the graph of temperature variations influence on the classification particle size -200 # based rate equation control chemical reactions or diffusion mechanism. The data in figure 6 has show a valueof R^2 = 0.9822 and R^2 = 0.9821, if seen from the amount of R^2, the biggest determining the nominal control of there action is diffusion, although the amount of R^2 not much different. While the temperature variation or the rate of temperature change is controlled by diffusion and chemical reactions or can be assumed intermediates because of the similarity of regression coefficients.

4. Conclusions

The process of baking phospate decomposition using solid NaOH is the stage of pre-treatment in mineral processing for the purpose of decomposing phosphate contained. On the baking process percent decompose phosphate contained optimum at the particle size classification -200 mesh and temperature of 400°C which is 91.40%. The particle size is very influential visible from experimental data on the classification of the biggest particle size -50+100 mesh phosphate decomposes 74.10% at a temperature of 400°C. Temperature also plays an important role in the decomposition of phosphate, the higher the temperature the greater the percentage of phosphate decomposes at temperatures of 500°C seen in the classification of particle size 200 mesh decompose phosphate highest 92.96%. The mechanism process of baking decomposition reaction is through determination of thermodynamic reaction obtained $\Delta G_o^{298}<0$, so that the reaction takes place spontaneously. Factors supporting the kinetics of the process after a reaction known as spontaneous, gained control in the reaction intermediates based on differences in R2 that is not too high values.

Acknowledgements

We would like to thank to Tri Purwanti for their hard working on the experimental research and to Center for Accelerator Science and Technology BATAN for financial support.

References

[1]. Archana K, Rekha P, Manis K J, Jin Y L, Kumar J R, Kumar V 2014 Thermal treatment for the separation of phosphate and recovery of rare earth metals (REMs) from Korean monazite Journal of Industrial and Engineering Chemistry Published by Elsevier BV on behalf of The Korean Society of Industrial and Engineering Chemistry
[2]. Habashi F 2013 Extractive Metallurgy of Rare Earth The Canadian Journal of Metallurgy and Materials Science **52 3**

[3]. Gupta C K and Krishnamurthy N 2005 Extractive metallurgy of rare earths CRC Press ISBN 0-415-33340-7
[4]. Gupta A and Yan S D 2006 Mineral Processing Design And Operation Perth Australiae Book ISBN9780080454610
[5]. Haxel G, Hedrick J and Orris J 2012 Rare Earth Elements Critical Resources for High Technology Reston (VA): United States Geological Survei USGS Fact Sheet 087-02
[6]. Lucas J 2015 Rare Earths Purification, Separation, Precipitation and Calcination, Science, Technology Production and Use Pages 69–91
[7]. Chrikst D E, Lobacheva O L and Berlinskii V 2010 *Russian Journal of Physical Chemistry A* **84** 12 2047-2050
[8]. Pecht, Michael G, Robert K E, Song Xin 2012 Rare Earth Materials Insight and Concerns Center for Energetic Concepts Development CALCE Espc Press University of Maryland College Park Maryland USA
[9]. Rekha P and Archana K 2014 J. of Industrial and Engineering Chemistry **2** 2035-2042
[10]. Kumari A, Ret P 2014 Thermal treatment for the separation of phosphate and recovery rare earth metals (REMs) from Korean MonaziteJournal of Industrial and Engineering Chemistry XXX
[11]. Chrikst D E, Lobacheva O L and Berlinskii V 2010 Gibbs Energies of Formation of Hydorxides of Lanthanides and Yttrium ISSN 00360244 Russian Journal of Physical Chemistry A **84 12** 2047–2050
[12]. Husein H B, Mulyasih Y and Anggraeni A 2011 Extraction and chromatographic studies on rare-earth elements (REEs) from their minerals: the prospect of REEs production in Indonesia Proceedings of the 2nd International Seminar on Chemistry 2011 421-430
[13]. Onyedika G, Christopher O N, Ayebatonworio O and Martin O 2013 Comparative Kinetics of Iron Ore Dissolution in Aqueous HCl-HNO3 System Journal of Minerals and Materials Characterization and Engineering **1** 153-159
[14]. Renata D, Abreu, Morais C A 2010 Purification of Rare Earth Elements from Monazite Sulphuric Acid Leach Liquor and The Production of High-Purity Ceric Oxide Minerals Engineering **23** 538 – 540.
[15]. Habashi F 2013 Extractive metallurgy of rare earths Canadian Metallurgical Quarterly 2013 **52 3**.
[16]. Onyedika G O, Christopher N, Ayebatonworio O and Martin O 2013 Comparative Kinetics of Iron Ore Dissolution in Aqueous HCl-HNO3 System Journal of Minerals and Materials Characterization and Engineering 1153-159
[17]. Gharabaghia M, Irannajadband M, Azadmehr A R 2013 Leaching kinetics of nickel extraction from hazardous waste by sulphuric acid and optimization dissolution conditions, chemical engineering research and design **91** 325–331.

IOP Conf. Series: Journal of Physics: Conf. Series **1091** (2018) 012013 doi:10.1088/1742-6596/1091/1/012013

Analysis of structural and functional groups on the magnetic composite rubber – Lanthanum Manganite modification

M Anggaravidya[1,a], Sudirman[2,b], and WA Adi[2,c]

[1]Centre for Industrial Assessment Process – Agency for the Assessment and Application of Technology, Puspiptek Area Serpong, Tangerang Selatan, Banten, Indonesia

[2]Centre for Science and Technology of Advanced Materials – National Nuclear Energy Agency, Kawasan Puspiptek Area Serpong, Tangerang Selatan, Banten, Indonesia

Corresponding author: mahendra_anggaravidya@bppt.go.id

Abstract. The synthesis and characterization of composite rubber – lanthanum manganite modification have been performed. The composite was prepared by rubber elastomer thermoplastics, stearic acid, ZnO, n-cyclohexyl-2-benzothiazyl sulphenamide (CBS), sulfur, and magnetic powder of $La_{0.8}Ba_{0.2}(Fe, Mn, Ti)O_3$. The raw material was mixed by open two-roll mixing for 32 min and then pressed at temperature of 140 °C for 30 min. The element analysis showed that the sample contained carbon, oxygen, sulfur, zinc, lathanum, barium, manganese, titanium and iron. The refinement results of x-ray diffraction pattern show that the sample issemicrystallinewith the crystallinity of 46%. The sample consist of the matrix is amorphose phase and the filler is crystalline phase. The filler consist of two phases, namely ZnO and $La_{0.8}Ba_{0.2}(Fe, Mn,Ti)O_3$phases. The functional groups analysis show that the sulfur addition modified the polymer by forming crosslink (bridges) between individual polymer chains and bonding between magnetic filler and rubber matrix. We concluded that this study has been successfully made a composite of rubber – $La_{0.8}Ba_{0.2}(Fe, Mn,Ti)O_3$ for microwave absorber sheet application.

1. Introduction

Elastomer Thermoplastic (ETP) is one of polymer that had the quality both elastic and thermoplastic. ETP is generally used in the automotive industry until household appliances. At this time the ETP is also a target for the electronics industry utilization. Recently introduced a product of electromagnetic absorbing can provide a relatively easy solution to reduce unwanted RF noise in electronic equipment, namely absorber sheet materials. Absorber sheet is a product of a composite material with magnetic particles embedded in the ETP. The composite has good noise attenuation performance from hundreds of MHz to several GHz.

The primary objective of this study is to develop a composite material based on natural rubber- that can be used to absorb the electromagnetic waves. Indonesia is one of the biggest rubber producers in the world with a total production of about 1.4 million tons per years. Natural rubber is an elastomer, it has a relatively high molecular weight and the potential for the manufacture of elastomer thermoplastic (ETP). Use of ETP from natural rubber-based is increasing every year because it has many advantages namely the process is easy and simple, the manufacturing process is faster and can be recycled [1]. The filler used magnetic material which was able to absorb the electromagnetic waves. Previous study has been carried out on the development of magnetic absorber material Ba-Srhexaferrite by substituting Fe ions with Mn and Ti [2]. Recent development of other magnetic material that serves as an absorber material is modification of hexaferrite systems [3-5] and ABO_3 perovskite [6-9]. The combination of both materials is expected to obtain superior composite materials that can be used for electromagnetic waves absorber and is of flexible nature.

However to make a good composite materials is not that simple and straight forward, because the process is influenced by several factors. The composite manufacturing process begins with the termination of the polymer chain. The step is called by mastication process to facilitate the mixing process between rubber, magnetic powder and other chemicals. The second step is vulcanization process to create a crosslink (bridges) between individual polymer chains. The filler of magnetic material must be homogeneous and blended with the polymer matrix. Temperature- and time of heating control must be precise. Therefore the result and discussions have focused on the synthesis result and structural characterization and functional groups analysis of the composite. So the aim of this study is to synthesize a composite material from a combination of rubber and lanthanum manganite modification method, and to understand the microstructure, the crystal structure, and the bonding mechanism of the composite.

2. Materials and Methods

The synthesis of $La_{0.8}Ba_{0.2}(Fe, Mn,Ti)O_3$ material was performed by using the solid reaction method. This magnetic material was prepared by oxide materials as starting materials, namely $BaCO_3$, Fe_2O_3, $MnCO_3$, TiO_2 and La_2O_3. The raw materials were obtained from Merck product with a purity of more than 99%. The raw materials are mixed by using a high-energy milling (HEM) apparatus of the type Spex 8000 for 10 hours at room temperature. The finely mixed powder was compacted at 5000 psi into pellets and sintered in the electric chamber furnace THERMOLYNE at 1000°C for 10 hour in the air at atmospheric pressure. The synthesis of composite material is started with the weighing of $La_{0.8}Ba_{0.2}(Fe, Mn,Ti)O_3$material and then the mixing process is carried out by using an open two-roll mixer (model: HF-2RM, series: 2394). The mixing process consists of mastication, vulcanization, pressing, and heat treatment. Vulcanization usually used sulfur and additive. These additives include accelerators, activators like zinc oxide and stearic acid and n-cyclohexyl-2-benzothiazyl sulphenamide (CBS). Accelerators can increase the rate of cure by catalyst the addition of sulfur chains to the rubber molecules. Formulation of the composite material is showed in Table 1.

Table 1. Formulation of the composite material

No.	Material	Fraction (per hundred rubber)
1.	Rubber (ETP)	100.0
2.	Stearic acid	2.0
3.	Zinc oxide (ZnO)	5.0
4.	$(La_{0.8}Ba_{0.2})Fe_{0.3}(Mn,Ti)_{0.35}O_3$	20.0
5.	n-cyclohexyl-2-benzothiazyl sulphenamide (CBS)	1.4
6.	Sulfur	2.5

3. Results and Discussion

The composite is prepared by rubber elastomer thermoplastics, stearic acid, ZnO, n-cyclohexyl-2-benzothiazyl sulphenamide (CBS), sulfur, and magnetic powder of $La_{0.8}Ba_{0.2}(Fe,Mn,Ti)O_3$. Fig. 1 is showing the results of the elementary analysis by using energy dispersive spectroscopy method on the composite material.

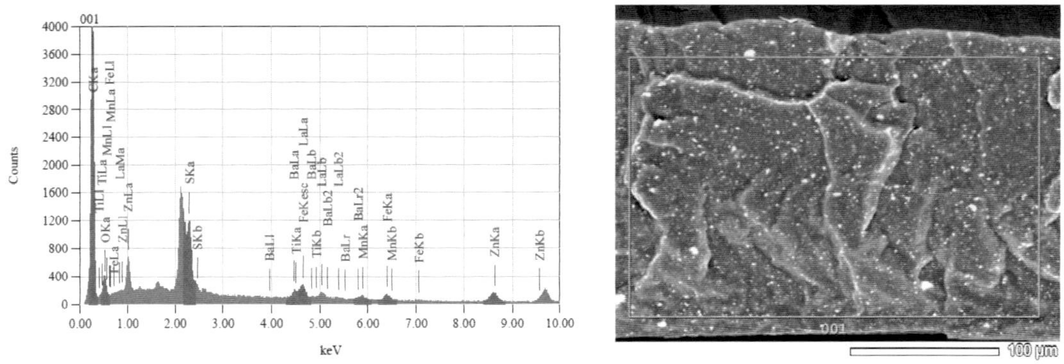

Fig.1. The elementary analysis of composite material by using energy dispersive spectroscopy

The energy dispersive spectrum as shown in Fig. 1 shows that the dominant elements are carbon (C), oxygen (O), sulfur (S), zinc (Zn), lathanum (La), barium (Ba), manganese (Mn), titanium (Ti) and iron (Fe). This means that the sample is iron rich. The detailed content of the sample is shown in Table 2.

Table 2. The result of elementer analysis of the iron sand sample

No.	Unsure	Content (wt%)
1.	Iron (Fe)	1.80 ± 0.06
2.	Titanium (Ti)	0.46 ± 0.07
3.	Manganese (Mn)	0.80 ± 0.09
4.	Lanthanum (La)	6.03 ± 1.85
5.	Barium (Ba)	1.42 ± 0.50
6.	Zinc (Zn)	5.60 ± 1.73
7.	Oxygen (O)	13.07 ± 1.65
8.	Sulfur (S)	3.00 ± 0.25
9.	Carbon (C)	67.82 ± 0.21

The SEM micrograms of the sample show that the filler of magnetic particles are embedded in the ETP matrix and are distributed homogeneously through out the sample's surface. And the results of elemental analysis have shown that all of the elements contained in the composite have emerged. So that, a further confirmation about phases formed in the sample is needed which is achieved by using XRD method as is shown in Fig. 2

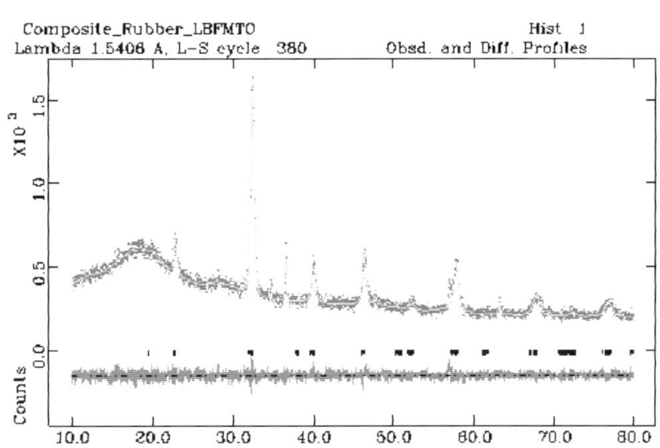

Fig. 2. The refinement result of XRD pattern on the sample.

The refinement result of XRD pattern on the sample has produced an excellent quality of fitting with the reliability factor R being very small. And the goodness of fit value χ^2 (chi-squared) is in good agreement with the standard value allowed in the rietveld analysis [10]. The sample is semicrystalline with the crystallinity of 46%. The sample consists of an amorphous phase matrix and a crystalline phase filler. The filler consist of two phases, namely ZnO and $La_{0.8}Ba_{0.2}(Fe,Mn,Ti)O_3$ phases. $La_{0.8}Ba_{0.2}(Fe,Mn,Ti)O_3$ phase had monoclinic structure, space group of I 1 2 / a 1, the lattice parameters of a = 5.543(1) Å, b = 5.587(1) Å and c = 7.822(1) Å, $\alpha = \gamma = 90^o$ and $\beta = 89.73(3)^o$, V = 242.3(1) $Å^3$, and ρ = 6.414 $gr.cm^{-3}$. ZnO phase had hexagonal structure, space group of P 63 m c, the lattice parameters of a = 3.2487(9) Å, b = 3.2487(9) Å and c = 5.204(1) Å, $\alpha = \beta = 90^o$ and $\gamma = 120^o$, V = 47.56(3) $Å^3$, and ρ = 5.467 $gr.cm^{-3}$. The phase identification referred to COD of No. 1001820 and No. 9011662 for $LaMnO_3$ and ZnO phases, respectively. This result is evidenced by the calculation results of the mass fraction as shown in Table 3.

Table 3. Mass fraction of the samples

Sample	Phase	Mass fraction (%)	wRp	χ^2 (chi-squared)
Composite	$La_{0.8}Ba_{0.2}$(Fe, Mn,Ti)O_3	94.20	5.84	1.196
	ZnO	5.80		

However, the XRD results could not explain the bonding mechanism,including the formation of crosslink (bridges) between individual polymer chains in the composite so that it is needed furtherconfirmation by using FTIR equipment as shown in Fig. 3.

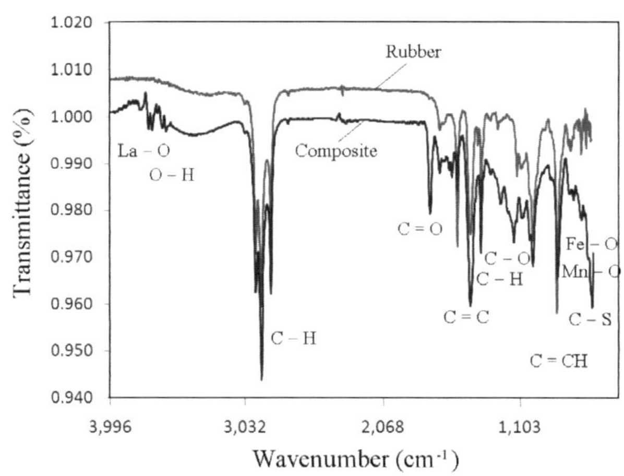

Fig. 3. The FTIR transmittance spectrum of the sample.

Fig. 4 represented the FTIR spectrum between 4000 to 400 cm^{-1} of the composite sample. Transmittance peaks of the FTIR spectra on the composite sample indicate the presence of vibration of La–O, O–H, C–H, C=C, C=O, C–O, C=CH, Mn–O, Fe–O, C–S bond [11-14]. Natural rubber consists of suitable polymers of the organic compound isoprene with minor impurities of other organic compounds and water. Forms of polyisoprene that are useful as natural rubbers are classified as elastomers. Chemical structure of cis-polyisoprene as the main constituent of natural rubber was showed in Fig. 4.

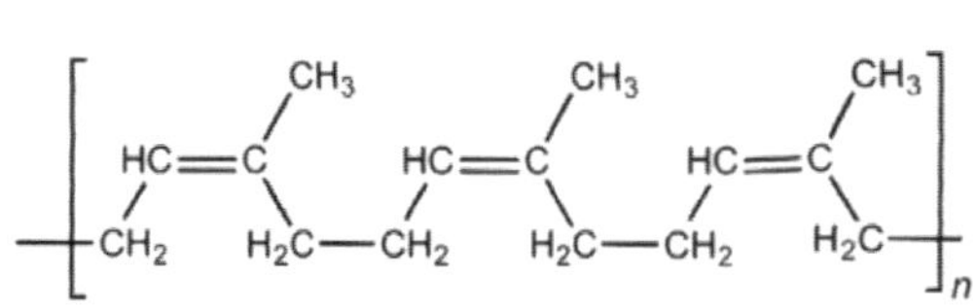

Figure 4. Chemical structure of cis-polyisoprene as the main constituent of natural rubber

The functional groups analysis results obtained by using the FTIR method, have shown that transmittance peaks of the rubber appear at wave numbers around 2849-2959, 1373 cm^{-1}, 1539 cm^{-1}, and 843 cm^{-1} that indicated the presence of the C–H, C=C, and C=CH bond vibrations, respectively. Meanwhile transmittance peaks of the $La_{0.8}Ba_{0.2}Fe_{0.3}(Mn,Ti)_{0.35}O_3$ appear at wave numbers of 3694 cm^{-1}, 613 cm^{-1} and 576 cm^{-1} attributed to the La–O, Mn–O, and Fe–O bond vibrations, respectively. The composite manufacturing process needs to be supplemented by a vulcanization process to create a crosslink (bridges) between individual polymer chains. A cross-link is a bond that links one polymer chain to another. They can be covalent bonds or ionic bonds. Vulcanization process usually used sulphur in the process. Both of these contain a sulphur atom in the molecule that initiates the reaction of the sulphur chains with the rubber. The chemical process of vulcanization is a type of cross-linking so that it is expected that the C-S bond would materialize as shown in Fig. 5.

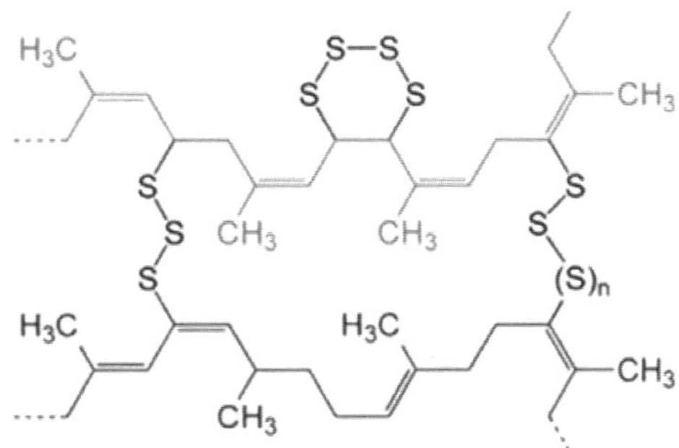

Figure 5. Schematic presentation of two polymer chains cross-linked after the vulcanization

The functional groups analysis showed that transmittance peaks of the crosslink appeared at a wave number of around 590 cm^{-1} that indicates the presence of the C–S bond vibrations. However, it has also been found that transmittance peaks of the samples appear at wave numbers around 1728 cm^{-1} and 1146 cm^{-1} that indicate the presence of C=O and C–O, bond vibrations, respectively. It is suspected that there was a bonding between the magnetic filler and the rubber matrix.

4. Conclusion

The synthesis and characterization of composite rubber – lanthanum manganite modification have been carried out using chemical process. Microstructure of the sample showed that the filler of magnetic particles are embedded in the ETP matrix and are distributed homogeneously on the sample's surface. The elemental analysis shows that the sample contains carbon, oxygen, sulfur, zinc, lathanum, barium, manganese, titanium and iron. The refinement results of x-ray diffraction pattern showed that the sample is semicrystalline with the crystallinity of 46%. The sample consist of the matrix was amorphous phase and the filler was crystalline phase. The filler consist of two phases, namely ZnO and $La_{0.8}Ba_{0.2}(Fe,Mn,Ti)O_3$ phases. The functional groups analysis showed that the sulfur addition modified the polymer by forming crosslink (bridges) between individual polymer chains and bonding between magnetic filler and rubber matrix. This study has been successfully synthesized composite rubber–lanthanum manganite modification and has been understood changes in the microstructure, crystal structure and bonding mechanism of the composite.

References

[1] Passador F R, Rojas G J A and Pessan L A 2013 Journal of Macromolecular Science Part B **52** **8** 1142-1157.
[2] Adi W A and Manaf A 2012 J. Basic. Appl. Sci. Res. **28** 7826-7834.
[3] Mehdipour M, Shokrollahi H 2013 J. Appl. Phys.**114** **4** 043906.
[4] Rai B K, Wang L, Mishra S R, Nguyen V V, Liu J P 2014 J. Nanosci. Nanotechnol. **14** 5272-

5277.
[5] Wang Y, Huang Y, Wang Q 2012 J. Mag. Magn. Mater. 324 19 3024-3028.
[6] Hazra S, Patra M K, Vadera S R, Ghosh N N 2012 J. Am. Ceram. Soc. 95 1 60-63.
[7] Zhang S, Cao Q 2012 Materials Science and Engineering B 177 678– 684.
[8] Adi W A and Manaf A 2013 Advanced Materials Research 789 97-100.
[9] Rochman N T and Adi W A 2013 International Journal of Academic Research Part A 5 4 91-96.
[10] Idris M S, Osman R A M 2013 Advanced Materials Research 795 479-482.
[11] Darmawan A, Smart S, Julbe A, Diniz da Costa, J C 2011 Materials 4 448–456.
[12] Hamzah R, Bakar M A, Khairuddean M, Mohammed I A and Adnan R 2012 Molecules 17 10974-10993.
[13] Ohnuki T 2015 International Polymer Science and Technology 42 8 39-45.
[14] Nariyal R K, Kothari P and Bisht B 2014 Chemical Science Transactions 3 3 1064-1066.

IOP Conf. Series: Journal of Physics: Conf. Series **1091** (2018) 012014 doi:10.1088/1742-6596/1091/1/012014

The effect of doping Ni to the resistivity and ratio magnetization on $La_{0.67}Ba_{0.33}Mn_{1-x}Ni_xO_3$ (x = 0.0; 0.1; 0.2; 0.4)

M Zulziar[1], B Kurniawan[2] and S A Saptari[3]

[1]Department of Industrial Engineering,Pamulang University, Pamulang 15417, Indonesia
[2]Department of Physics, University of Indonesia, Depok 16424, Indonesia.
[3]Department of Physics, State Islamic University, Ciputat 15412, Indonesia

E-mail : marjuki.zulziar@gmail.com

Abstract. $La_{0.67}Ba_{0.33}Mn_{1-x}Ni_xO_3$ with x = 0.0; 0.1; 0.2; 0.4 were prepared from powder of La_2O_3, $BaCO_3$, $MnCO_3$, and NiO by the solid-state reaction method. All the powder is mixed by using Planetary Ball Milling for 25 hours. After that, all the samples are compacted, calcinatedat 800°C for 10 hours and sintered 1350°C for 20 hours. Phase identification was carried out by using XRD spectrometer and refined by using High Scorer Plus. The results showed that $La_{0.67}Ba_{0.33}Mn_{1-x}Ni_xO_3$ were single phases for all x compositions. The crystal structureswereorthorhombic. The resistivity and magnetoresistance (MR) properties were evaluated by using FPP (Four Points Probe) method, while magnetization was measured by using permagraph spectrometer. From the magnetoresistance results, we observed that the samples showed different sensitivity's responds for each x value. By increasing of x values, the magnetoresistances tend to change from negative magnetoresistance to positive magnetoresistance. Besides, the sensitivity's responds of magnetoresistance were increased by increasing the x values. The highest sensitivity was for x=0.4. The permeagraph results showed that all the samples were paramagnetic at room temperature due to substitution of Ni. We concluded that the doping of Ni gives effect to the resistivity and magnetoresistance of the samples and also to the sensitivity's responds as a candidate of magnetoresistance sensor's material.

1. Introduction

Hole-doped perovskite manganite of the form $R_{1-x}AMnO_3$ (R is the rare-earth ion, A is alkaline earth divalent ion) have drawn considerable interest in recent years, due to their unusual magnetic and electronic properties. Various attempts have been extensively done in this manganite[1][2][3][4]. The magnetoresistance (MR) effect has been substantial importance technologically, especially in connection to read-out heads for magnetic disks and as sensors of magnetic fields[5][6]. For a long time, hole doped manganite $La_{1-x}A_xMnO_3$ (A = Ca, Sr, Ba, Pb) with perovskite structure were known as an interesting class of materials[7]. Proposed a new phenomenological model that quantitatively explains the observation of unusual transport and magnetoresistance (MR) phenomena near TC, and illuminates the origin of the intrinsic CMR. In addition, the discovery of low field magnetoresistance (LFMR) stimulates the investigations on the granular systems [8][9]. Although several research works had been performed on $La_{0.67}Ba_{0.33}Mn_{1-x}Ni_xO_3$, the position of Ni dopants (with various concentrations) in the $La_{0.67}Ba_{0.33}Mn_{1-x}Ni_xO_3$host lattice have not been well clarified yet. Based on x-ray absorption studies, the authors have pointed out that the Ni dopants in $La_{0.67}Ba_{0.33}Mn_{1-x}Ni_xO_3$ are in fact Ni^{2+} ions, which substitute for Mn ions in the perovskite structure. The chemical shift of the main absorption edge towards higher energy as the increase of the Ni doping concentration does substantiate the increase in concentration of Mn^{4+} ions. It is observed that the doping of Ni does affect the resistivity and magnetoresistance of the samples and the sensitivity's response as a candidate of magnetoresistance sensor's material.

2. Experimental

$La_{0.67}Ba_{0.33}Mn_{1-x}Ni_xO_3$ with x = 0.0; 0.1; 0.2; 0.4 were prepared from powder of La_2O_3, $BaCO_3$, $MnCO_3$, and NiO by the solid-state reaction method. All the powder is mixed by using Planetary Ball Milling for 25 hours. Afterward, all the samples are compacted, calcinated at 800°C for 10 hours and sintered 1350°C for 20 hours for all x compositions. Phase identification was carried out by using XRD. The resistivity and magnetoresistance (MR) properties were evaluated by using FPP (Four Points Probe) method, while magnetization was measured by using a permagraph spectrometer.

3. Result and Discussion

Figure 1 presents the XRD pattern of the $La_{0.67}Ba_{0.33}Mn_{1-x}Ni_xO_3$ with x = 0.0; 0.1; 0.2; 0.4 sample. The diffraction pattern of the sample shows a single phase and it is free from impurities. Crystal structure was further refined by using High Scorer Plus Software. Refinement results of the $La_{0.67}Ba_{0.33}Mn_{1-x}Ni_xO_3$ sample can be seen in Table 1.The table shows that the lattice parameter of the variation of composition *x* in $La_{0.67}Ba_{0.33}Mn_{1-x}Ni_xO_3$ were single phases for all x compositions. Thus, the crystal structure remains unchanged although with different composition x, the crystal structure was orthorhombic. The X-ray diffraction spectra for samples with x = 0; 0.1; 0.2 and 0.4 have the same diffraction pattern and that the maximum reflection peak's intensity is at the 2θ diffraction angle of 32°.

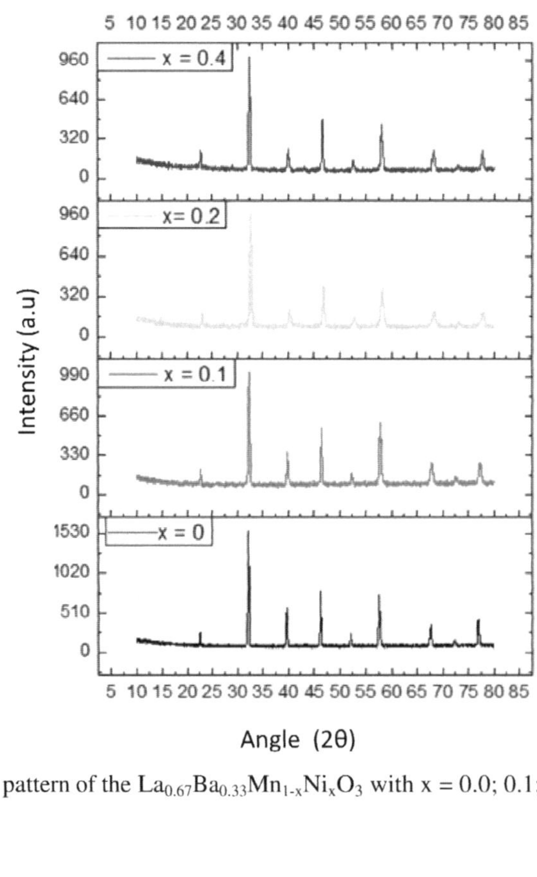

Figure 1. XRD pattern of the $La_{0.67}Ba_{0.33}Mn_{1-x}Ni_xO_3$ with x = 0.0; 0.1; 0.2; 0.4 sample.

The magnetic properties of sample $La_{0.67}Ba_{0.33}Mn_{1-x}Ni_xO_3$ with x = 0.0; 0.1; 0.2; 0.4 are characterizedby using a permagraph spectrometer. Magnetic materials can be divided into hard magnetic and soft magnetic. For a material having a large coercivity(above 10 kA/m) is called hard magnetic, while a material having small coercivity (below 1 kA/m) is called soft magnetic. Figure 2 show the comparation of the hysteresis curve between the four samples. The permeagraph results showed that all the samples were paramagnetic at room temperature due to subtitution of Ni.

Table 1.Refinement results of $La_{0.67}Ba_{0.33}Mn_{1-x}Ni_xO_3$.

x	a(Å)	b(Å)	c(Å)	Volume (Å^3)	χ^2	W_{Rp}
0	5.51	7.80	5.53	238.49	1.82	13.82
0.1	5.49	7.79	5.52	236.61	1.80	12.99
0.2	5.48	7.76	5.53	235.83	1.62	12.75
0.4	5.46	7.74	5.49	232.60	1.51	12.79

A negative magnetoresistance ratio at$La_{0.67}Ba_{0.33}Mn_{1-x}Ni_xO_3$samples decreased with increasing Ni doping at Mn site. The largest negative value magnetoresistance there when x = 0.1 which is-1.54 %. The same behavior has been reported by other researchers [10], which discusses the changes in the temperature of the magnetoresistance in samples $La_{0,67}Ba_{0,33}Mn_{1-x}Ti_x/Bi_xO_3$, when granting small Ti doping xbelow 0.1 (x = 0; 0.1 ; 0.02 and 0.05)but for x above 0.1, negative magnetoresistance values decreased more sharply, and become to be a positive magnetoresistance . This is happened due to a decreasing valueof the Curie temperature (T_c) in the system [10].

From the magnetoresistance resultsin figure 3 and table 2, we observed that the samples showed different sensitivity's responds for each x value. By increasing of x values, the magnetoresistances tend to change from negative magnetoresistance to positive magnetoresistance. For x = 0 and 0.1 the sampel is negatif magnetoresistance (–MR) and for x = 0.2 and 0,4 the sampel is positif magnetoresistance(+MR). Besides, the sensitivity's responds of magnetoresistance were increased by increasing the x values. The highest sensitivity was for x=0.4. Therefore it could be concluded that the doping of Ni does affect the resistivity, the magnetoresistance and the sensitivity responseof the samples. These indications support the samples as a good magnetoresistance sensor material candidate.

Table 2.Δρ and MR of $La_{0.67}Ba_{0.33}Mn_{1-x}Ni_xO_3$

x	Δ ρ	MR
0	-0.94	-15.61 %
0.1	-0.51	-1.54 %
0.2	2.32	5.99 %
0.4	8.16	12.59 %

4. Conclusion

The diffraction pattern of the sample shows that the samples are single phase and are free from impurities for all values of x compositions. The crystalstructure is orthorhombic.The permagraph results showed that all the samples were paramagnetic at room temperature due to substitution of Ni.The resistivity was increased, and the conductivity was decreased while increasing Ni dopant. By increasing of x values, the magnetoresistances tend to change from negative magnetoresistance to positive magnetoresistance. The sensitivity response of magnetoresistance was increased by increasing the x values. The highest sensitivity was for x=0.4. Finally, it could be concluded that the doping of Ni does affect the resistivity, the magnetoresistance and the sensitivity responseof the samples.These indications support the samples as a good magnetoresistance sensor material candidate.

Acknowledgement
This work was supported by University of Indonesia Material Magnetic and Dielectric Research Group No 1861/ UN2.R12/HKP 05.00/2015.

References

[1] Altintas S P, Amira A, Varilci A and Terzioglu C 2012 Influence of Gd-doping in $La_{0.7}Ca_{0.3}MnO_3$ on its structural and magneto-electrical properties *J. Magn. Magn. Mater.* **324** **7** 1331–1336.

[2] Ayadi F, Regaieg Y, Cheikhrouhou-koubaa W, Koubaa M, and Cheikhrouhou A 2015 Preparation of nanostructured $La_{0.7}Ca_{0.3-x}Ba_xMnO_3$ ceramics by combined sol – gel and spark plasma sintering route and resulting magnetocaloric properties *J. Electron. Mater.* **381** 215–219.

[3] Creel T F 2013 Structural and magnetic properties of $La_{0.7}Sr_{0.3}Mn_{1-x}Ni_xO_3$ ($x \leq 0.4$) *J. Appl. Phys.***114** **1** 013911.

[4] Cui K, Cheng Y, Dai J and Liu J 2013 Synthesis , characterization and microwave absorption properties of $La_{0.6}Sr_{0.4}MnO_3$/polyaniline composite *Mater. Chem. Phys.* **138** 810–816.

[5] Widyaiswari U, Kurniawan B, Imaduddin A and Saptari S A 2016 Study of $La_{0.67}Sr_{0.33}MnO_3$ Structure and Resistivity Synthesized by Using Sol Gel Method *Journal of Physics: Conference Series* **776** 012041.

[6] Saptari S A, Manaf A and Kurniawan B 2014 The Influence of Ni-Doping on Structure and Magnetic Properties of $La_{0.67}Ba_{0.33}MnO_3$ *Adv. Mater. Res.* **896** 406–409.

[7] Winarsih S 2016 Investigation of structural and electrical properties of $La_{0.7}(Ba_{1-x}Ca_x)_{0.3}MnO_3$ compounds by sol-gel method *AIP conference proceedings* **1746** 020004–1–02004–5.

[8] Pratama R 2016 Effect of Various Sintering Temperature on Resistivity *AIP conference proceedings* **1729** 020028–1–020028–4.

[9] Phan M, Yu S, Hur N H and Jeong Y 2004 Large magnetocaloric effect in a $La_{0.7}Ca_{0.3}MnO_3$ single crystal *J. Appl. Phys.* **96** **2** 1154–1158.

[10] Yuan X, Liu Y, Hung B, Chen W, Wang C J and Zhang R 2005 Enhancement of Magnetoresistance in Bi or Ti doped $La_{0.67}Ba_{0.33}MnO_3$ *Mater. Sci. Eng. B*, **117** **1** 93–97.

Influence of Fe_3O_4 Addition in TiO_2 Catalyst on The Degradation Process of Methylene Blue

Siti Wardiayati and Sari Hasnah Dewi
Center for Science and Technology of Advanced Materials – National Nuclear Energy Agency. Kawasan Puspiptek Serpong. 15314. Indonesia

Email : siti-war@batan.go.id

Abstract. Synthesis of composite titanium dioxide with iron oxide magnetic material (Fe_3O_4/TiO_2 composite) has been successfully carried out by precipitation method. Photocatalytic performance test of this composite Fe_3O_4/TiO_2 was performed for degradation process of the methylene blue solution with the experiment parameters consist of the solution pH, lighting time, type of light, amount of catalyst, and the concentration of methylene blue. The experimental results showed that the Fe_3O_4 addition in TiO_2 catalyst can simplify recovery process of the catalyst, and able to broaden the absorption area up to visible light, so that the degradation process can be carried out with UV and visible light. Degradation of methylene blue by Fe_3O_4/TiO_2 composite in under sunlight more effectively compared UV light.

1. Introduction

Titatium dioxide nanoparticles (TiO_2) of anatase phase are a catalyst material that has proven to be very efficient in the degradation of organic compounds because it has a relatively large energy gap of about 3.2 eV [1]. However, there are obstacles in its use in the recovery process of the catalyst material after the degradation process. It is given that the catalyst material was dispersed into the solution so that the separation process needs to use centrifuges. The separation process using the centrifuges needed a lengthy period of time to finish. Therefore, we need further development namely combines these TiO_2 catalyst materials with magnetic material to form a composite material to facilitate the recovery process of TiO_2 catalyst after degradation process. This composite development is expected to shorten the separation process so that it becomes more effective and economical

Magnetic materials which are suitable for use as magnetic composite photocatalyst, are of the superparamagnetic type. Because this magnetic material has no spontaneous magnetic moment, it can be well dispersed in the water, and also easier to be collected back because it can produce a maximum magnetic response when given an external magnetic field. The best iron oxide-based superparamagnetic materials to be used as a carrier magnetis Fe_3O_4 and γ-Fe_2O_3 because it has a fairly high magnetization. In addition, synthesis of this magnetic material is much simpler and non-toxic so that is safe for the environment.

Photocatalysts magnetic composite of Fe_3O_4/TiO_2 can be prepared by various methods, such as sol-gel[2], plasma spray [3], precipitation [4-5], solvothermal [5], sonochemical [7-8], and micro-emulsion [9]. In addition, various forms morphology of Fe_3O_4/TiO_2 catalyst have been successfully made also include core/shell [6], the nanotube core/shell [7,8] and hollow [6,10].

In the previous study, TiO_2-based catalyst material with good performance has been successfully synthesized and it is able to degrade almost 100% methylene blue with a degradation time of 150 minutes using ultraviolet light [12]. However, the process of recovery of the catalyst material is still using the centrifuge equipment with a relatively limited capacity. Based on these results, in the present research a composite of Fe_3O_4/TiO_2 catalyst will be developed,it is to be synthesized by precipitation method using the precursors $FeCl_2.4H_2O$, $FeCl_3.6H_2O$, and titanium ortho titanate. The photocatalyst performance of the Fe_3O_4/TiO_2 composite catalyst was done through the degradation process of organic compounds (methylene blue), and then will be compared with the performance capabilities of the result of previous studies (TiO_2 photocatalyst). Therefore, the objective of this study was to observe the effect of Fe_3O_4 addition in TiO_2 catalyst in the recovery processes of methylene blue degradation with research parameters include pH, lighting time, type of light, the amount of catalyst, and the concentration of methylene blue.

2. Materials and Methods

Materials

TiO_2 and Fe_3O_4 synthesis results of pre-vious experiment with specification of anatase (TiO_2) and particle size of ± 10 nm, methylene blue, NaOH, H_2SO_4, NH_4OH, tetra bythyl ortho titanate (TBOT), and demineralization water.

Method

The photocatalytic performance test was carried out in a solution of methylene blue in a catalytic reactor with magnetic stirrer and UV-pen light (wavelength 356 nm). While the experimental procedure was performed as follows; the Fe_3O_4/TiO_2 powder was incorporated into the catalytic reactor containing methylene blue solution whose pH level has been set. The suspension was stirred in accordance with the desired time, and then after the photocatalytic process the catalyst material was retrieved by using a magnet. After that the solution was analyzed using a UV-Vis spectrophotometer to determine the remaining concentration of methylene blue. The photocatalytic test result of catalyst composite Fe_3O_4/TiO_2 was compared with the TiO_2 catalyst (the result of previous study). The study parameters consist of the solution pH (2, 4, 6, 8, 10, and 12); lighting time (30, 45, 60.75, 90, 120, and 150 minutes); concentration of methylene blue (10, 20, 30, 40, and 50 mg/L) and the amount of TiO_2 catalyst (10, 25, 50, 100, 150, and 200 mg).

3. Result and Discussion

Effect of pH solution on the efficiency of methylene blue degradation

In the photocatalytic process, the pH scale of the solution has a profound influence on the efficiency of methylene blue degradation, because the presence of forming free radicals that contribute to the oxidization of organic compounds into simpler compounds that are not harmful to the environment depends on the pH value of the surrounding solution.

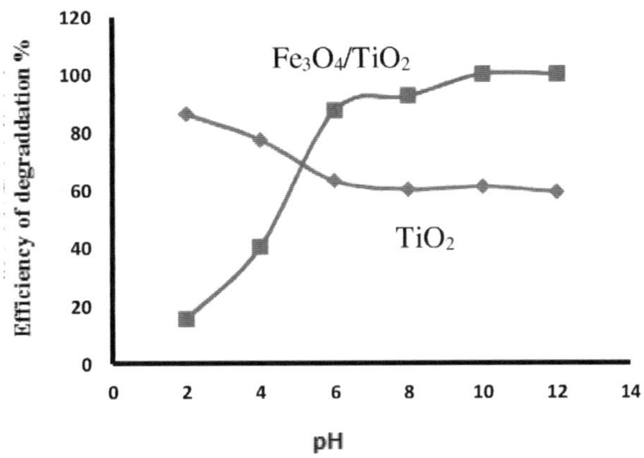

Figure 1. Effect of pH on efficiency degradation of methylene blue (methylene blue solution of 100 mL [10 mg/L], amount of catalyst of 100 mg, irradiation time of 120 min)

The reaction of free radical formation in the process of photocatalytic TiO2 by UV light has been discussed in the previous study [12]. Effect of pH values on the efficiency of methylene blue degradation photocatalytic composite Fe_3O_4/TiO_2 is shown in Figure 1. Figure 1 shows that the effect of photocatalytic Fe_3O_4/TiO_2 on the methylene blue degradation is very effective and it occurs at bases pH of about 10-12. In these case the degradation process nearly perfect (100%). While the methylene blue degradation has not occurred at acidic pH (pH $\leq$ 4.0). This is due to Fe3O4 magnetic composites with TiO_2 will change the properties of the catalyst material. Fe_3O_4 at acidic pH will not be stable, because Fe_3O_4 under acidic conditions will react with the acid around it as shown in the following reaction:

$$Fe_3O_4 + HNO_3 \rightarrow 3\ Fe(NO_3)_3 + NO_2 + 5\ H_2O$$

With the reduction of Fe_3O_4 in the catalyst material,it becomes very influential on the degradation efficiency, because Fe_3O_4 was absorbing the methylene blue. The absorption properties appeared from the experimental results of methylene blue degradation in the dark atmosphere (without light) as shown in Figure 2. Experimentsl of methylene blue degradation in the dark atmosphere was conducted to show that the degradation process occurs not because of the photocatalytic process but because of the absorption process. In Figure 2 it is clearly shown that the occurrence of the decrease in the concentration of methylene blue in the dark atmosphere is relatively high with an efficiency of 76%, whereas when using TiO_2 catalyst in the same condition it only reaches an efficiency of about 19%.

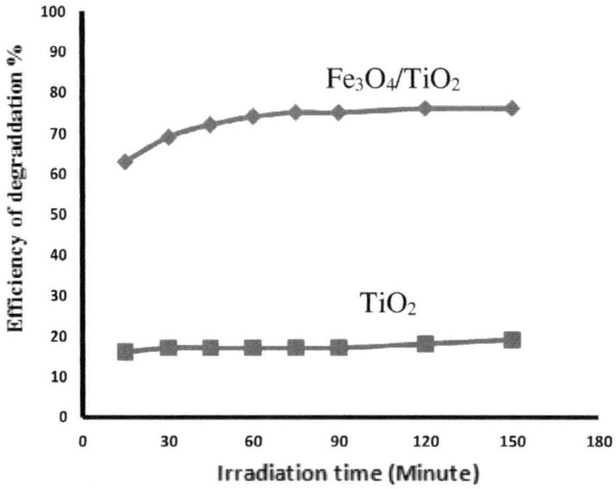

Figure 2. . Efficiency of degradation of methylene blue by TiO_2 and Fe_3O_4/TiO_2 in the dark situation (without light) (vol. methylene blue solution [10 mg/L] of 100 mL, amount of catalyst of 100 mg, pH solution of 10)

Effect of the addition of Fe_3O_4 on the performance of TiO_2 catalyst

Degradation process of the photocatalytic method is strongly influenced by the strength of light, light distance and lighting time. The longer it takes for the lighting degradation process to finish, the more effective it will be. In the photocatalyst method, degradation will occur when the photon energy of UV-Vis absorbed by the catalyst material is greater than the threshold energy between electrons on the valence band and the conduction band. In addition, the presence of other compounds in this material will also affect the performance of the catalyst material.

Investigation of the effect of the addition of Fe_3O_4 on the performance of TiO_2 catalyst materials was carried out with a variety of lighting experiments using UV light and sunlight in the duration of 15-150 min as shown in Figure 3.

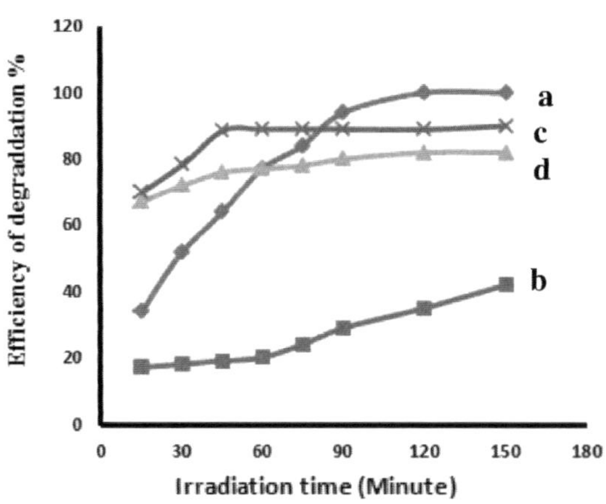

Figure 3. Efficiency of methylene blue degradation by various catalyst and irradiation source a. TiO_2 with UV light, b. TiO_2 with sunlight, c. Fe_3O_4/TiO_2 with UV light and d. Fe_3O_4/TiO_2 with sunlight

As shown in Figure 3, the performance of TiO_2 catalyst is very low when using sunlight, because TiO_2 only absorb the UV light, while the UV light content in the sun ray spectrum was only around 4-5% 9[13]. However, when using UV light from a UV lamp of pen models with a wavelength of 356 nm, there is an indication that the photocatalytic performance is very good. The degradation process of methylene blue by Fe_3O_4/TiO_2 using sunlight is more effective compared with using UV light, this is due to the Fe_3O_4 addition in the TiO_2 catalysts will cause a widening of the absorption region from the UV light into the visible light region. This is evident from the measurement results of absorbance TiO_2 Fe_3O_4/TiO_2 using reflectance of UV-Vis Spectrophotometer as shown in Figure 4.

In Figure 4 it appears that the addition of Fe_3O_4 in the TiO_2 catalyst can broaden the light absorption from UV light (wavelength $\leq$ 400 nm) up to the visible light region with wavelength > 400 nm. In another work, the authors have successfully synthesized a photocatalyst Fe_3O_4/TiO_2 by sol-gel method and come up with the same result that the efficiency degradation occurs up to the visible light region, but the concentration of methylene blue in this study can be three times higher [14]. When compared with the Fe_3O_4/TiO_2 research results reported by Chen-Yu Hsieh and Yung-Hsu, 2014, the authors have used precursor tetra orthotitanate isopropanol and have shown that the optimum photocatalytic performance only occurs in the UV light region whereas in this study it occurs up to the visible light region.

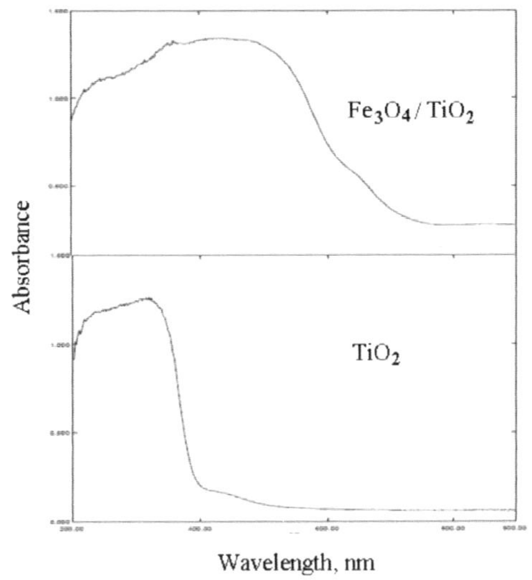

Figure 4. Measurement of light absorption TiO_2 and Fe_3O_4/TiO_2 catalyst material by UV-Visible Diffuse Reflectance spectroscopic

Effect of the amount of catalyst on the effectiveness of photocatalytic

Based on the photocatalytic method, the amount of catalyst does affect the effectiveness of the organic compounds degradation. The higher the amount of catalyst, the more amount of free radicals will be formed in a manner that the decomposition of organic compounds will be more effective, but it is limited by the maximum amount of catalyst because these catalyst can hinder the incoming light [12, 14-15]. Therefore the amount of catalyst has influence on the effectiveness of methylene blue degradation as shown in Figure 5.

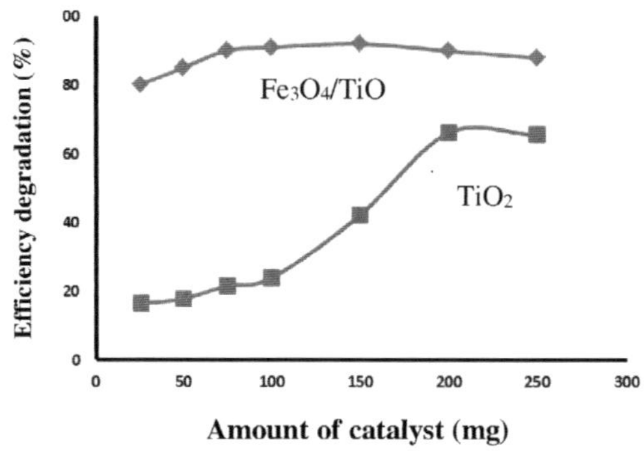

Figure 5. Effect of efficiency degradation on the amount of catalyst material (methylene blue solution of 100 mL, and irradiation time of 150 min)

Figure 5 shows that the degradation efficiency of Fe_3O_4/TiO_2 initially increases by increasing the amount of catalyst, but after reaching the optimal value up to 50 mg, the reaction rate does not change and even decreased. This is due to the active surface area of the catalyst has increased so that the reaction becomes faster. However, after the amount of catalyst becomes greater than 50 mg, it is suspected thatagglomerationhas occured amongthe particles, and causes the incoming light to be blocked. With obstruction of light on the catalyst material will reduce the formation of free radicals so that the process of degradation decreases.This also occurs in the use of TiO_2 catalyst as shown in Figure 5. At the same amount of Fe_3O_4/TiO_2 and TiO_2 catalyst, the efficiency degradation with Fe3O4/TiO2is much higher than with TiO_2, because Fe_3O_4/TiO_2 have adsorption properties and wider absorbency regions.

Effect of the methylene blue concentration on the efficiency of degradation by Fe_3O_4/TiO_2 catalyst material

The concentration of methylene blue will affect the degradation efficiency. The higher concentration of methylene blue would affect a decrease in the degradation efficiency. This is due to the higher concentration of methylene blue, the available load of Fe_3O_4/TiO_2 catalyst to degrade the organic compounds is also higher. The influence of the methylene blue concentration on the residual methylene blue was shown in Figure 6.

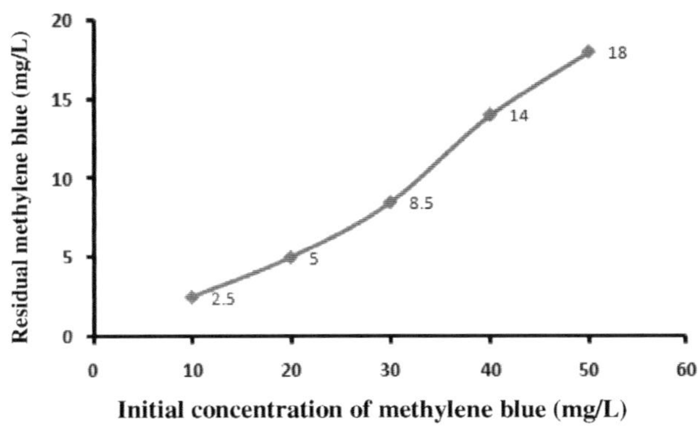

Figure 6. Effect of the methylene blue concentration on the efficiency degradation by Fe_3O_4/TiO_2 catalyst material (Fe_3O_4/TiO_2 of 50 mg, methylene blue solution of 100 mL, pH of 10, and irradiation time of 45 min)

Figure 6 is show that with the increase in the concentration of methylene blue, the efficiency degradation decreased. It is consistent with the results of research Salmin S. Al-Shamali et al [13] who said that with the increase of methylene blue will increase the number of molecules methylene blue to be adsorbed on the surface of TiO_2, so would inhibit light which will penetrate TiO_2 and resulted in the formation of free radicals is reduced so that the efficiency of degradation of methylene blue decreasing.The success rate of the photocatalytic properties can be measured by the repetition of use of catalyst to degrade of dyes. From the experimental results showed that repetition Fe_3O_4/TiO_2 synthesized have very good photocatalytic properties. In this researche of repetition of degradation of methylene blue by Fe_3O_4/TiO_2 done untill 6 times still give high relative degradation efficiency of about 88 %.

The experimental results presented by the authors have demonstrated that the composite catalyst Fe_3O_4/TiO_2 can shorten the process of degradation, because for the recovery of the catalyst material it was sufficient to just use permanent magnets as shown in Figure 7. Beside that addition of Fe_3O_4 on TiO2 catalyst can widen the absorption of the UV light up to visible light region, so that the degradation process could be achieved by the sunlight.

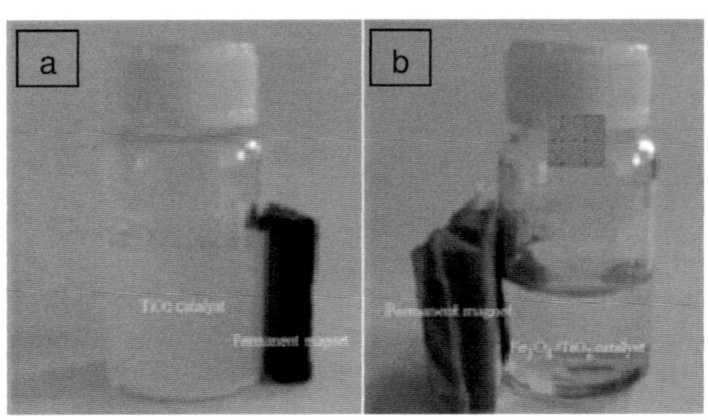

Figure 7. The recovery of catalyst after methylene blue degradation processing.

Figure 7 (a) shows the TiO_2 catalyst prior to the addition of Fe_3O_4-shaped suspension in the water and is not attracted by a magnet, therefore in this case it is necessary for the recovery process to use a centrifuge. After the addition of Fe_3O_4 in the TiO_2 catalyst (Figure 7 (b)), these catalyst material is easily attracted by a magnet, so it no longer needsa centrifugation (separation process becomes much simpler).

4. Conclusion

In this study the composite Fe_3O_4/TiO_2 catalyst has been successfully synthesized by using precipitation method for application in photocatalytic materials. The experimental results showed that the Fe_3O_4 addition in TiO_2 catalyst can simplify recovery process of the catalyst, and able to broaden the absorption area up to visible light, so that the degradation process can be more effectively carried out by using sunlight compared with only using the ultra violet light. The composite Fe_3O_4/TiO_2 catalyst is very appropriately used for waste treatment industry, especially the textile industry.

Acknowledgement
The author would like to thank all those who have helpfully in this research. The author is grateful that this research is fully supported by a grant DIPA of Center for Science and Technology of Advanced Materials – National Nuclear Energy Agency Republic of Indonesia.

References

[1] Scanlon D O, Dunnill C W, Buckeridge J, Shevlin S A, Logsdail A, Woodley S M , Catlow R A, Powell M J, Palgrave R G, Parkin I P, Watson G W, Keal Thomas W, Sherwood P, Walsh A and . Sokol A A. 2013 Nature Materials **12** 798–801
[2] Ma W F, Zhang Y, Li L L, You L J, Zhang P , Zhang Y T , Li J M, Yu M, Guo J, Lu H J, and Wang C C 2012 ACSNANO **6** 3179 –388
[3] Ctibor P Stengl V and Pala Z 2013 American Chemical Science Journal **3** 38-400
[4] Chang C Y and Hsieh Y H 2012 Journal of Nanomaterial **Article ID 13542** (1-6)
[5] Xuan S , Jiang W, Gong X , Hu, Y , and Chen Z 2009 Journal of Physic and chemitry C **113** 553-558
[6] Jiang W, Zhang X, Gong X, Yan F, Zhang Z. 2010 International Journal of Smart and nan Material **1** 278-287
[7] He Q, Zhang Z, Xiong J, Xiong Y and Xiao H 2008 A. Optical materials **31** 380-384
[8] Shi Z L, Dum C and Yao S H 2011 Journal of The Taiwan Institute of Chemical Enggineers **42** 652-657
[9] Kontos, Likodimos V, Stergiopoulos T, Tsoukleris D S, and Falaras P 2009 Chemical Material **21** 662-762
[10] Agrawal M, Gupta S, Pich A. Zafeiropoulus N E, Rubio-Retama J, Jehnicchen, D and Stammm 2010 Langmuir **26** 17649-17655
[11] Zaleska A 2008 Doped –TiO_2 : A Review, Recent Patents on Engineering **2** 157-164
[12] Fisli A, Saridewi R, Dewi S H and Gunlazuardi J 2013 Advavced Materials Research **629** 131-137
[13] Al-ShamalI S S 2013 Australian Journal of Basic and Applied Sciences 7 **4** 172-176
[14] Wardiyati S 2014 Prosiding Kimia dalam Pembangunan **XVII**
[15] Zhu C L, Zhang M , Qiao Y J, Xiao G, Zhang F, and Chen Y J 2010 Journal of Physic and Chemistry C **114** 16229-16235

IOP Conf. Series: Journal of Physics: Conf. Series **1091** (2018) 012016 doi:10.1088/1742-6596/1091/1/012016

Removal study of Cu(II), Fe(II) and Ni(II) ions from wastewater using polymer-coated cobalt ferrite ($CoFe_2O_4$) magnetic nanoparticles adsorbent

N S Asri[1], F A Nurdila[1], T Kato[2], S Iwata[3], and E Suharyadi[1]

[1]Department of Physic, Gadjah Mada University, Sekip Utara BLS.21 Yogyakarta, Indonesia 55281

[2]Department of Electrical Engineering and Computer Science, Nagoya University, Furocho 1 Chikusa-ku, Nagoya Japan 464-8601

[3]EcoTopia Science Institute, Nagoya University, Furocho 1 Chikusa-ku, Nagoya Japan 464-8601

E-mail : esuharyadi@ugm.ac.id

Abstract. Now days, heavy metal pollution has become one of the most serious environmental problems. The magnetic nanoparticles (MNPs) have received great attention to solving the problems as adsorbent materials due to their unique physical and chemical properties. $CoFe_2O_4$ MNPs was prepared by co-precipitation method, and its application for removal of Cu(II), Fe(II) and Ni(II) ions from wastewater. The various structural as well as morphological properties were characterized by x-ray diffraction (XRD), transmission electron microscopy (TEM), and vibrating sample magnetometer (VSM) analysis. Factor of affecting adsorption including temperature, contact time, particles size, magnetic properties and additional Polyethylene glycol (PEG-4000) as coating agent of adsorbent were further investigated. The results show that maximum Fe(II) was adsorbed at all of various parameters. Adsorption process of heavy metal ions from wastewater was known as increase as temperature range from RT to 60°C, but decrease at 120°C for Cu(II) and Ni(II). The maximum adsorption of Cu(II) and Ni(II) also occurs at the smallest particle size, 4nm. On the other hand, percentage of removing metal ions using PEG-coated MNPs decrease comparing to Co-MNPs without coating, for Cu(II), Fe(II), and Ni(II) is 97.1%, 100%, and 99.6%, respectively. The ability of Co-MNPs for removing metal ions decrease after coated using PEG-4000 maybe due to the small surface area and existence of polymer as nonmagnetic. Furthermore, the result of this experiment revealed that $CoFe_2O_4$ MNPs can be used as the effective adsorbent to remove metal ions in the wastewater.

1. Introduction

The presence of heavy metals as contaminant in our environment poses a serious problem due to human activity. Among these heavy metals are such as Hg [1], Pb[1-3,22], Cu and Zn [3-5], Cr [6], Sr[2], As [7],and Ni [4] which were released into the environment, and are now becoming a worldwide concern due to its toxic and bioaccumulation. In order to solve the problems, many experimental techniques have been applied to remove the contaminant as the colour, odour, organic and inorganic pollutant. There are several methods which have been developed such as ion exchange[4], nanofiltration [8], solvent extraction, electrowining, electrocoagulation and adsorptionmethod [3,5]. Compared with the other methods, the adsorption method is highly efficient, easy in application and is inexpensive [3].

Adsorption method depend on the ability of the surface of adsorbent to adsorb contaminants. The commonly employed adsorbents used for heavy metal removal are graphene oxide [2,9], active carbon [10],Chitosan modified [11], magnetite [12], etc. More recently, much attention has been paid to nanoparticles adsorbent, such as Fe_3O_4, $MnFe_2O_4$, etc., due to their basic properties, extremely small size, high surface area to volume ratio, and unique morphology, which provide better kinetics for the adsorption of metal ions from wastewater [13].

The one of magnetic nanoadsorbent which has attracted attention is a ferrite which chemical formula is $MeFe_2O_4$ (Me = Co, Ni, Zn, Mg, etc.). Among ferrites, $CoFe_2O_4$ is an interesting magnetic material due to its moderate saturation magnetization, excellent chemical stability, and mechanical hardness [14]. Compared with the other ferrites of the same size, cobalt ferrite exhibits better magnetic response [15]. In order to further improve the dispersion of nanoparticles as well as to prevent their agglomeration, polyethylene glycol (PEG) is widely used as the coating agent. In addition to these applications, PEG can also support magnetic nanoparticles due to its pore forming ability and its hydrophilic characteristicwhich can improve the adsorption capacity rate [16].

In this study, $CoFe_2O_4$ magnetic adsorbent are used for Cu(II), Fe(II) and Ni(II) metal ions removal. We focus on various study under control parameters such as temperature, contact time, particles size, magnetic properties, and modification of $CoFe_2O_4$ using PEG as a coating agent.

2. Material and methods

2.1. Preparation of magnetic nanoparticles $CoFe_2O_4$

$CoFe_2O_4$ nanoparticles were prepared by the co-precipitation of an aqueous solution of $FeCl_3.6H_2O$ (2 mol; 14.35 mL) and $CoCl_2.6H_2O$ (1mol; 7.9 mL). The mixture of the solution is then dispensed into a NaOH solution drop by drop at a temperature of 80°C. The solution is stirred with a 450 rpm stirring speed for 3 hours. The synthesized materials were rinsed and repeated for ten times to remove salt and impurities particles. In producing nanoscale, subsequent furnace at 80°C for 4 h was needed.

2.2. Characterization

The characterization of magnetic nanoparticles was carried out using the X-ray diffraction (XRD) (Shimadzu model XD-3H) method for crystal identification. The instrument was equipped with a copper anode generating Cu-Kα radiation (λ = 1.5406 Å). A bright field TEM apparatus (JEOL JEM-1400) was used for size investigation and morphology of magnetic nanoparticles and the VSM method (Riken Denshi Co., Ltd) was used for magnetic behavior analysis.

2.3. Adsorption studies Batch

A batch of experiments have been conducted to examine heavy metal removal from artificial wastewater by $CoFe_2O_4$. The process of making artificial waste using basic materials $CuSO_4.5H_2O$, $FeSO_4$. H_2O, and $NiSO_4.6H_2O$, respectively, in which 1.5 grams of material is dissolved in 1000mL of distilled water. The solution was stirred by amagnetic stirrer (room temperature) for 4 hours and then HCl (chloride acid) was added drop wise into the solution for as many as 20 drops.

The adsorption process in wastewater is carried out with the mixture 0.4 g of $CoFe_2O_4$ in 80 mL waste water solution, blended thoroughly using a magnetic stirrer.NH_4OH added to the solution in order to control the pH level of the solution at level 9.The mixture solutionis then shaken for 3 hours at room temperature. The solution was separated with the pollutant subsequent filtering. The authors then carried out similar process for various parameters, such as temperature (room temperature, 60, 90, 120)°C, contact time (60, 180, 300) min, particles size (4, 8, 9, 10) nm, magnetic properties (8,18, 37) emu/g and adding polymer PEG-4000 (polyethylene glycol) as coating agent with mass ratio 1 : 1.

The amount of adsorbed Cu(II), Fe(II) and Ni(II) ions was calculated according to the following equations :

$$\textit{metal ion removal efficiency } Re = \frac{(C_0 - C_f)}{C_0} \times 100(\%) \quad (0.1)$$

where C_0 and C_f are the initial and the final concentration of Cu(II)/Fe(II)/Ni(II) ions in the solutions, respectively.

3. Result and discussion

3.1. Characterization of magnetic nanoparticles

The XRD spectra of $CoFe_2O_4$ and $CoFe_2O_4$+PEG-4000are shown in figure 1, and $CoFe_2O_4$ sample for various particles are shown in figure 2. It indicates that the $CoFe_2O_4$ phase has been formed in the sample.The indication is based upon the main reflection peak (311) which appears at the 2θ diffraction angle of 35.379°, which is consistent with the typical characteristic of $CoFe_2O_4$. The other reflection peaks (220), (400), (511) and (440) for $CoFe_2O_4$ (JCPDS 22-1086) are also clearly observed. The obtained lattice parameter is (8,408±0,040 Å). This result is close to the literature value for the lattice bulk of $CoFe_2O_4$ (8,395±0,005Å) [17]. The addition of PEG as coating agent into $CoFe_2O_4$ samples has resulted in the formation of impurities; α-Fe_2O_3 (hematite), α–FeO(OH),andγ-FeO(OH). The particles size values were obtained by using the Scherrer formula 8.8 nm for $CoFe_2O_4$and 10.6 nm $CoFe_2O_4$+PEG-4000, respectively.

1
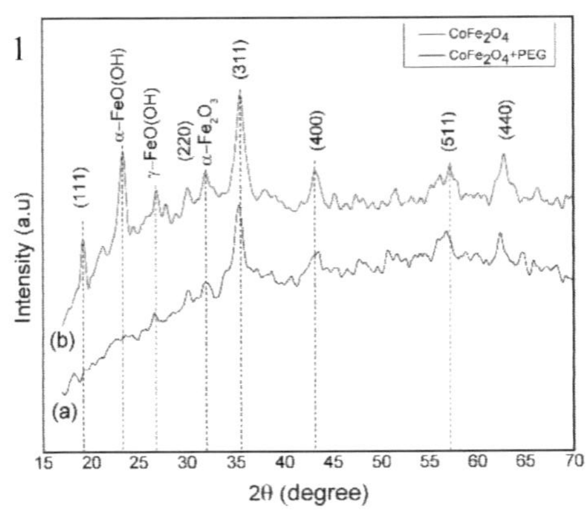

Figure 1. XRD spectrum; (a) $CoFe_2O_4$, (b)$CoFe_2O_4$+ PEG-4000

2
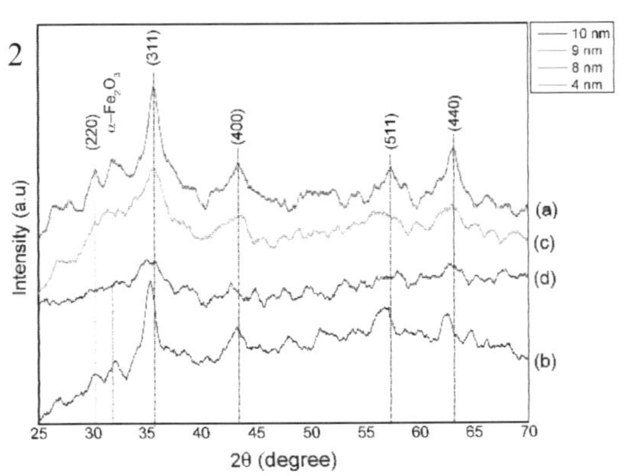

Figure 2. XRD spectrum of $CoFe_2O_4$ at various particles size; (a) 10 nm, (b) 9 nm, (c) 8nm and (d) 4nm.

The TEM images of the prepared $CoFe_2O_4$ particles were obtained as shown in figure 3(a).$CoFe_2O_4$ surface morphology analysis showed the agglomeration of many ultrafine particles with a diameter of about 8.4 nm. This is consistent with Scherrer analysis on the XRD diagram, which is about 8.8 nm. The pattern of dash ring showed the diffraction of particles samples and the Miller indices that could be identified were (220), (311), (400), (511),and (440).

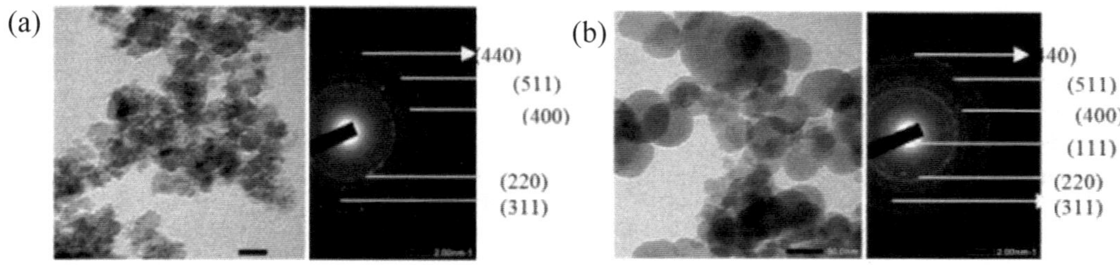

Figure 3. The result of TEM analysis of sample (a) $CoFe_2O_4$ and (b) $CoFe_2O_4$ coated by PEG-4000

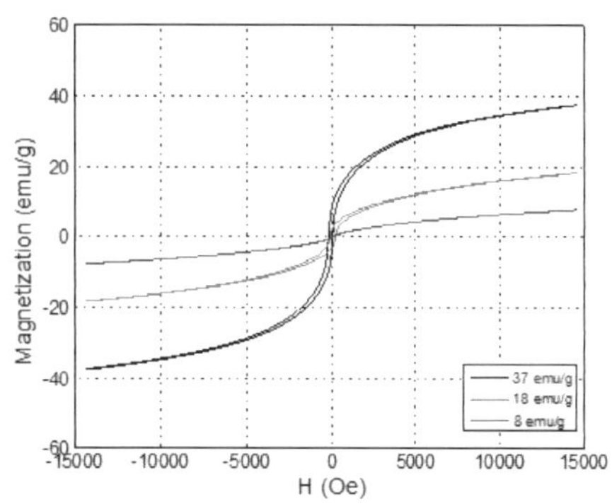

Figure 4. Magnetic moment vs. applied magnetic field for various sample of $CoFe_2O_4$ adsorbent.

$CoFe_2O_4$+PEG sample has spherical form and is homogeneous as shown in figure 3(b). The particle size after PEG coating is larger than that of pure $CoFe_2O_4$. Miller indices from this sample can be identified as (111),(220), (311), (400), (511),and (440).The particles size distribution obtained from TEM analysis is around 10nm.

It is known that magnetic particles which less than 30 nm diameter size will exhibit superparamagnetism[6]. Therefore, in this experimental work, the authors would like to claim that the $CoFe_2O_4$ sample has superparamagnetic properties. It has been verified by the magnetization curve measured by VSM method, as shown in figure 4. The magnetic samples are expected to respond well to the magnetic field without any permanent magnetization.

3.2. Adsorption Studies

3.2.1. Effect of temperature

To investigate the effect of temperature on the adsorption capacity, the experiment was carried out for initial metal uptake concentration of 5 g/L at four different temperature of 30°C, 60°C, 90°C and 120°C, respectively. It has been observed that for the increasing temperature, the percentage of metal uptake would decrease. It arises due to the weakness of bonding energy between the adsorbent and pollutant [18].

Table 1 shows the adsorption of Fe(II) maximum, below 0.02 ppm (limit detection). In this case, temperature did not have any significant effect on the adsorption. The sorption of Ni(II) increases proportional to the increasing temperature. On the other hand, the sorption of Cu(II) reaches a maximum at 60°C, and decreases at 120°C. The decreases of metal uptake due to binding of Cu(II) and adsorbent have desorption at the temperature [19]

Table 1. The effect of temperature on removal efficiency.

Parameter	Raw material	Removal efficiency	Temperature (°C)			
			30	60	90	120
Cu	292.144	(ppm)	2.73	0.87	1.19	6.17
		(%)	99.07	99.70	99.59	98.28
Fe	306.240	(ppm)	no detection	no detection	no detection	no detection
		(%)	100	100	100	100
Ni	358.940	(ppm)	0.34	0.23	0.19	2.45
		(%)	99.91	99.94	99.95	99.34

Table 2. The effect of contact time on removal efficiency.

Parameter	Raw material	Removal efficiency	Contact time (min)		
			60	180	300
Cu	292.144	(ppm)	1.64	2.73	1.30
		(%)	99.44	99.07	99.55
Fe	306.240	(ppm)	no detection	no detection	no detection
		(%)	100	100	100
Ni	358.940	(ppm)	2.13	0.34	0.52
		(%)	99.41	99.91	99.85

The adsorption by the system is of the physics adsorption typeso the temperature is a major contributor tothe major effect on exothermic process of pollutants. The adsorption occurs due to the Van der Waals interaction at the exothermic condition.Van der Waals interaction is the weak interaction between molecules and it releases low energy. It means that at low temperatures, molecules could attract adsorbents. The increase in temperature will affect an increase in the energy and ionic reactivity. As with this increase, ions will be enabled to overcome the energy barrier and to chemically interact with the active sites on the surface.

Since the ionic reactivity increases, then the number of diffusingions at the adsorbentsurface also increases. Therefore, ions will be adsorbed more at the surface; and as one could see, the metal Cu(II) removal increases as the temperatureincreases, from 30°C to 60°C. Equilibrium temperature occurs at 60°C. Above that, ionic reactivity could break the other bond, however due to a very weak Van der Waals interaction,a desorption might occur instead.

3.2.2. Effect of contact time

Contact time is another factor in batch adsorption process. In order to establish an equilibration time for the maximum uptake, the adsorptionof heavy metal was studied as a function of contact time.For this purpose 0.4 g of $CoFe_2O_4$ in 80 mL of wastewater was shaken for a period of 60-, 180- and 300 minutes. Table 2 shows the variation or the removal efficiency of the Cu(II), Fe(II) and Ni(II). The adsorption rate was very fast and adsorption was achieved by Fe(II) within 60 minutes. On the other hand, removal of Cu(II) decreased at the 180 minutes time, and increased when the contact time run into 300 minutes. The decrease in the of Cu(II) removal is caused by a desorption process. The desorption process occurs at weak interaction between Cu(II) ions and $CoFe_2O_4$as adsorbent. Reverse phenomenon occurs at adsorption study of Ni(II) ions. The maximum removal of Ni(II) took180 minutes but it decreased after 300 minutes. In the adsorption process, any interaction andcollision occurs between metal ions and adsorbent. The increase in contact time means an increase in the interaction that resulted in the molecular bond being broken. This provided the chance for the active sites to adsorb each metal ions. In this case, the adsorption is known to be reversible.

Table 3. The effect of particle size of $CoFe_2O_4$ on removal efficiency.

Parameter	Raw material	Removal efficiency	Particle size (nm)			
			10 nm	9 nm	8 nm	4 nm
Cu	292.144	(ppm)	3.49	2.73	3.16	2.30
		(%)	99.08	99.28	99.17	99.40
Fe	306.240	(ppm)	no detection	no detection	no detection	no detection
		(%)	100	100	100	100
Ni	358.940	(ppm)	0.89	0.34	0.74	0.80
		(%)	99.77	99.91	99.81	99.80

Table 4. The effect of magnetic properties on removal efficiency.

Parameter	Raw material	Removal efficiency	Magnetic properties (emu/g)		
			37	18	8.0
Cu	292.144	(ppm)	2.73	2.19	2.30
		(%)	99.28	99.42	99.40
Fe	306.240	(ppm)	no detection	no detection	no detection
		(%)	100	100	100
Ni	358.940	(ppm)	0.34	0.74	0.77
		(%)	99.91	99.81	99.80

Table 5. The effect of polymer-coated $CoFe_2O_4$ on removal efficiency.

Parameter	Raw material	Removal efficiency	Nanoadsorbent	
			$CoFe_2O_4$	$CoFe_2O_4$ + PEG
Cu	292.144	(ppm)	2.73	8.47
		(%)	99.07	97.10
Fe	306.240	(ppm)	no detection	no detection
		(%)	100	100
Ni	358.940	(ppm)	0.34	1.31
		(%)	99.91	99.63

3.2.3. Effect of particle size

The authors prepared $CoFe_2O_4$ samples,each with a different particles size, and used them as adsorbent to adsorb heavy metal in wastewater. Table 3 shows result of particles size effect on the metal uptake of Cu(II), Fe(II) and Ni(II). The maximum removal occurs with the smallest particles size. The adsorbent of small particles size has large surface area to adsorb the heavier metals [20]. Surface area besides the particles size, depends on the adsorbentsmorphology. However, in case of the 8 nm particle size samples, the capacity of adsorption decreased. In addition, the phenomenon occurred due to magnetic properties of metal pollutant (volume of metal ions) and solubility of metal in water.

3.2.4. Effect of magnetic properties
The heavy metal adsorption efficiency by different magnetic properties show some particular trend but not very significant. The maximum adsorption efficiency of Ni(II) by $CoFe_2O_4$ would occur at the utmost magnetic properties. Moreover, the magnetic properties of Ni(II) pollutant contribute to nano adsorbent $CoFe_2O_4$. In addition, due to the large saturation magnetization values, they can be separated from the solution with a permanent magnet in a relatively short time [21]. On the other hand, the adsorption efficiency of Cu(II) has reversed from metal uptake of Ni(II), as seen from table 4. The sample of $CoFe_2O_4$nano adsorbent with the lowest magnetic properties has the maximum adsorption to Cu(II). The possible explanation of this phenomenon is the role played by some other factors such as surface area, particle morphology, and available active sites [6].

3.2.5. Effect of PEG-coating
PEG (Polyethylene glycol) is used to coat the $CoFe_2O_4$ nano adsorbent in order to stablize it against oxidation, corrosion, and spontaneous aggregation to increase their physic-chemical stability and to provide a functionalized surface [21]. Furthermore, the functions of PEG is being nanoparticles template to control the size and pores structures of particles. However, the particle size of nano adsorbent which was coated with PEG is larger than before coating. The particle size of nano adsorbent is related to surface area of the particles. It means that the particles with the small size have larger surface area than the bigger one. The effect of that is the particles have many active sites to adsorb heavy metal in wastewater. Therefore, the PEG coated $CoFe_2O_4$ nano adsorbent would adsorb the heavy metal Cu(II) and Ni(II) less than $CoFe_2O_4$ without coating, which can be seen in table 5.

4. Conclusion
$CoFe_2O_4$ magnetic nanoparticles (MNPs) is effectively used as adsorbent to remove heavy metal Cu(II), Fe(II) and Ni(II) ions from wastewater. A variety of parameters were then introduced in the experiment in order to complete this study. The percentage of metal ions removal is not significantly affected by the change of temperature and contact time. Particles size gives the major contribution on large surface area, so that metal uptake has a maximum value. Magnetic properties of $CoFe_2O_4$ contributes to the separation of the pollutants within the solution. However, the additional polymer polyethylene glycol as agent of adsorbent coating decrease the ability of $CoFe_2O_4$ MNPs to remove metal ions. The possible explanation is that the particles size was increased after coating by polymer, and consequently the surface area become small.

Acknowledgement
1. HibahKompetensi (Hikom) KementerianRistekDikti, Indonesia, 2015.
2. Indonesia Endowment Fund for Education, Ministry of Finance, Republic of Indonesia, 2013-2015.

References
[1] Pastora J G, Bringsas E, Ortiz I 2014 *Chemical Engineering Journal* **256** 187-204
[2] Zhang Y, Yan T, Yan L, Guo X, Cui L, Wei Q, Du B 2014 *J. of Molecular Liquids* **198** 381-387
[3] Ge F, Li M -M, Ye H, Zhao B -X 2012 *Journal of Hazardous Materials* **211– 212** 366– 372
[4] Rudnicki P, Hubicki Z, Kołodynska D 2014 *Chemical Engineering Journal* **252**362–373
[5] Foroughia F, Tabrizib S A H, Amighiana J, Teluri A S 2015 *Ceramics International* **41** 6844–6850
[6] Li X, Niitsoo O, Couiz A 2014 *J. of Colloid and Interface Science* **420** 50-56
[7] Dey A, Singh R, Purkait M K 2014 *J. of Water Process Engineering* **3** 1-9

[8] Zha C, Du S, Wang T, Zhang J, Luan Z 2012 *Adv. Chem. Eng. Sci.* **2** 366-371
[9] Deng J H, Zhang X R, Zeng G M, Gong J L, Niu Q Y, Liang 2013 *J. Chemical Engineering Journal* **226** 189–200
[10] Fu L, Liu F, Ma Y, Tao X, Ling C, Li A, Shuang C, Li 2015 *Chemical Engineering J* .**263** 83-91
[11] Ren Y, Abbood H A, He F, Peng H, Huang K 2013 *Chemical Engineering Journal* **226** 300–311
[12] Iwahori K, Wanatabe J, Tani Y, Seyama H, Miyata N 2014 *J. of Bioscience and Bioengineering* **117** 333-335
[13] Zhao L, Li X, Zhao Q, Qu Z, Yuan D, Hu X, Chen G 2010 *J. of Hazardous Materials* **184** 704-709
[14] Huang Y, Wang W, Feng Q, Dong F 2013 *J. of Saudi Chemical Society*
[15] Shipley, H J, Yean S, Kan A T, Tomson M B 2010 *Environmental Science and Pollution Research* **17** 1053-1062
[16] Coey J M D 2010 *Magnetism and Magnetic Materials* United States of America, Cambridge University Press.
[17] Hu J, Lo, M C I, Chen G 2007 *Separation and Purification Tecnology* **56** 249-256
[18] Baldi G, Bonacchi D, Innocenti C, Lorenzi G, dan Sangregorio 2007 *J. of Magnetism and Magnetic Material* **311** 10-16
[19] Liu X M, Fu S Y, Zhu L P 2007 *J. Solid State Chemical* **180** 461-466
[20] Ai L, Zhou Y, Jiang J 2011 *Desalination* **266** 72-77
[21] Satyro S, Race M,Marotta R, Dezotti M, Spasiano D, Mancini G, Febbricino M 2014 *J. of Environmental Chemical Engineering* **2** 1969-1979
[22] Zhou L, Ji L, Cheng M, Shao, Y P, Zhang H, Gao W, Li Y 2014 *Journal of Hazardous Materials* **265** 104– 114

IOP Conf. Series: Journal of Physics: Conf. Series **1091** (2018) 012017 doi:10.1088/1742-6596/1091/1/012017

The effect of dry and wet milling on microstructure and properties of bonded NdFeB magnets

P Sebayang[1*)], T Sudiro[1], D Aryanto[1], D Hasibuan[2], Z Sitorus[2]

[1]Research Center for Physics-LIPI, Kawasan Puspiptek Serpong Gd. 440-442, Tangerang Selatan 15314, Indonesia
[2]Physics Department, Faculty of Mathematics and Natural Science, Universitas Sumatera Utara, Medan 20155, Indonesia

E-mail: [*)]sebayang.perdamean@gmail.com

Abstract. The preparation of NdFeB has been conducted in a dry and wet milling condition using a planetary ball mill (PBM) for ½, 1, 2, 3, 4, and 5 hours. The dry and wet milling was performed in an argon gas and toluent, respectively with ball and powder ratio of 10:1. The milled powder was dried in the vacuum condition of 11 mBar at the temperature of 60°C for 10 hours. 5wt.%Celuna (WE-518) was used as binder, mixed with magnetic powder and anisotropically pressed at 25 kgf/cm^2. Curing process was conducted in a vacuum chamber of 11 mBar at temperature of 180°C for 1 hour. The particle diameter and the true density of NdFeB powder were measured by using PSA and Pycnometer, respectively. The morphology and phase composition were characterized using SEM and XRD, respectively. The obtained results show that the optimum milling time of NdFeB powder is 3 hours using wet milling method. The results also concluded that the powder true density of 7.79 g/cm^3, the bulk density of 5.98 g/cm^3 and the mean particle diameter of 9.59 μm can be achieved. The SEM image confirmed that the milling process has an effect in reducing the particles size. But the milling process lead to a damage in the crystal structure of the material, resulting in decreasing the magnetic flux density. The half hour dry milling is capable to produce a maximum magnetic flux density of 1816 Gauss.

1. Introduction

The demand for magnetic materials continues to rise in line with advances in the electronics technology, computerization and transportation [1]. It was begun after the discovery of the rare earth-based permanent magnets, such as NdFeB, RECo, and REFeB [2,3]. Currently, permanent magnet materials in a generator/electric motor have also been used to convert the electrical energy into the mechanical energy or vice versa [2,4]. The advantageous of a permanent magnet is the fact that the energy produced is quite high and environmentally friendly [5,6]. In general, the main qualities of permanent magnet are characterized by several parameters, such asmagnetic remanence (Br), magnetic coercivity (Hc), the maximum energy product (BH_{max}), magnetic flux density, Curie temperature, etc [7,8].

The NdFeB permanent magnets have been widely used, especially in the electronic equipment, electrical generators, sensors transducer, automotive, petrochemical industries, medical equipment products and other applications. Currently, permanent magnet is also used in the automotive field as electric power steering system, sensor rpm/velocity and electric motor. The numbers continue to increase rapidly every year [6,7].

Currently, permanent magnet based on $Nd_2Fe_{14}B$ phase is the best type of permanent magnet which has a maximum energy product (BH_{max}) that reaches up to 27 MGOe[9]. Therefore, small size $Nd_2Fe_{14}B$ permanent magnet has high capability [10]. NdFeB magnets have also been used to replace samarium cobalt magnets in some applications, in particular for the application at the temperatures less than 80°C [5]. The disadvantage of this type of magnet is its low curie temperature, easily oxidized and difficult to be applied at high temperatures condition [11,12].

In order to support the development of NdFeB magnets, the research is focusing on the effect of dry and wet milling process on fabrication, microstructure and properties of NdFeB magnets. In this study, NdFeB powder (type MQP-B+ 10118-70) is selected as the main raw material, Planetary Ball Mill (PBM) is used for milling process, argon and toluent as milling media, and celuna (WE-518) as an adhesive medium. The compaction and curing process was conducted at 25 kgf/cm^2 in the magnetic field press and at 180°C in vacuum (11 mBar) for 1 hour. The powder mean diameter and true density, bulk density, morphology, microstructure, and magnetic flux density was characterized, measured and discussed.

2. Experimental Method

The raw material powders used in this study is NdFeB powder type MQP-B+10118-70 (technical raw materials). The specification of NdFeB powder MQP-B+ are specified in the following: residual induction (Br) = 8.95-9.15 kG, intrinsic coercivity (HCI) = 9.0-10.5 kOe, coercive force (H_c) = 6.8 kOe, energy product (BH_{max}) =15.8-16.8 MGOe, Curie temperature (T_c) =360°C, maximum process temperature = 200°C and maximum operating temperature =120-160°C. The wet and dry milling was done by using a Planetary Ball Mill (PBM) that has powder and ball ratio of 1:10 for variation of time: ½, 1, 2, 3, 4 and 5 hours. In the wet milling process, toluene was used as milling medium and in the dry milling process; argon gas was injected in the crucible PBM. The ball milled powder was dried in a vacuum oven of 11 mBar at 60°C for 10 hours.

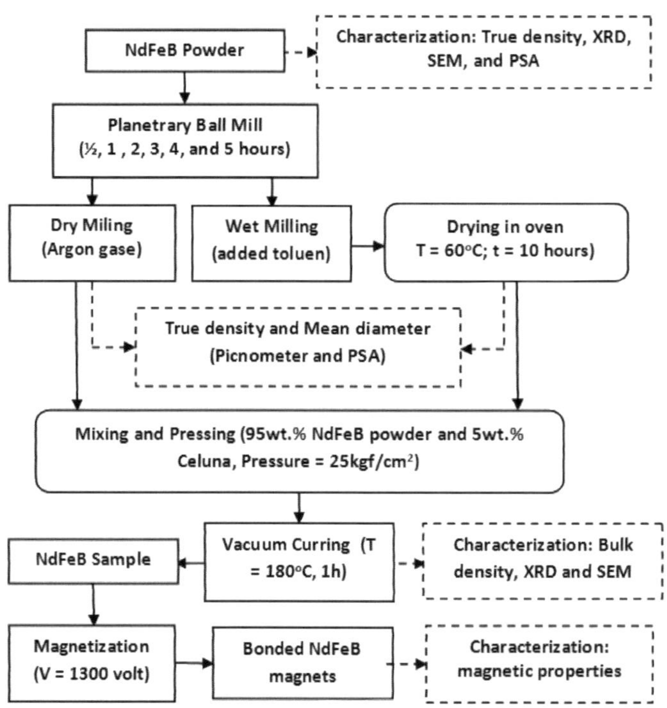

Figure 1. Flow diagram of preparation and characterization of bonded NdFeB permanent magnets.

For compaction, 3 grams of NdFeB powder was mixed with 5wt.%Celuna (WE - 518) and pressed in a magnetic field press (anisotropy) with a pressure of 25 kgf/cm^2. The compact sample was then cured in a vacuum chamber of 11 mBar at temperature of 180°C for 1 hour. The specimen magnetization was conducted by using Impulse Magnetizer (voltage of 1300 volts). The specimen characterization tested are the mean particle diameter, true density, bulk density, morphology, microstructure, and magnetic flux density by using PSA, pycnometer, Archimedes principle, SEM, XRD and Gaussmeter, respectively. The flowchart of the fabrication process and the characterization of bonded NdFeB permanent magnets are shown in figure 1.

3. Results and Discussion

The cumulative diameter of NdFeB powders after dry and wet milling for 0.5; 3 and 5 hours were measured by using a Particle Size Analyzer (PSA) as shown in figure 2.

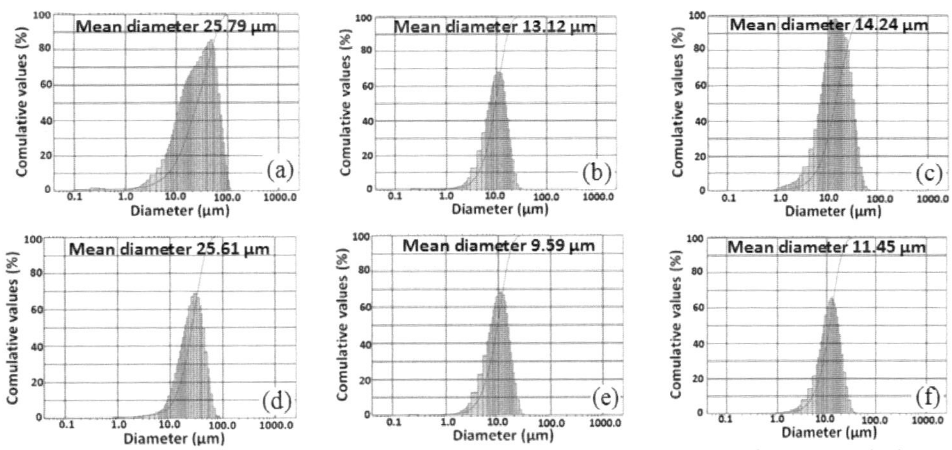

Figure 2. Cumulative value versus diameter of NdFeB powders after dry milling for ; (a). 0.5h;(b). 3.0 h;(c). 5.0h and wet milling for ; (d). 0.5 h;(e). 3.0 h and (f). 5.0 h.

The diameter of NdFeB powder as a function of milling time prepared by dry and wet milling are shown in figure 3. It can be seen that the mean particle diameter decreased at longer milling time until it reaches the minimum conditions. However, when the milling time was extended, the grain size was not decreased due to the powder agglomeration. The wet milling is a relatively better technique compared with the dry milling. This is probably due to that the collisions between the ball and grains are more appropriately milled. The optimum milling time is achieved at 3 hours with a mean diameter of 13.12 µm for dry milling and 9.59 µm for wet milling.

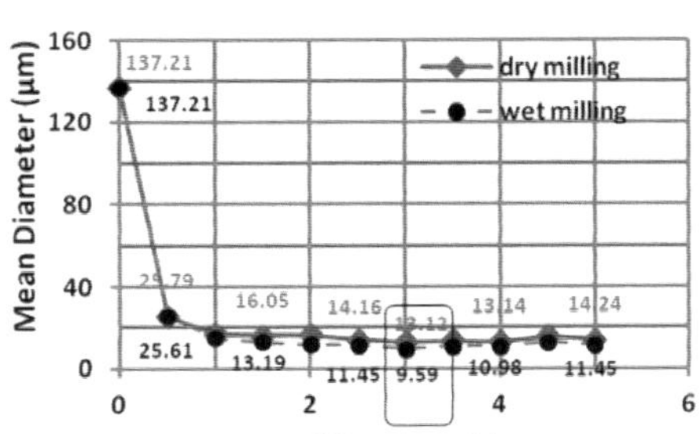

Figure 3. The diameter of NdFeB powders after dry and wet milling as function of milling time.

Figure 4 shows the true density of NdFeB powder at various milling time. The highest true density was obtained after 3 hours milling time. This means that as the grain size decreased, the true density increased. The true density of wet and dry milled powders after 3 hours milling time is 7.79 g/cm^3 and 7.75 g/cm^3, respectively.

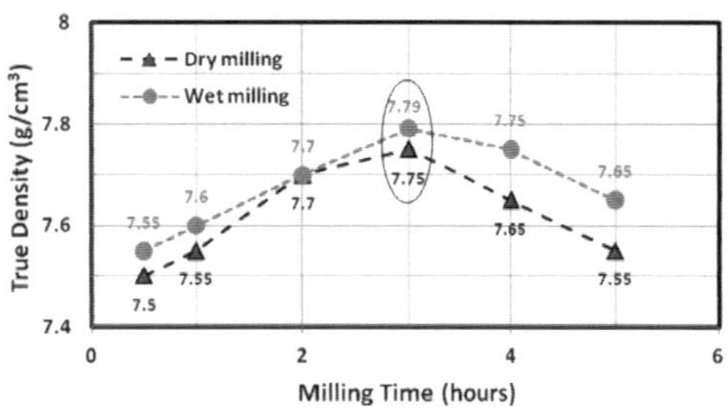

Figure 4. True density of NdFeB powders as a function of milling time after dry andwet milling process.

The SEM images of NdFeB powder in different condition of milling and time are shown in figure 5. All SEM photographs of NdFeB powder was taken with a magnification of 500X. Based on the results, it can be seen that the NdFeB powder before milling is relatively has larger diameter and randomly distributed. The SEM image also clearly shows the effect of milling in reducing the diameter of the particle. The smaller particles result in higher level of density and lower level of porosity. But, there are no significant changes in SEM images.Figure 6 shows that the bulk density of bonded NdFeB magnets at various milling time. The results show that the bulk density of bonded NdFeB magnet increased with the increase in milling time for up to 3 hours and then decreased at extended milling time. The optimum bulk density of the specimen is 5.98 g/cm^3 for wet milling and 5.90 g/cm^3 for dry milling condition. The bulk density of NdFeB samples prepared by wet milling is relatively higher than the dry milling samples. At longer milling time (>3h), the bulk density of the sample is likely to decrease. This is due to the occurrence of grain or agglomeration which results in the presence of pores or voids in the sample.

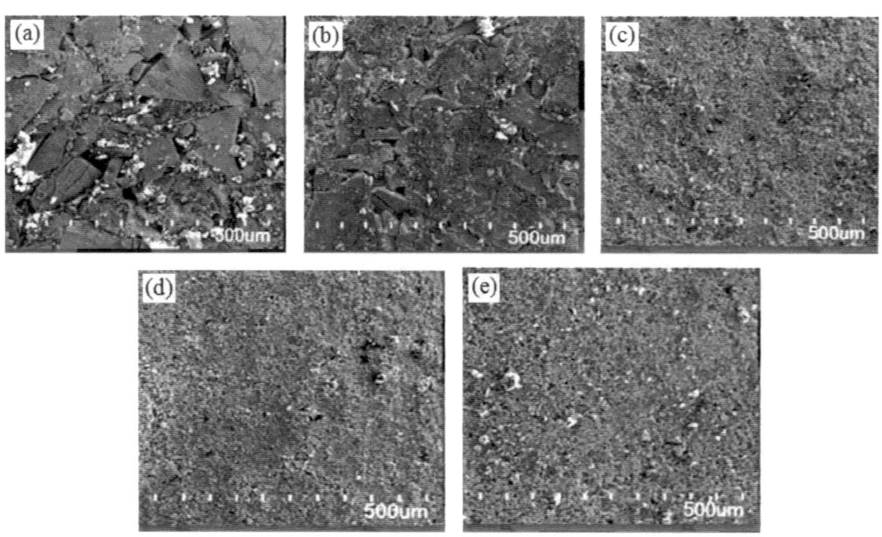

Figure 5. SEM images of bonded NdFeB magnets (a). without milling ; dry milling at : (b). ½ h, and (c). at 5 h ; wet milling at (d). ½ h and (e). 5 h.

The relationship between magnetic flux density of bonded NdFeB magnets and the milling time is shown in figure 7. The optimum magnetic flux density of bonded NdFeB magnets is obtained at half-hour milling time. The bulk density of dry milling process is relatively better than that of wet milling process. At half-hour milling time, the bulk density of the specimen is 1816 Gauss for dry milling and 1434 Gauss for wet milling. The magnetic flux density of bonded NdFeB magnet decreased with the increase in milling time. This suggests that the milling process can lower the magnetic properties of bonded NdFeB samples. This is probably due to the oxidation of the raw material during the milling or handling process.

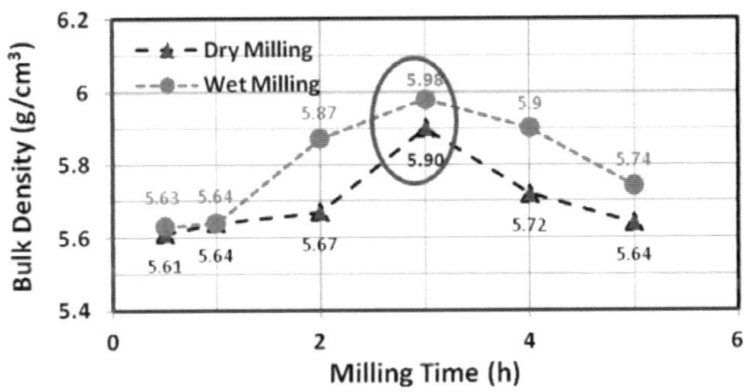

Figure 6. Bulk density of bonded NdFeB magnets at varying milling time.

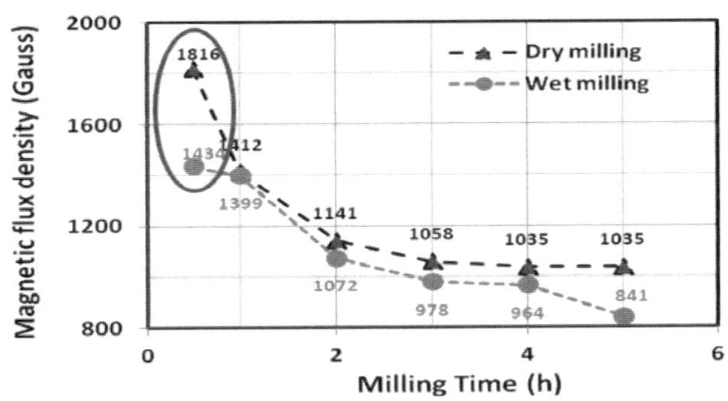

Figure 7. Magnetic flux density of bonded NdFeB magnet at varying milling time.

Figure 8 shows the XRD patterns of original powder, dry and wet milled powder (half and five hours). Apparently, $Nd_2Fe_{14}B$ phase was detected in the original powder, but it disappeared in the dry milled and wet milled powder. The ball milled powder was turned into an amorphous phase. This means that the milling process has a detrimental effect on the structure of the material. The powder is likely oxidized to form the oxide layer during the milling process.

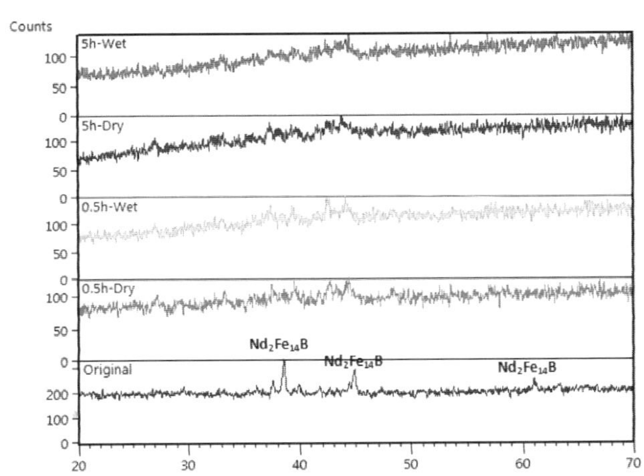

Figure 8. XRD pattern of (a) original powder, dry milled powder at : (b). ½ h and (c). 5 h; wet milled powder at : (d). ½ h and (e). 5 h.

4. Conclusion

Based on the above results, it can be concluded that the NdFeB powder that have been prepared using wet milling is relatively better compared to dry milling condition. The optimum milling time was achieved for 3 hours. These results show that a mean particle diameter of 9.59 μm and the true density of 7.79 g/cm^3 can be achieved. The optimum bulk density of bonded NdFeB magnets is 5.98 g/cm^3 that have been obtained using wet milling for three hours. The obtained results show that the milling process can reduce the size of the particles, but the milling process can also lead to damages in the crystal structure of $Nd_2Fe_{14}B$. Consequently, the magnetic flux density of bonded NdFeB magnet was decreased. The half-hour dry milling process was only able to reach a maximum magnetic flux density of 1816 Gauss.

References

[1] Pèrigo E A, deCampos M F, Faria R N and Landgraf F J G 2012 *Powder Technology* **224** 291-296
[2] Kramer M J, Mccallum R W, Anderson I A, And Constantinides S. 2012 *The Journal of The Minerals, Metals & Materials Society* **64** 7
[3] Li J, Liu Y, Gao S J, Li M, Wang Y Q, and Tu M J 2006 *J. Magn. and Magn.Mat.* **299** 195-204
[4] Liu Z W, Huang H Y, Gao X X, Yu H Y, Zhong X C, Zhu J and Zeng D C 2011 *J. Phys. D: Appl. Phys.* **44 025003** 1-10.
[5] Hamada N, Mishima C, Mitarai H and Honkura Y 2003 *IEEE Trans. on Magn.* **39 5** 2953-2955
[6] Zhang X H, Xiong W H, Li Y F, Song N 2009 *Materials and Design* **30** 1386-1390
[7] Zhai F, Sun A, Yuan D, Wang J, Wu S, Volinsky A A and Wang Z 2011 *J. Alloys and Compounds* **509** 687-690
[8] Sugimoto S 2011 J. Phys D: Appl Phys **44** 1-11
[9] Vuong N V, Khanh N V and Thuy D M 2003 *Physica B* **327** 349-351
[10] Jurkovic S, Rahman K, Bae B, Patel N and Savagian P 2015 *Energy Conversion Congress and Exposition (ECCE)* (ISBN: 978-1-4673-7151-3)
[11] Sebayang P, Muljadi, Sardjono P, Yuswitasari A and Adi W A 2013 *AIP Conference Proceedings* **40** 1555
[12] Nakamura H, Hirota K, Shimao M, Minowa T, Honshima M 2005 *IEEE Trans. Magn. Magn.* **41** 3844-3846

IOP Conf. Series: Journal of Physics: Conf. Series **1091** (2018) 012018 doi:10.1088/1742-6596/1091/1/012018

Study of magnetic properties of sintered alloy Fe-Cr ODS using VSM

R Salam, A Dimyati, Mujamilah and M Silalahi

Center for Science and Technology of Advanced Materials – BATAN, Kawasan Puspiptek Serpong, Banten, INDONESIA

E-mail : arbi.dimyati@gmail.com

Abstract. In this experimental work Vibration Sample Magnetometer (VSM) was used to study the magnetization property of ultrasonic assisted sinter material ODS Fe-Cr containing dispersoid particles Y_2O_3. The results of magnetization diagram analysis showed an anisotropy with saturation values 150 and 200 emu/gr for the angles 270 and 360°, respectively. Although higher saturation level, the alloy exhibits exceptionally small hysteresis loop with very low remanence and coercivity. As powder Fe-Cr mixture before sintering has maximum saturation 200 emu/gr. Properties combination of anisotropy, the absent of hysteresis loop and high saturation value has been opened new research opportunity on application of Fe-Cr based ODS material in field of sensors and magnetic assisted electro mechanic parts such as transformers and motors.

1. Introduction

The search for a combination of materials to obtain a more powerful magnet is still a major topic in all parts of the world. Among them is the research conducted at the Center for Science and Technology of Advanced Materials (PSTBM) BATAN, in the framework of the investigation process of the formation of micro alloys, the Center has created a Fe-Cr ODS sintered alloy material by using ultrasonic preliminary process[1]. In publications that have also been presented in the same seminar, it was reported that this material has the properties of a magnetoresistance alloy. Measurements showed that the value of the materials resistance decreased with the effect of magnetic fields up to 3-fold. The purpose of this study was to determine the magnetic properties of sintered alloy material Fe-Cr ODS containing Y_2O_3 dispersed particles with measurement techniques using Vibrating Sample Magnetometer (VSM). VSM is a standard tool used for testing the magnetic properties of a material. In addition to the components and their functions VSM, observation characteristics Vibrating Sample Magnetometer (VSM) measurement results become the focus of discussion conducted in this paper.

2. Experimental Method

2.1. Material

The material used in this study is the ODS Fe-Cr alloy containing Y_2O_3. The chemical composition of the material is shown in table 1. Alloys are prepared as follows: first the powder starting materials Feand Crare prepared using the ultrasonic method for 50 hours at the frequency of 20kHz and 750 Watts power. Then the mixture was compacted by isostatic pressing method with a manual press machineat a pressureof 6000 psi using molds with diameter of 10mm. After ward, the pelletized solid sample is transfered into an ampoule made of a quartz glass materials and then positioned inside the vacuum chamber. The ampoule and the sample is thensinteredin an oven at 1000°C and 1300°C gradually and without interruption for one hour each. After the sintering process, the pelletized samples are taken out of the ampoule in a way to solve it, followed by polishing using sand paper to get rid of the passivation layer which may grow during the sintering process with residual air left in the ampoule in the vacuum process.

Table 1. The chemical composition ODS Fe-Cr alloy.

No	material	contents[weight%]
1	Fe	87
2	Cr	12
3	Y_2O_3	1

2.2. VSM Components

All the VSM tool known today in principle has the same main components, namely as intended schematically in figure 1.

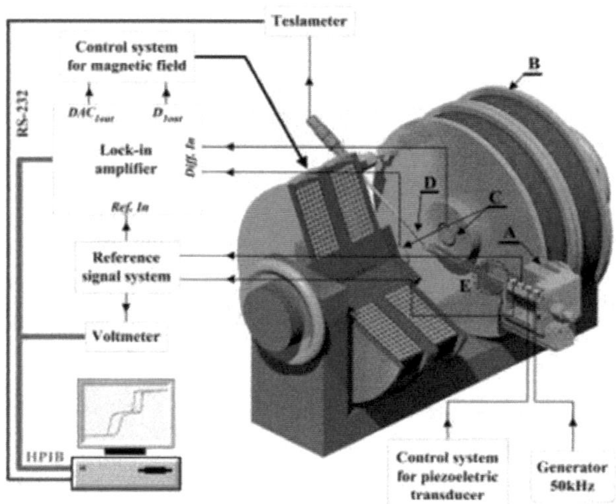

Figure 1. Test equipment vibrating sample magnetometer (VSM).

As shown in the figure can be broken down components of VSM following functions in sequence respectively as follows;

1. Head Generator is a place holder vibrator transmitted towards the sample in the sample holder or sample holder through a piezoelectric transducer.
2. The electromagnet or coil Hemholtz function generates a magnetic field to magnetize the sample then the process vibration transforms into electrical current that can be measured by the data processing system.
3. Pickup Coil transmit electrical signals from the sample into the amplifier. The signal will be transferred by the results of the induction coil pickup in the differential input to the lock-in amplifier. The signal from the pick-up coil is detected by a lock-in amplifier is noted as a function of magnetic field that allows us to get a hysteresis loop of the samples tested. For harmonic oscillation of the sample, the signal (e) induced in the pick-up coil is proportional to the amplitude of oscillation (K), the oscillation frequency of the sample (ω) and magnetic moment (m) of the samples tested.
4. The Hall Sensor is used to convert and transfer the magnetic field energy into voltage that will produce electric current. Hall sensors are also used to measure the flow without disrupting the flow

of currents that exist on the conductor. The current measurement sensor connects hall with teslameter.

5. Sensor capacity serves to provide a signal proportional to the amplitude of oscillation of samples and supplies voltage to the electronic system which generates a reference signal. The next signal will be given to the reference input of the lock-in amplifier. Digital converter output will be sent to analog (DAC1out) and digital output (D1out) of the lock-in amplifier controls the current flowing through the electromagnet and show direction respectively.

Detailed information about the VSM has been published [2]–[10].

2.3. VSM Measurement

To prepare for VSM testing, a sample is cut to the size of 2x2x2 mm and then weighed. Weight of the sample was less than 10 mg, is the weight for optimum function of the sensor system in the VSM. In the next step of the process the samples were put into a small container and covered with wax to ensure that the position has not changed during the vibration measurement. After that the container together with the samples are inserted into the sample holder rod and fastened using a sealing tape, as shown in figure 2.

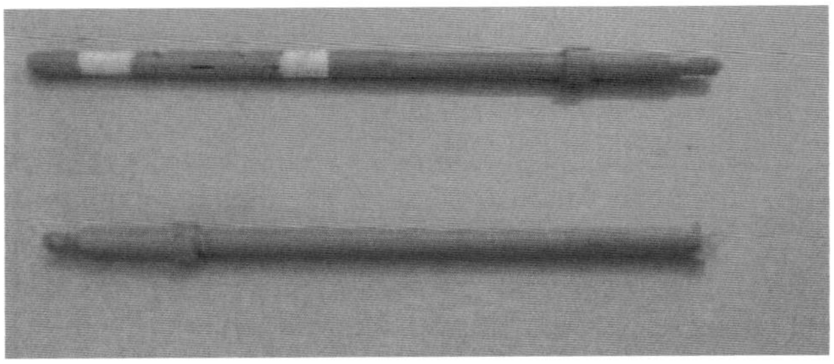

Figure 2. The placement of the sample in the sample holder stem of the VSM test equipment.

The necessary steps in the testing procedure are outlined below and visually shown in figure 3;

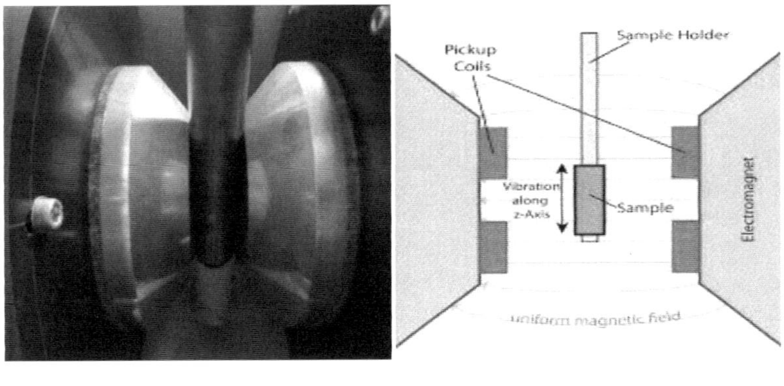

Figure 3. Working Principle vibrating sample magnetometer.

1. Put the sample into the prepared medium. Before running the VSM tool functions, the initial step that must be done is to place the sample holder between the poles of the magnetic field source.
2. Induction of the sample's dipole moments. Samples that have been positioned will be conditioned by a uniform magnetic field. This happens because the magnetic induction performed by the edge of the VSM wall is magnetizing the pickups coil.

3. Measure the sample standard signal. Once the magnetic induction has completed the magnetization process, samples will show the signal in the form of vibrations with a sinusoidal motion in the medium pickup coil. This signal has the same frequency, where the sample will be proportional to the vibration amplitude and magnetic materials.
4. Output unit vibration magnetometer. Signals sent from the pickup coil system will be forwarded to the differential amplifier contained in the vibration unit. The output of the differential amplifier is then processed in an amplifier which receives a reference signal. And the end result of the process of identification of this signal will be given by the magnetometer in the form of a proportional DC signal which provides information magnetic moment of the sample being tested.

3. Results and Discussion

Figures 4 to 6 show the results of VSM testing in the form of charts showing magnetization as a function of magnetic field applied to the sample powder and solid samples taken from different directions with different angle of 90°. The results showed that the magnetization value of the material (as read off the chart), either in powder or solid state magnetism, is a very high value compared to the values in ferromagnetic materials or permanent magnets. The maximum value of the measured saturation is 200 emu/g at 1 Tesla magnetic field strength. Besides all the characteristics of a graph showing the results of magnetization with a slim hysteresis loop. It shows that all the samples do not have a remanence value or coercivity which indicates that the influence of even a small or weak external magnetic field will cause the magnetic properties of materials to immediately return to its original state. In other words, the material is inherent magnetic properties will not immediately disappear when the influence of the magnetic properties of the external magnetic field is removed. These properties are a very important quality to have for the components used in devices whose performance or operation depends upon the rapid changes in the magnetic field such as motors, transformers and relays.

As shown in figure 4, the powder sample has a saturation value of about 200 emu/g. This value is approximately equal to the results of measurements on solid samples and was taken at an angle of 270° (see figure5). This shows that the state of the sample, whether as a powder or a sintered material has no effect on the value of magnetism, at least for one side of the sample. This could be an indication that the influence of magnetism is apparently caused by the ultrasonic process only, and not by the compaction process. However, because the ultrasonic process in this study was done on time, energy and the same frequency, so in this work, the authors can not be certain what effects would occur when the ultrasonic treatment parameters were changed. In order to able to clarify these aspects,the authors would have to do further research.

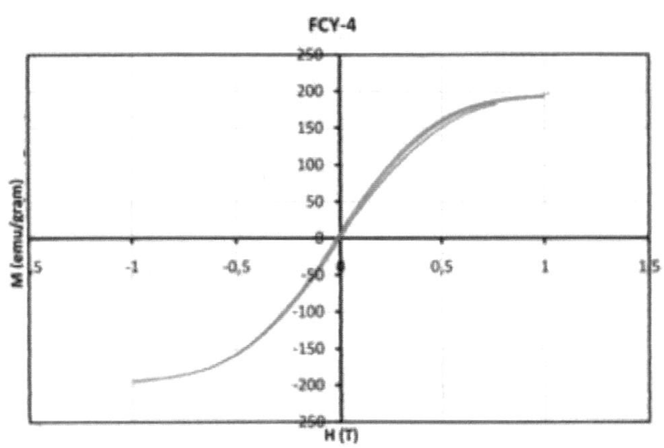

Figure 4. Graph magnetization VSM testing on samples of powder.

IOP Conf. Series: Journal of Physics: Conf. Series **1091** (2018) 012018 doi:10.1088/1742-6596/1091/1/012018

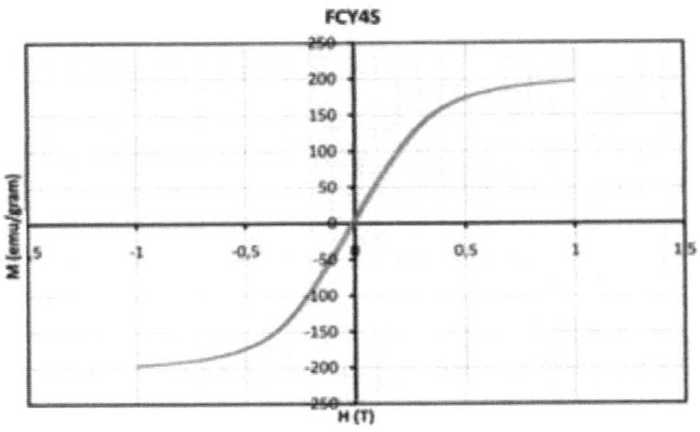

Figure 5. Graph magnetization VSM results on alloy samples taken in 270°.

VSM measurement results in solid samples taken from different points of 90° or perpendicular to the previous measurement are displayed in figure 5, showing the value of saturation at 1 Tesla magnetic field recorded at 160 emu/g (figure 6). This value is lower, 40 emu/g.if measured at the angle of 270°. Analogous to what is commonly found in metallographic, this phenomenon shows the texture of magnetism, where the sample has a tendency toward a certain type of magnetism. Causes of what can be called a magnetic anisotropy have not been found in this study. There is a possible relation to the physical processes that occur during the mechanical pressing procedure. When pressed grains of powder materials change shape and become elongated in the horizontal direction and retracts in the axial direction, which metallographically speaking would be the cause of crystalline texture to arise in the material. However, the physical processes that occur in it is not the object of discussion in this study. It requirs further research.

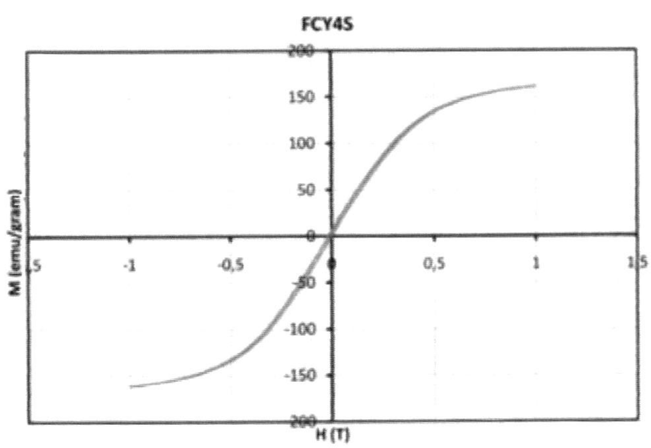

Figure 6. Graph magnetization VSM results on powder samples taken in 360°.

4. Conclusion

Based on the present study, it can be concluded the following: ODS sintered alloy Fe-Cr has a maximum saturation of 200 emu/g. The graph shows the magnetization of the material an isotropy on the angle of 90° with a sleek hysteresis loop shows instability in the magnetization of small magnetic field changes. This has opened up the possibility of new research linking objects between metallurgical and research-oriented magnetic sensor product in the form of materials or substances that work-based mechanical components magnetic.

References

[1] Silalahi M, Dimyati A, Harjanto S, Untoro , and Suharno B 2014 Microalloying of Fe-Cr by using ultrasonic irradiation *Int. J. Technol.* **5** 2.

[2] Mujamilah 2000 Vibrating sample magnetometer (VSM) tipe oxford vsm1.2h *Pros. Semin. Nas. Bahan Magn. I* 77–81.

[3] Case W E and Harrington R D 1966 *Calibration of vibrating-sample magnetometers* **70 4** 255–262.

[4] Ciprian R, Carbucicchio M and Palombarini G 2010 Interface phenomena and magnetic properties of Fe/Cr multilayers *IEEE Trans. Magn.* **46 2** 432–435.

[5] Žák T 2008 Testing New Vibrating Sample Magnetometer 5 K – 1273 K **59 7** 44–47.

[6] Žák T and Parma J 2006 Upgrading the Vibrating Sample Magnetometer with New HW and SW **57 8** 88–91.

[7] Raut A R and Murhekar G H 2015 Vibrating Sample Magnetometery Studies Modified and Doped Polyvinyl Alcohol Conjugates **4** 446–454.

[8] Kirupakar B R, Vishwanath B A, Sree M P and Deenadayalan 2016 Vibrating sample magnetometer and its application in characterisation of magnetic property of the anti cancer drug magnetic microspheres *Int. J. Pharm. Drug Anal.* **4 5** 227–233.

[9] Stamenov P and Coey J M D 2006 Vector vibrating-sample magnetometer with permanent magnet flux source *J. Appl. Phys.* **99 8** 2005–2007.

[10] Burgei W, Pechan M J and Jaeger H 2003 A simple vibrating sample magnetometer for use in a materials physics course *Am. J. Phys.* **71 8** 825–828.

IOP Conf. Series: Journal of Physics: Conf. Series **1091** (2018) 012019 doi:10.1088/1742-6596/1091/1/012019

Study of structure and resistivity of electron-doped superconducting cuprates $Eu_{1.85}Ce_{0.15}Cu_{1-y}Zn_yO_{4+\alpha-\delta}$

Risdiana, M. Manawan, L Safriani, T Saragi, A Aprilia, N Syakir, S Hidayat, A Bahtiar, Fitlilawati, and R E Siregar

Department of Physics, Padjadjaran University, Jl. Raya Bandung-Sumedang Km.21 Jatinangor, Sumedang 45363, Indonesia

E-mail : risdiana@phys.unpad.ac.id

Abstract. Partially Zn-substituted for Cu in electron-doped superconducting cuprates of $Eu_{1.85}Ce_{0.15}Cu_{1-y}Zn_yO_{4+\alpha-\delta}$ (ECCZO) with y = 0, 0.01, 0.02, 0.05 and various δ values have been studied in order to elucidate the effect of nonmagnetic impurity of Zn to their structure and resistivity properties in the electron-doped cuprates. It has been found that the main peaks of T' structures are clearly observed in all samples. Superconductivity disappeared with Zn substitution for Cu in electron-doped system of ECCZO.

1. Introduction

Mechanism of high T_c superconducting cuprates (HTSC) has been one of interesting topic to be studied in order to develop these materials for several applications. Based on charge carrier, there are two system of HTSC, namely, hole-doped system and electron-doped system. When conducting planes in HTSC lose one electron, one mobile hole remains in the planes, leading to the formation of a hole-doped system of HTSC. On the other hand, when conducting planes get an excess electron, the system forms an electron-doped HTSC. For hole-doped system especially La-based 214 cuprates, some reports were published to explore various properties such as magnetic properties, impurity effect, electronic phase diagram and conductivity [1-8]. Different with hole-doped system, only limited reports were published for electron-doped system [9-12] so that some properties in electron-doped system are still needed to be explored to understand mechanism of HTSC in electron-doped system.

For the crystal structure, both systems have common structures consist of both the conduction layer and the charge reservoir layer. The structure of 214 cuprates is basically tetragonal as shown in Figure 1. Each cooper ion is strongly bonded to four oxygen. There are different configurations of conducting layer between hole-doped and electron-doped system. In hole-doped system of $La_{2-x}Sr_xCuO_4$, cooper ion and oxygen form octahedral configuration. The structure is called the T structure. Oxygen atoms are located above and below each Cu atom, called apical oxygen as shown in Figure 1(a). In electron-doped system of $Ln_{2-x}Ce_xCuO_4$ with Ln is rare earth elements of Nd, Pr, Sm or Eu, cooper ion and oxygen form planar configuration. The structure is called the T' structure as shown in Figure 1(b). Superconductivity in electron-doped system is obtained after removing the oxygen excess at apical site that exist in the sample during heating or sintering process. This is one of the reason for difficulty in preparing high quality electron-doped superconductivity.

The study of relationship between some physical properties and superconductivity in hole-doped and electron-doped system via the impurity effect has also been intensively studied [5,7,10,13]. It is found that the superconducting volume fraction in hole-doped system of $La_{2-x}Sr_xCuO_4$ is more significantly decreased by a small amount of nonmagnetic impurity of Zn for Cu than that by magnetic impurity of Ni [8]. In electron-doped system, the superconductivity is suppressed through the magnetic impurity of Ni for Cu more markedly than through the Zn substitution [14].

The study of structure and crystallite size associated with the effect of nonmagnetic impurity of Zn for Cu in electron-doped system is also very limited. Here, we reported the study of structure and transport properties in electron-doped system of $Eu_{2-x}Ce_xCuO_4$ in order to elucidate the effect of nonmagnetic impurity of Zn to their structure and resistivity properties in the electron-doped cuprates.

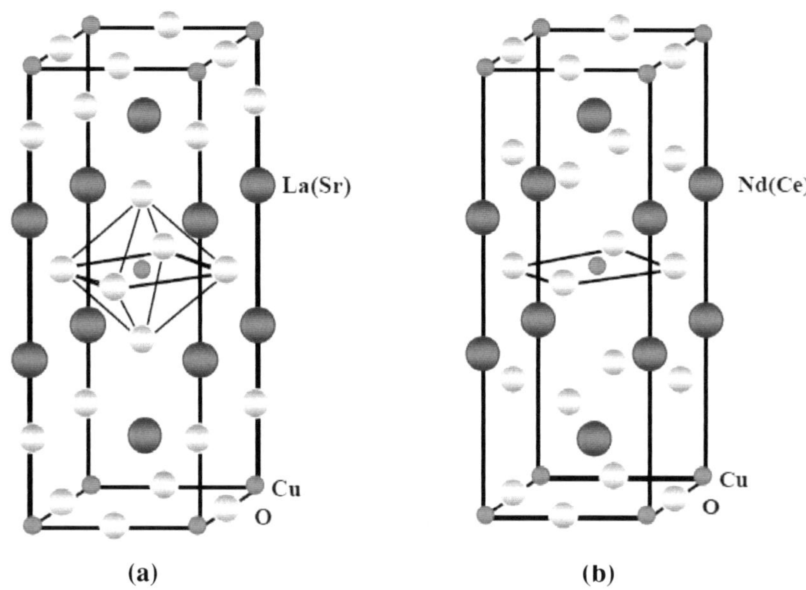

Figure 1. Crystal structure of hole-doped superconducting cuprates of $La_{2-x}Sr_xCuO_4$ (a), and electron-doped cuprates of $Nd_{2-x}Ce_xCuO_4$ (b).

2. Experimental

Electron-doped cuprates of $Eu_{1.85}Ce_{0.15}Cu_{1-y}Zn_yO_{4+\alpha-\delta}$ with y = 0 - 0.05 were prepared by the ordinary solid-state reaction method. As-grown samples were obtained after sintering process at 1050°C for 16 h. The oxygen excess (α) during sintering process was removed by flowing Ar gas at various temperatures in a range of 900°C – 950°C for 10 h – 15 h. Values of removed oxygen by annealing (δ) were estimated from the weight change before and after annealing. All of the samples were characterized by the powder x-ray diffraction (XRD) measurements and electrical resistivity measurements at Graduate School of Engineering, Tohoku University, Japan.

3. Results and Discussion

XRD pattern of electron-doped system of $Eu_{1.85}Ce_{0.15}Cu_{1-y}Zn_yO_{4+\alpha-\delta}$ (ECCZO) with $y = 0$, $\delta = 0$ (a) and $y = 0$, 0.01, 0.02, 0.05 in various δ values of 0.041 (b), 0.090 (c), 0.107 (d), 0.045 (e), respectively, were shown in Figure 2. For reference, the theoretical positions of XRD pattern of $Eu_{1.85}Ce_{0.15}CuO_4$ with selected main peaks from ICSD (code: 72249) [15] were also displayed (f). Main peaks of T' structures were clearly observed in all samples.

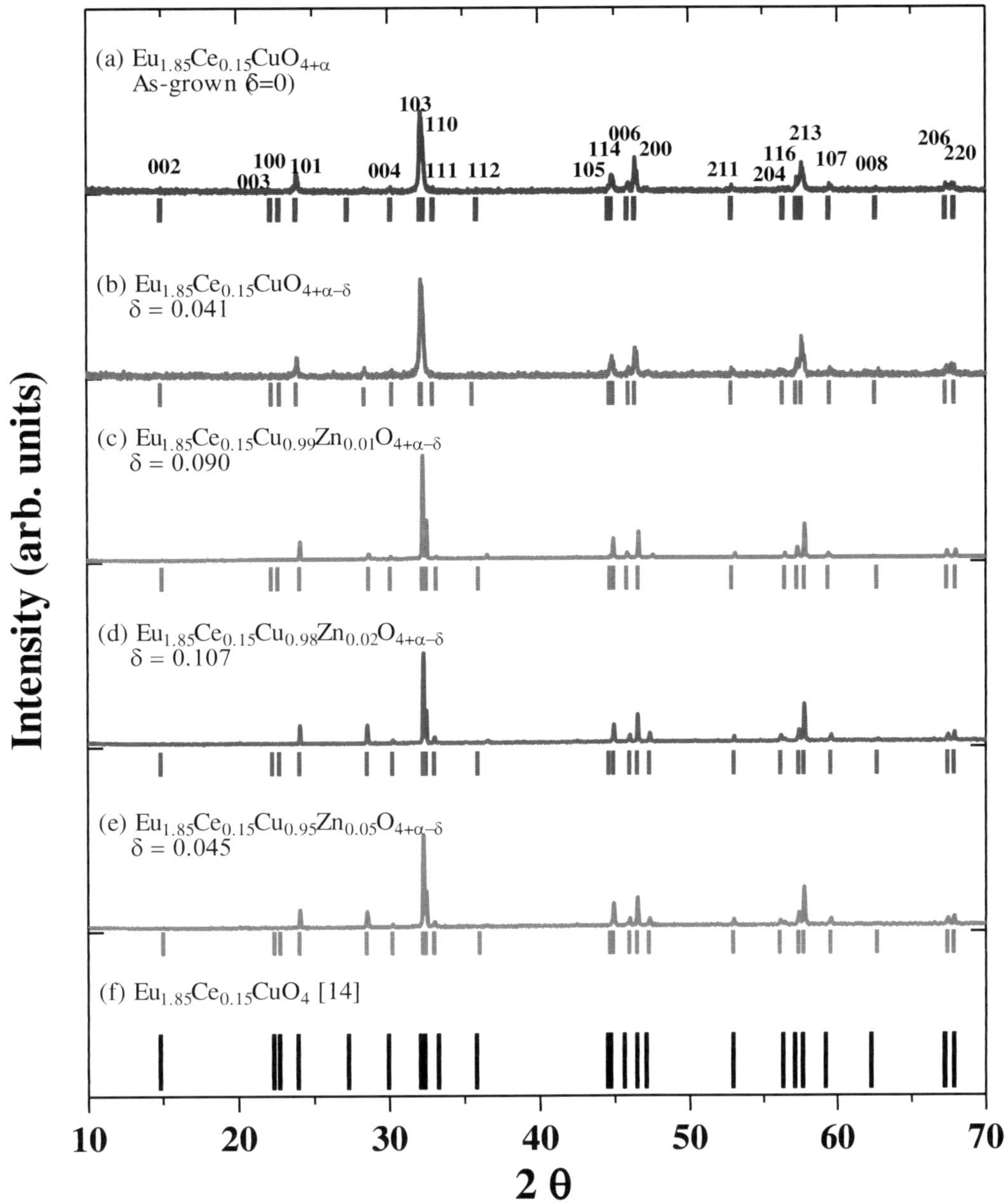

Figure 2. XRD pattern of $Eu_{1.85}Ce_{0.15}Cu_{1-y}Zn_yO_{4+\alpha-\delta}$ with $y = 0$, $\delta=0$ (a) and $y = 0$ (b), 0.01 (c), 0.02 (d), 0.05 (e) in various δ values. For reference, XRD pattern of $Eu_{1.85}Ce_{0.15}CuO_4$ data with ICSD code 72249 [15] (f) are also displayed.

Figure 3 shows the typical lattice parameters in the *a*-axis, *c*-axis and volume of crystal as function of Zn concentration of y. The value of *a*-axis stay almost unchanged as the *y* value varies. One of the reason is the absence of apical Cu-O bonds makes no Jahn-Teller effect in ECCZO samples. Jahn-Teller distortion in the hole-doped system which has apical oxygen ions makes the apical Cu-O bonds longer than the Cu-O bonds in the CuO2 plane, while substitution of Zn for Cu has tendency to equalize the bond lengths between the apical Cu-O bonds and the Cu-O bond in the CuO2 plane. The value of *c*-axis and volume of crystal slightly increased with very small changed as increasing *y* which is probably due to the slightly different size between Zn and Cu.

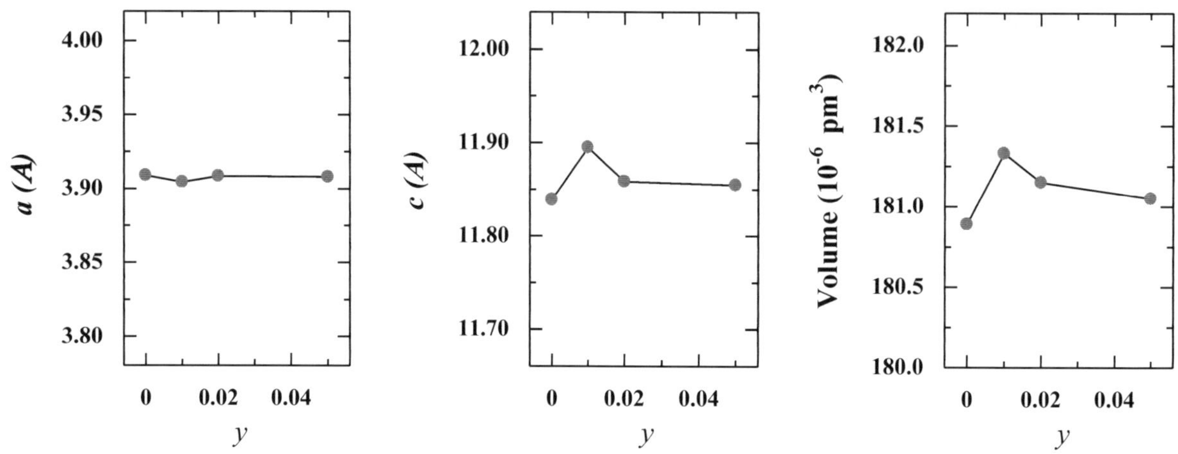

Figure 3. Lattice parameters in the *a*- and *c*-axis as function of Zn concentration of *y* in $Eu_{1.85}Ce_{0.15}Cu_{1-y}Zn_yO_{4+\alpha-\delta}$.

Figure 4 shows temperature dependence of resistivity of $Eu_{1.85}Ce_{0.15}Cu_{1-y}Zn_yO_{4+\alpha-\delta}$ with $y = 0.01, 0.02, 0.05$ with various δ values. For $y = 0$, the trace of superconductivity in temperature dependence of resistivity was reported [16]. For $y = 0.01$, 0.02, and 0.05, resistivity increased with decreasing temperature. It is found that Zn substitution for Cu was affected in destruction of superconductivity in electron-doped system of ECCZO which is due to the effect of Zn in localized charge carriers in electron-doped system.

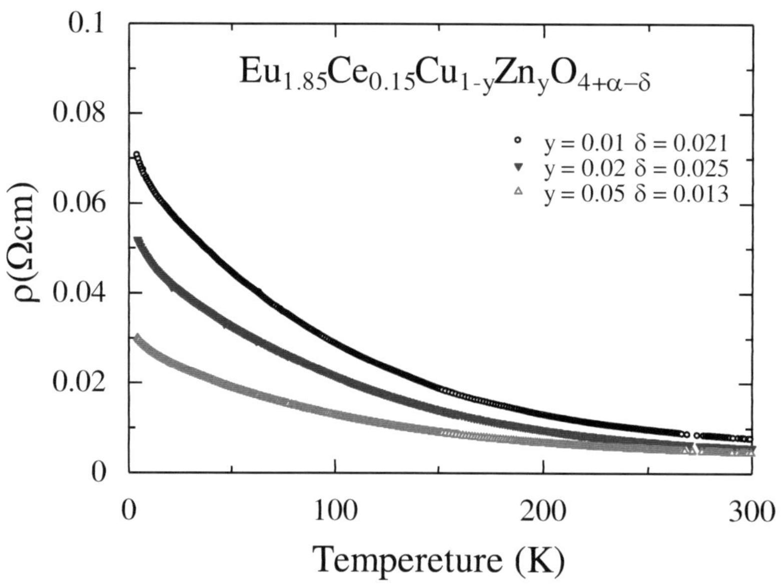

Figure 4. Temperature dependence of resistivity of $Eu_{1.85}Ce_{0.15}Cu_{1-y}Zn_yO_{4+\alpha-\delta}$ with $y = 0.01, 0.02, 0.05$ with various δ values.

4. Summary

Polycrystalline samples of electron-doped system of $Eu_{1.85}Ce_{0.15}Cu_{1-y}ZnyO_4$ with $y = 0$, 0.01, 0.02, 0.05 and various δ values have studied in order to study effect of Zn impurity for Cu to its structure and resistivities. From XRD pattern, it is found that all the main peaks of T' structures are observed.

From resistivity measurements, it is found that Zn substitution for Cu was affected in destruction of superconductivity in electron-doped system of ECCZO, which is due to the effect of Zn in localized charge carriers in electron-doped system.

Acknowledgements
We would like to thank Y. Koike, T. Adachi, M. Kato, T. Noji, T. Kawamata and H. Sato for their technical support in the x-ray diffraction and electrical-resistivity measurements at Tohoku University, Japan. These works were supported by Academic Leadership Grant of Padjadjaran University. Part of this work was also supported by research grant of International Research Collaboration and Scientific Publication No: 393/UN6.R/PL/2015, 431/UN6.3.1/PL/2016, and 718/UN6.3.1/PL/2017.

References

[1] K. M. Suzuki, T. Adachi, H. Sato, I Watanabe, Y. Koike, 2016, *Journal of the Physical Society of Japan* **85** 124705
[2] Y. Tanabe, T. Adachi, K. Suzuki, T. Kawamata, Risdiana, T. Suzuki, T. Suzuki, I. Watanabe, Y. Koike. 2011 *Physical Review* B **83** 144521
[3] K. Suzuki, T. Adachi, Y. Tanabe,Y. Koike, T. Kawamata, Risdiana, T. Suzuki, I. Watanabe. 2010 *Phys. Rev. B* **82** 054519
[4] Y. Tanabe, T. Adachi, Risdiana, , T. Kawamata, T. Suzuki, I. Watanabe, Y. Koike. 2009 *Physica B* **404** 717
[5] Risdiana, T. Adachi, N. Oki, S. Yairi, Y. Tanabe, K. Omori, T. Suzuki, I. Watanabe, A. Koda, W. Higemoto and Y. Koike. 2008 *Physical Review B* **77** 054516
[6] T. Adachi, N. Oki, Risdiana, S. Yairi, Y. Koike, I. Watanabe. 2008 *Physical Review B* **78** 134515
[7] Y. Koike, T. Adachi, N. Oki, Risdiana, M. Yamazaki, T. Kawamata, T. Noji, K. Kudo, N. Kobayashi, I. Watanabe, K. Nagamine. 2005 *Physica C* **426-431** 189
[8] T. Adachi, N. Oki, Risdiana, S. Yairi, Y. Koike, I. Watanabe. 2007 *Physica C* **460-462** 1172-1173
[9] Risdiana, T. Adachi, N. Oki, Y. Koike, T. Suzuki, I. Watanabe. 2010 *Physical Review B* **82** 014506
[10] Risdiana, L. Safriani, W. A. Somantri, T. Saragi, T. Adachi, I. Kawasaki, I. Watanabe, Y. Koike. 2014 *Advanced Materials Research* **896** 354
[11] Z. Guguchia, T. Adachi, Z. Shermadini, T. Ohgi, J. Chang, E. S. Bozin, F. Von Rohr, A. M. Dos Santos, j. J. Molaison, R. Boehler, Y. Koike, A. R. Wieteska, B. A. Frandsen, E. Morenzoni, A. Amato, S. J. L. Billinge, Y. J. Uemura, R. Khasanov 2017 *Physical Review B* **96** 094515
[12] Risdiana, M. Saputri, M. F. Sobari, A. I. Hanifah, W. A. Somantri, T. Saragi. 2017 *IOP Conf. Series: Materials Science and Engineering* **196** 012012
[13] K. Suzuki, T. Adachi, Y. Tanabe, H. Sato, Risdiana, Y. Ishii, T. Suzuki, I. Watanabe, Y. Koike. 2012 *Physics Procedia* **30**, 275
[14] J. M. Tarascon, E. Wang, S. Kievelson, B. G. Bagley, G. W. Hull, R. Ramesh. 1990 *Physical Review B* **42**, 218
[15] T. Uzumaki, K. Hashimoto, N. Kamehara. 1992 *Physica C* **202** 175-187
[16] Risdiana, S. Pratiwi, D. Suhendar, W. A. Somantri, L. Safriani, T. Saragi. 2016 *AIP Conf. Proc.* **1712** 050020

IOP Conf. Series: Journal of Physics: Conf. Series **1091** (2018) 012020 doi:10.1088/1742-6596/1091/1/012020

Ni-Substitution to $La_{0.67}Ba_{0.33}Mn_{1-x}Ni_xO_3$ compounds (x = 0; 0.01; 0.02; 0.03; 0.04; and 0.05) andits effect to resistance under magnetic fields

Romadhon Nuzuly[1], Budhy Kurniawan[1,2] and Sitti Ahmiatri Saptari*[3]

[1]Material Science Study Programme, Department of Physics, University of Indonesia, Depok 16424
[2]Department of Physics, University of Indonesia, Depok 16424
[3]Department of Physics, State Islamic University, Ciputat 15412

Email: sitti.ahmiatri@uinjkt.ac.id

Abstract. Ni-substituted to perovskites $La_{0.67}Ba_{0.33}Mn_{1-x}Ni_xO_3$ with x = 0; 0.01; 0.02; 0.03; 0.04; and 0.05, have been done and investigated by x-ray diffraction spectrometer, permagraph spectrometer, and four-point probe (FPP) measurements. X-ray diffraction spectrometer result refined by high score software. All samples show an orthorhombic structure at room temperature. At room temperature the cell parameter a, b and c of the samples seem to be constant although the doping of Ni was increased. The permagraph spectrometer results showed that all the samples were paramagnet at room temperature since magnetic remanence is zero for all samples. The four-point probe (FPP) results negative magnetoresistance and positive magnetoresistance. The negative magnetoresistance appears at x = 0 with ratio 15.625% and the positive magnetoresistance appears at x = 0.01 until 0.05 with maximum ratio 18.341% for x = 0.01. Theseevidences showed that highest sensitivity's respond for resistivity change before and after applying external magnetic fields to perform magnetoresistance phenomena is for x = 0.01.

1. Introduction

The $A_{1-x}B_xMnO_3$ perovskite are among the most studied magnetoresistance materials in condensed matter physics. In 1950, Jonker and Van Santenhave studied perovskite and found that theCurie temperature value for (La,Ba)MnO_3 compounds for $x \approx 0.33$ was about 340 K[1]. This is anindication that the compound is still ferromagnetic at room temperature and therefore other authors will perform further study on magnetoresistance in perovskites. Zener's double exchange (DE) interaction between Mn^{3+} and Mn^{4+} ions in the oxygen orbital 2p was introduced inorder to explain the coupling of magnetic and electronics properties in these compounds[2]. Undoped $LaMnO_3$ is an A-type antiferromagnetic insulator. By substitution of La^{3+} with a divalent cation, $LaMnO_3$ can be driven into a metallic and ferromagnetic state [3]. There have been some experimental works carried out to study the phenomenon of magnetoresistance in perovskites. $LaMnO_3$ was studied by Sukirman*et al*and the resulting magnetoresistance (MR) ratio was about -7% in room temperature when a 0.76 T magnetic induction was applied[4]. Next, La^{3+} site in $LaMnO_3$ doped with Ba^{2+}[5-6], Ca^{2+}[7-8] or Sr^{2+}[9] and MR ratio for $LaMnO_3$ which doped by Ba^{2+} is about -30% in temperature room when a 6 T magnetic induction was applied. On the other hand, Xiao Bo Yuan *et al* have doped the Mn site with Bi and Ti[10]. Theyobtained an MR ratio of approximately-10.7% for x = 0.02 at room temperature when a 1 T strong magnetic induction was applied. Therefore in this work the authors investigated the $La_{0.67}Ba_{0.33}MnO_3$ compound which was doped with Ni in order to gain information onthe MR ratio and the Ni ferromagnetic ordering in the compound;and whether the Ni doped compound would likely to yield the same results as the Fe doped compound because the ordering isalso ferromagnetic.

IOP Conf. Series: Journal of Physics: Conf. Series **1091** (2018) 012020 doi:10.1088/1742-6596/1091/1/012020

2. Experimental and Method

La_2O_3, $BaCO_3$, $MnCO_3$, and NiO powder were prepared by mechanical alloying to prepare $La_{0.67}Ba_{0.33}MnO_3$ doped with Ni with x = 0; 0.01; 0.02; 0.03; 0.04; and 0.05 in 10 H. After that, all samples were calcinated at 800°C for 8 hours. Before being sinteredat 1350°C for 24 hours, all samples were pressed into pellet. The structure of the samples was determined by the X-ray diffraction (XRD) method. Magnetic properties were measured by a permagraph instrument. Electrical measurements were carried out using the four-point probe (FPP) method.

3. Results and discussion

XRD result indicates that all samples are in the single phase. XRD patterns for all samples are shown in figure1.For refinement process, we use the High Score Plus (HSP)application software because it is simple to use. The refinement results are shown in table1. All samples are in good quality because the magnitude of the goodness of fit or chi square (χ^2) values turned out to be less than four. As for the crystal structure, all samples have exhibitedthe orthorhombic space group symmetry with the the proper lattice parameters and lattice constant values.

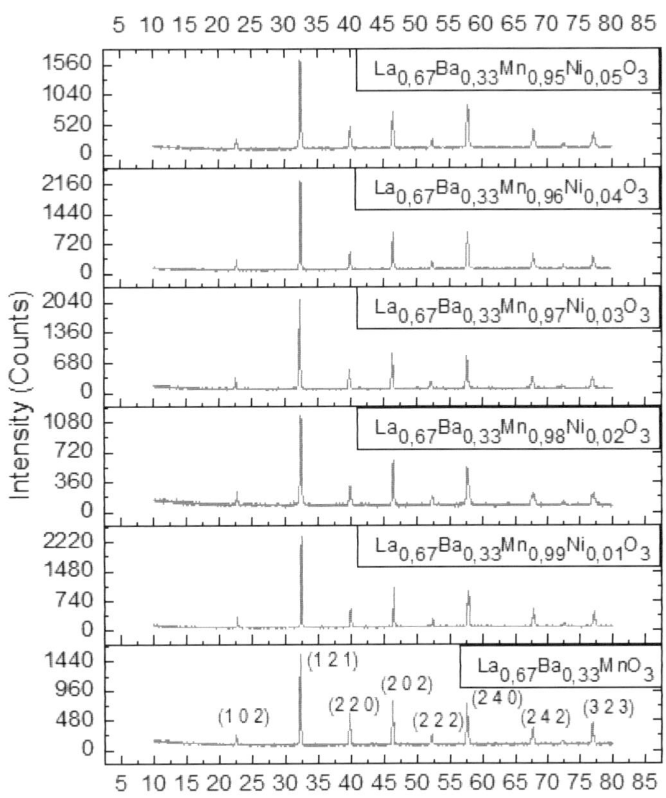

Figure. 1. X-ray diffraction pattern for all samples $La_{0.67}Ba_{0.33}Mn_{1-x}Ni_xO_3$

IOP Conf. Series: Journal of Physics: Conf. Series **1091** (2018) 012020 doi:10.1088/1742-6596/1091/1/012020

Table 1. Refinement result for all samples $La_{0.67}Ba_{0.33}Mn_{1-x}Ni_xO_3$

Sample	GoF = χ^2	$\alpha = \beta = \gamma$	Lattice Parameter (Å)		
			a	b	c
x = 0	2.148	90°	5.518	7.811	5.534
x = 0.01	2.105	90°	5.516	7.807	5.534
x = 0.02	1.559	90°	5.510	7.814	5.545
x = 0.03	1.584	90°	5.513	7.801	5.532
x = 0.04	1.930	90°	5.515	7.806	5.533
x = 0.05	1.759	90°	5.514	7.799	5.534

After measurement with the XRD, the authors measured all samples with the permagraph instruments at room temperature to investigate their magnetic properties. Permagraph results are shown in figure 2,all samples are in the paramagnetic state

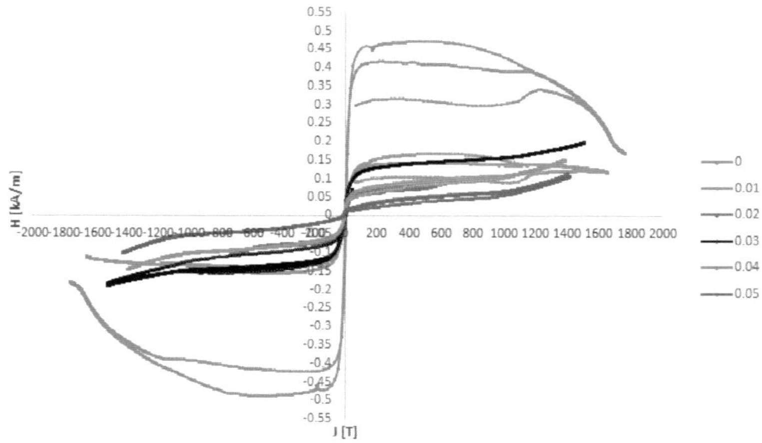

Figure 2. Permagraph measurement for all samples $La_{0.67}Ba_{0.33}Mn_{1-x}Ni_xO_3$

In our discussion, the XRD resultshaveindicatedthe orthorhombic crystalline space-group symmetry for all samples, which then allowed the authors to conduct the next experiments, namely the permagraphand the four-point probe measurements. In permagraph method, all samples have paramagnetic properties because magnetic remanence of all samples is zero. This is an indicatethat all samples become paramagnetic at room temperature while the Ni values are increasing.

Figure3 shows the MR ratio versus H curves at room temperature. This figure gives the curve of MR = $\frac{R(H)-R(0)}{R(0)} x\,100\%$ versus x, where the magnetic field varies from 0 to 0.76 T. For x = 0 the MR ratio is -15.625%. When the amount of the Ni-dopant increases, the MR ratio also rises and this in turn directs the MR to have a positive magnetoresistance which then tend to drop-off toward the x axis. The Ni-dopant value of x=0.01 would then affect the biggest positive magnetoresistance value in the sample as shown in table 2.

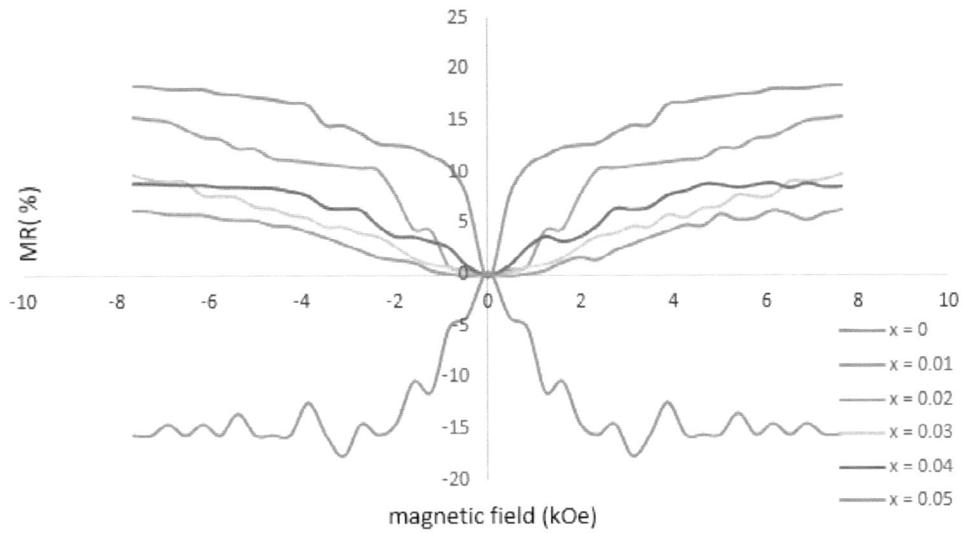

Figure 3. Curve of MR ratio versus magnetic field

In the next measurements, employing the four-point probemethod, the results show both negative and positive magnetoresistance. The negative magnetoresistance has been recorded in the $La_{0.67}Ba_{0.33}Mn_{1-x}Ni_xO_3$ sample for x = 0 witha ratio of 15.625%. The positive magnetoresistance value has been found in the other sampleswith a maximum ratio of about 18.341% for the $La_{0.67}Ba_{0.33}Mn_{1-x}Ni_xO_3$ sample with x = 0.01.

For the samples with a doping concentration of x = 0.01, the value 0.33 in Ba^{2+} makes double exchange interaction possible. Because Ba-dopantsat the La^{3+} site will produce Mn^{4+} and Mn^{3+} as in the stoichiometry $Ba^{2+}_{0,33}(Mn^{3+}_{0,67}Mn^{4+}_{0,33})O^{2-}_{3}$. Double exchange interaction will set the sample in the ferromagnetic condition and make the MR ratio in negative and this is displayed in table 2. Besides Ba^{2+} ions which were doped to the La^{3+} sites, the Mn sites in this work were doped with Ni^{2+} ions. Ni-doped to the sample sets the sample in a paramagnetic condition as shown in the figure2. Then, it will make the MR ratio become positive as shown in the figure3.

Table 2. The MR ratio for all samples

No	Sample $La_{0.67}Ba_{0.33}Mn_{1-x}Ni_xO_3$	ΔR (Ohm)	MR Ratio (%)
1	x = 0.00	-0.942	-15.625
2	x= 0.01	103.871	18.341
3	x = 0.02	50.240	15.414
4	x = 0.03	6.945	9.867
5	x = 0.04	4.170	8.541
6	x = 0.05	2.443	6.302

4. Conclusion

The magnetic and magnetoresistance properties for the $La_{0.67}Ba_{0.33}Mn_{1-x}Ni_xO_3$ samples with x = 0; 0.01; 0.02; 0.03; 0.04; and 0.05 have been studied. The results have shown that it is the doping which turnsthe samples into a paramagnetic state and to posses interesting magnetoresistance behaviors. Undoped samples have shown negative MR values with with a ratio of -15.625%. When the samples were doped with Ni, it turns the samples to have a positive MR. The biggest positive MR is at x = 0.01 sample with a ratio of 18.641%. These results are recommended for the studied application.

Acknowledgement
The author gratefully acknowledge the support of Departement of Physics, University of Indonesia for the research facilities. We are also thankful for the financial support provided by the university under the research grant contract no. 1861/UN2.RI1/HKP 05.00/2015.

References
[1] Jonker G H and Van Santen J H 1950 Ferromagnetic compounds of manganese with perovskite structure *Physica* **16 3** 337–349.
[2] Nasri M, Triki M, Dhahri E, Hussein M, Lachkar P and Hlil E K 2013 Investigation of structural, magnetocaloric and electrical properties of $La_{0.6}Ca_{0.4-x}Sr_xMnO_3$ compounds *Phys. B Condens. Matter* 408 1 104–109.
[3] Altintas S P, Amira A, Varilci A and Terzioglu C 2012 Influence of Gd-doping in $La_{0.7}Ca_{0.3}MnO_3$ on its structural and magneto-electrical properties," *J. Magn. Magn. Mater.* **324 7** 1331–1336.
[4] Sukirman E and Purwamargapratala Y 2010 Analisis Struktur Kristal $LaMnO_3$ dengan Teknik Difraksi Sinar-X dan Metode Rietvield **12 1** 70–74.
[5] Mansuri I and Varshney D 2012 Structure and electrical resistivity of $La_{1-x}Ba_xMnO_3$ (0.25 < x < 0.35) perovskites *J. Alloys Compd.* **513** 256–265.
[6] Pratama R 2016 Effect of Various Sintering Temperature on Resistivity *AIP conference proceedings* **1729** 020028–1–020028–4.
[7] Ramadhan M R 2016 Sintering Temperature Effect on Electric Transport *AIP conference proceedings* **1729** 020025–1–020025–4.
[8] Aguirre M H 2012 Structure , microstructure , and high-temperature transport properties of $La_{1-x}Ca_xMnO_{3-\delta}$ thin films and polycrystalline bulk materials *Phys. B Phys. Condens. Matter* **013703 2008** 1–5.
[9] Widyaiswari U, Kurniawan B, Imaduddin A and Saptari S A 2016 Study of $La_{0.67}Sr_{0.33}MnO_3$ Structure and Resistivity Synthesized by Using Sol Gel Method *Journal of Physics: Conference Series* **776** 012041.
[10] Yuan X, Liu Y, Hung B, Chen W, Wang C J and Zhang R 2005 Enhancement of Magnetoresistance in Bi or Ti doped $La_{0.67}Ba_{0.33}MnO_3$ *Mater. Sci. Eng. B* **117 1** 93–97.

IOP Conf. Series: Journal of Physics: Conf. Series **1091** (2018) 012021 doi:10.1088/1742-6596/1091/1/012021

Synthesis and characterization of high purity Fe_3O_4 and α - Fe_2O_3 from local iron sand

Sari Hasnah Dewi and Wisnu Ari Adi
Centre for Science and Technology of Advanced Materials, Kawasan Puspiptek Serpong, Tangerang Selatan 15314, Indonesia

Email: hasyarri@batan.go.id and dwisnuaa@batan.go.id

Abstract. The synthesis and characterization of high purity of Fe_3O_4 and α-Fe_2O_3 from iron sand has been carried out. Iron sand samples retrieved from Banten, Indonesia. The iron sand powder was milled for 10 hours at room temperature by using high energy milling (HEM) and then was separated according to its type using magnetic separator. The iron sand powders are dissolved in acid chloride solution so that is formed a solution of iron chloride, and this solution is sprinkled with sodium hydroxide to obtain fine powders. The fine powders which formed were washed again with de-ionized water and sintered in the electric chamber furnace at 750 °C in the air at atmosphere pressure for 5 hours. The elemental analysis results using neutron activation analysis (NAA) is obtained that the fine powders contain the dominant of iron (Fe). The refinement results of X-ray diffraction pattern shows that the fine powders have a single phase of Fe_3O_4, and then after sintering changed to α-Fe_2O_3 single phase. The microstructure analysis showed that the particle of Fe_3O_4 and α-Fe_2O_3 shaped respectively like spherical and polygonal with the similarly particle sizes and homogeneously on the surface of the samples. We concluded that this study has successfully performed the process of purifying iron sand to obtain fine powders of Fe_3O_4 and α-Fe_2O_3 with high purity. The first section in your paper

1. Introduction

Indonesia is an archipelago which has a long coastline is endowed with rich natural resources, especially iron sand. The iron sand is generally to be found around the coastal area and which is also rich in various minerals such as iron and titanium. The iron sand in Indonesia has a huge potential with a proven reserve of up to 83 million tons with an average Fe content of around 59% is are dispersed almost throughout the coastlines of Indonesia, especially in Java, Sumatra and Nusa Tenggara. One of the iron sand reserve with a huge potential and having a very high Fe content is the iron sand found on the coast of Banten. According to the previous research results which have been carried out by Wisnu et al. there are some indications that based on the elemental test results of the iron sand by using neutron activation analysis method, the iron content in the iron sand of Banten is very high as shown in table 1 [1].

Table 1. The result of elementer analysis of the iron sand from Banten [1]

No.	Element	Content (μg/g)	
		Concentration	Uncertainty
1.	Fe	585,223.43	103.85
2.	Ti	70,051.89	4,250.25
		M (impurity) < 1,000	

From table 1 shows that the dominant elemental constituents are iron and titanium with around 58.52 and 7.01 wt% respectively which also implies that the iron sand has an economic potential to be processed in the future.

One of iron mineral compound comes in the form of iron oxide. Iron oxide is a popular material because of its wide-ranging applications such as absorber materials [2-4], in sensors [5], as catalysts [6], and in biomedical applications such as imaging- and drug delivery devices [7]. In addition Iron oxide has many advantages, such as: oxidation stability, non-toxicity and compatible liquid system. There are four variety of commonly known crystalline iron oxides, namely α-Fe_2O_3 also known as hematite, β- Fe_2O_3 and γ- Fe_2O_3 also known as maghemit, ε- Fe_2O_3 and Fe_3O_4 also known as as iron (II, III) crystalline [4].

Many methods of synthesis have been carried out to obtain iron oxides and the type of procedures chosen will determine the quality and quantity of the product material. Synthesis of iron oxides utilizing different processes will produce compounds with different magnetic and electrical properties.

Several synthesis methods of iron oxide powders have been reported, among them the hydrothermal [8] the sol gel [9], the micro-emulsion [10], the co-precipitation [11], and the ultrasonic method [12]. In this work, the authors will focus on the extraction of iron oxide from local raw materials of iron sand through two stages. The first stage (top-down method) is iron sand particles that are large will be broken down into smaller sizes by using mechanical, chemical, or with other energy forms. Techniques commonly used in the top-down method is mechanical milling, followed by magnetic separation process with the aim to separate the powder of magnetic and non magnetic (impurities) of iron sand.

At this stage the iron oxide produced still has the size and shape of the particles that have not yet become homogeneous and their purity level is still low [1]. Therefore it is now necessary to undertakethe second step, namely the bottom-up method. This second method is the synthesis by way of homogeneous nucleation of the liquid, namely binding by chemical reaction. The advantages of this method is homogenization of all the particles□ size, shape or morphology, the chemical composition and the intended crystal structure as was originally desired, such as the composition and the same surface, individual dispersionor mono-dispersion, and no agglomeration. Among all the bottom-up techniquessofar, chemical co-precipitation is the most superior method in the synthesis of iron oxide particles because the procedure is straight forward and uncomplicated. The main objectives of this study is to obtain high purity Fe_3O_4 and α- Fe_2O_3 oxide samples by processing and purifying iron sand into iron oxides. The discussion will be focused on the results of processing and purifying the iron sand, the phase analysis includes structure parameters of the synthesized Fe_3O_4 and α- Fe_2O_3, their particle morphology and the elemental content of the synthesized oxides.

2. Materials and Methods

The iron sand raw material was extractedin Banten province Indonesia. The purifying process of iron sand into iron oxide product in the form of Fe_3O_4 and α- Fe_2O_3 with high purity is shown in figure 1.

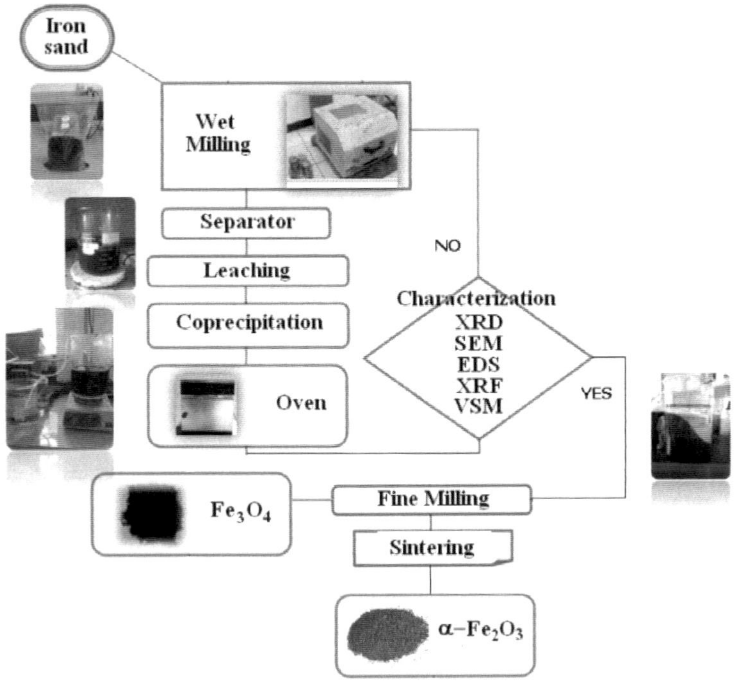

Figure 1. The purifying process of iron sand into iron oxide product

During the first stage, iron sand was milled by using high energy milling (HEM). Then proceed with the separation of iron sand using a magnetic separator with weak magnetic fields in order to obtain fine powder of type of titanomagnetite which has a high Fe content. In the second stage the fine powder of titanomagnetite was dissolved in a technical grade chloride acid (HCl) solution with the ratio of 0.25Kg/L HCl to produce a solution of iron chloride. Next, technicalgrade Sodium hydroxide (NaOH 5M) was added slowly to a solution of iron chloride while stirring continuously using a magnetic stirrer. Stirring was performed at 80 °C for 1 hour. The precipitate was washed with demineralization water until reaching the neutral pH state,in order to removeresidual salts and other impuritiesfrom the sample. The precipitate is then separated from the solvent using permanent magnets (Nd2Fe14B magnet) and after that dried in an oven at 100 °C. The dry powder sample was milled by using an agate mortar to homogenize the particle size of the sample and then heated at 750 °C in a furnace for 5 hours. Finally, the sample was milled again by using an agate mortal.

The iron sand purification result was analyzed by X-ray diffractometer (XRD) method using a Miniflex Rigakuapparatus with CuKα radiation (λ = 1.5406 Å). The qualitative and quantitative phaseanalysis was performed by applying a GSAS program using Rietveld method. The morphology of the powder samples was observed by the scanning electron microscope (SEM) method employing the JED 2300 JEOL scanning electron microscope. Finally, the elemental composition analysis was performed by using neutron activation analysis (NAA) method. The standard reference material from NIST is usedas the internal quality control standardand Al-0.1% Au alloyobtained from the Institute for Reference Materials and Measurement (IRMM) is used as flux monitorstandard. The irradiation has been performed at thermal neutron flux about 1013 $n.cm^{-2}.s^{-1}$ in the irradiation facility of Multi Purpose Reactor GA. Siwabessy in Serpong, Indonesia.

3. Result and Discussion

Fine powder of titanomagnetite type which has a high Fe content from the milling process and magnetic separator has been obtained from the first stage process. Then in the second stage, the powder is carried through the leaching- and sintering process in order to obtain Fe_3O_4 and α-Fe_2O_3 fine powder as shown in figure 2.

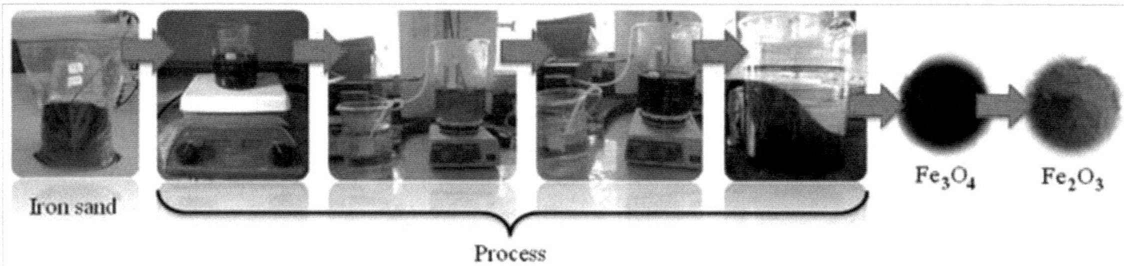

Figure 2. The iron sand powder after purifying process.

The fine powder of titanomagnetite type with high Fe content is characterized by XRD method and compared with the Fe_3O_4 and α-Fe_2O_3 fine powder resulting from the purification process. In figure 3 are shown the results of measurements of XRD reflection pattern of the third of powders both before and after the purification process.

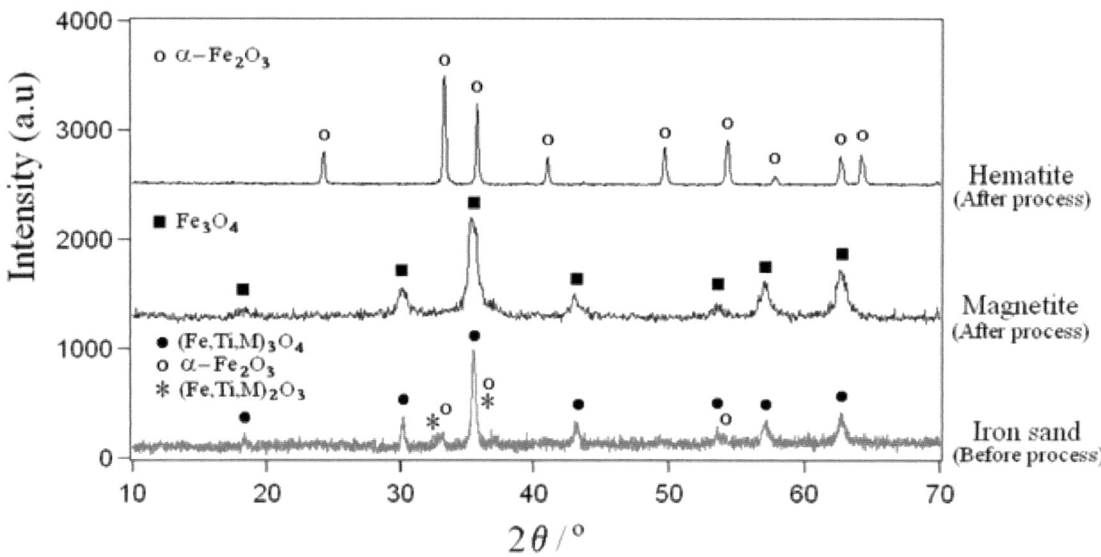

Figure 3. The XRD pattern of the sample.

The results of phase identification and analysis of titanomagnetite type samples have been reported in previous studies [1]. Based on the refinement result of the X-ray diffraction pattern,the mass fraction of phase for iron sand of titanomagnetite type is obtained as shown in table 2.

Table 2. The mass fraction of phases contents on the iron sand kind of titanomagnetite [1]

No.	Mineral name	Phase	Mass fraction (%)
1.	Ilmenite	$(Fe,Ti,M)_2O_3$	11.67 ± 0.01
2.	Hematite	α-Fe_2O_3	15.96 ± 0.01
3.	Titanomagnetite	$(Fe,Ti,M)_3O_4$	72.37 ± 0.04

M : Impurities (Na, Sc, Cr, Co, Zn, Ga, As, La, Sm, Eu, Ho, Yb, Hf, W, Pt, U, Sb, Th, Ce, Mg, Mn, V, Al, and Ta)

IOP Conf. Series: Journal of Physics: Conf. Series **1091** (2018) 012021 doi:10.1088/1742-6596/1091/1/012021

Table 1 indicates that the phases contained in the iron sand is rich in titanomagnetite phase, moreover there are still ilmenite and hematite phases left over in the sample. However, in both the titanomagnetite- and ilmenite phase some minute (in ppm) amount of impurity elements M are also found.

The results of X-ray diffraction pattern refinement of the Fe_3O_4 sample obtained from the purification process are shown in figure 4.

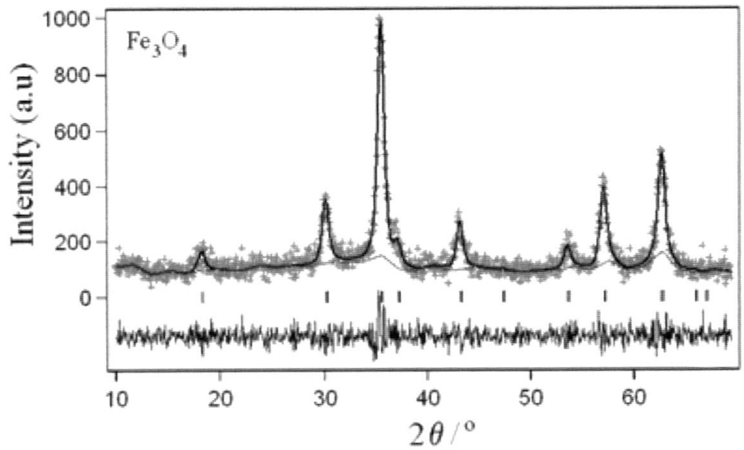

Figure 4. The refinement result of XRD profiles on the Fe3O4 sample after purifying process

Phase identification of Fe_3O_4 sample employs the crystallography open database (COD: 9013529) as a reference. According to the COD data the results obtained could be identified as the single phase of magnetite. Since the Hanawalt table can only be used for qualitative analysis, then the rietveld method (GSAS program) will be used for quantitative analysis.

The refinement result obtained are of excellent quality, the fitting converges to smallreliability R factor and with a goodness of fit value $\chi2$ (chi-squared) in good agreement with the standard value set by rietveld method. The structure parameter, factor R and goodness of fit ($\chi2$) are shown in table 3.

Table 3. Structure parameter, factor R, *goodness of fit* (χ^2).

Sample after purifying process			
Fe_3O_4 phase			
Space group : F d -3 m (227), crystal system : Cubic a = 8.380(3) Å, b = 8.380(3) Å, c = 8.380(3) Å, α=β=γ=90° V = 588.5(7) Å^3 and ρ = 5.226 gr.cm^{-3}			
R factor	wRp = 17.12	Rp = 13.26	χ^2 (chi-squared) = 1.245

Based on the obtained results of XRD refinement, it could be safely established that Fe_3O_4 sample synthesized using the purification process is a single magnetite phase.

In the next step, the Fe_3O_4 sample is set to undergo heat treatment to obtain a sample of α- Fe_2O_3. In figure 5 are shown the refinement results of X-ray diffraction pattern of the α- Fe_2O_3 sample after the heat treatment process.

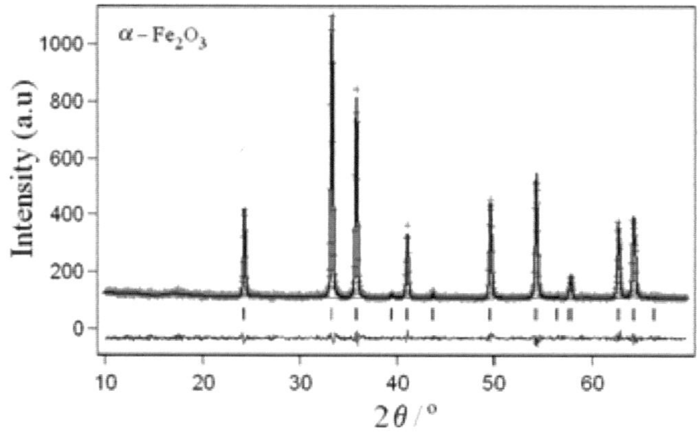

Figure 5. The refinement result of XRD profiles on the sample after purifying and sintering

The refinement result has also produced excellent quality of fitting. The phase identification refers to the crystallography open database (COD: 9000139) for the hematite phase. Figure 5 shows that the sample is a hematite single phase. The structure parameter, factor R and goodness of fit ($\chi 2$) are shown in table 4.

Table 4. Structure parameter, factor R, *goodness of fit* (χ^2).

Sample after purifying and sintering			
α-Fe_2O_3 phase			
Space group : R -3 c (167), crystal system : Hexagonal a = 5.0323(3) Å, b = 5.0323(3) Å, c = 13.730(1) Å, α=β= 90° and γ= 120° V = 301.14(4) Å^3 and ρ = 5.283 gr.cm^{-3}			
R factor	wRp = 18.26	Rp = 11.67	χ^2 (chi-squared) = 1.106

The SEM micrograms of the sample before and after purifying process are shown in figure 6.

(a) Iron sand after milling by HEM

(b) After separating by magnetic separator

(c) Fe_3O_4 powder of purifying result (a) α-Fe_2O_3 powder of purifying result

Figure 6. Surface morphology of samples before and after purifying process

In figure 6(a) is shown the particle morphology of the (mechanically) milled iron sand sample. The collision of the milling steel balls causes the particles of iron sand to be flattened and thus the fracture forming smaller particles. Figure 6(b) showsthe fine powder of iron sand from separation process results using a magnetic separator, so the particles seem to appear in the shape of aggregates and are almost uniform. Figure 6(c) is a particle morphology of Fe_3O_4 from purification process results and the particles appear in the form of spherical and are uniform. Spherical shape is thought to originate from the droplets NaOH at low speed. Figure 6(d) is a particle morphology of α- Fe_2O_3 with uniform polygonal shape. Specific form and nature of α-Fe_2O_3 particles follow the shape of a hexagonal structure. Uniform shape indicates that microscopically the samples have similar phase. These results are consistent with the results of refinement X-ray diffraction pattern of the sample of Fe_3O_4 and α-Fe_2O_3 obtained from purification process.

Thus, further analysis of the elemental content in the samples by using neutron activation analysis (NAA is required). NAA is both a quantitative and a qualitative method with high efficiency for the precise determination of a number of main-components and trace elements in different types of samples. NAA method is princippally based on the nuclear reaction between neutrons and target nuclei. It is a useful method for the simultaneous determination of about 25-30 major, minor and trace elements of geological, environmental, biological samples in ppb-ppm range without or with chemical separation [1].

Figure 7 shows the gamma-ray spectrum of neutron activation analysis (NAA) measurement result on the sample after the purification process.

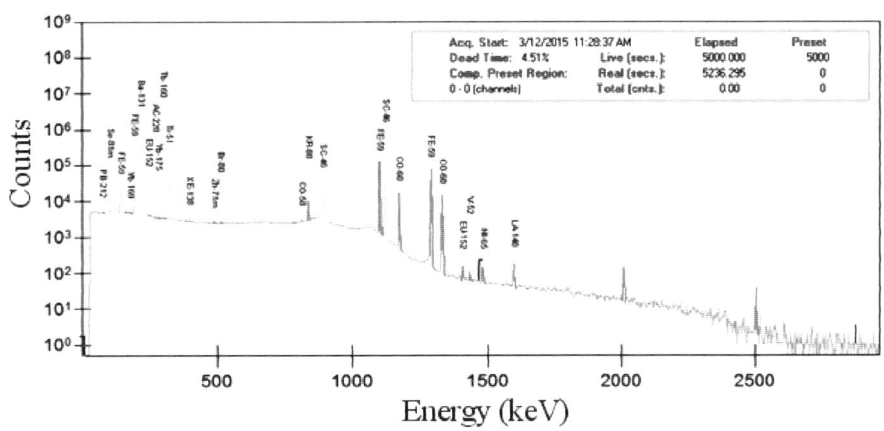

Figure 7. The elementary analysis of sample after purifying process by NAA

IOP Conf. Series: Journal of Physics: Conf. Series **1091** (2018) 012021 doi:10.1088/1742-6596/1091/1/012021

The spectrum of gamma energy as Figure 7 shows that the dominant elements are iron (Fe-59) at 1099.25 keV, this means that the sample is rich in content of iron. The detailed information on concentration of elemental content in the sample is given in Table 5.

Table 5. The result of elementer analysis of the Iron sand kind of titanomagnetite samples

No.	Element	Content (μg/g) Iron sand kind of titanomagnetite	
		Concentration	Uncertainty
1.	Fe	580,677,03	2,773,38
		M* (impurity)	
No.	M* (impurity)	Content (□g/g)	
		Concentration	Uncertainty
1.	Ce	19,44	0,53
2.	Yb	0,95	0,15
3.	Th	0,70	0,07
4.	La	8,14	0,37
5.	Eu	0,39	0,03
6.	Zn	561,65	21,08
7.	Co	144,78	1,48
8.	Sc	13,11	0,06
9.	Cr	652,08	7,53
10.	Sm	2,61	0,07
11.	Sc	14,94	0,88
12.	Na	536,13	10,18
13.	La	9,36	0,40
14.	Mg	7559,73	2760,42
15.	V	3154,05	83,31
16.	Mn	4534,03	40,14
17.	Ti	4249,37	512,42

According to the results of the elemental content analysis of the Fe_3O_4 sample, it is also a very complex sample. Table 5 displays the iron content in the sample of processing result, which in this case is around 58.07 wt%. It means that this sample has a very high iron content. These results show that in this work the processing of iron sand from local resources into high purity of Fe_3O_4 and α-Fe_2O_3 has successfully been carried out and because of its economic potential it is strongly recommended to continue with the production works.

4. Conclussion

This study has successfully performed the purifying process of iron sand from local resources to obtain fine powders of Fe_3O_4 and α- Fe_2O_3 with high purity. The co-precipitation method is a simple and an effective technique for preparing fine powder of Fe_3O_4 and α- Fe_2O_3. The refinement results of x-ray diffraction pattern shows that both Fe_3O_4 and α-Fe_2O_3 have a single phase. Fine powder of Fe_3O_4 has a structure of cubic (F d -3 m) with lattice parameters a = b = c = 8.380(3) Å, α= β= γ = 90o, V = 588.5(7) Å3 and ρ= 5.226 gr.cm-3. Meanwhile fine powder of α- Fe_2O_3 has a structure of hexagonal (R -3 c) with lattice parameters a = b = 5.0323(3) Å, c = 13.730(1) Å, α= β= 90o and γ= 120o, V = 588.5(7) Å3 and ρ= 5.226 gr.cm-3. SEM micrograms of the sample show that the particle of Fe_3O_4 and α- Fe_2O_3 are shaped like spheres and polygons respectively with similar particle sizes and homogeneously on the surface of the samples. The spectrum of gamma energy shows that the dominant elements are iron (Fe-59) at 1099.25 keV around 58.07 wt%., this means that the sample is rich in content of iron. The purifying process used is economical and environmentally friendly, as it involves technical quality materials, which make Fe_3O_4 and α- Fe_2O_3 as a promising material for industrial applications.

Acknowledgements
This work is supported by Research and Development of Smart Magnetic Program for DIPA 2015 grant of Center for Science and Technology of Advanced Materials, National Nuclear Energy Agency Republic of Indonesia.

References

[1] Adi W A and Rina M 2015 Composition analysis of titanomagnetite and ilmenite from iron sand using neutron activation analysis as preliminary study for producing iron oxide and titanium dioxide *International Journal of Academic Research Part A* **7** 3.
[2] Fadzidah M I, Hashim M, Ismayadi I, Idza I R, Norailiana A R, Rodziah N, Hapishah A N, Shafie M S E, Bahmanrokh G H, Masni M, Masoudi M 2014 Synthesis and characterization of barium-hexaferrite nanoparticles for microwave absorption, *Solid state science and technology* **22** **1-2** 12-20
[3] Rochman N T and Adi W A 2013 Preliminary study of the development of absorber electromagnetic wave materials by modifying iron sand *International Journal of Academic Research Part A* **5** **4** 91-96.
[4] Adi W A and Manaf A 2013 Microstructure and phase analysis of $La_{0.8}Ba_{0.2}Ti_xMn_{(1-x)}O_3$ system for microwave absorber material (x = 0 – 0.7) *Journal of Advanced Materials Research* **789** 97-100.
[5] Zhou J, Gan N, Li T H, Zhou H K, Li X, Cao Y T, Wang L H, Sang W G, Hu F T 2013 Ultratrace detection of C-reactive protein by a piezoelectric immunosensor based on Fe_3O_4@SiO_2 magnetic capture nanoprobes and HRP-antibody co-immobilized gold as signal tags *Sens. Actuators B* **178** 494-500.
[6] Gawande M B, Branco P S, Varma R S 2013 Nano-magnetic (Fe_3O_4) as a support for recycled catalyst in the development of sustainable methodologies *Chem. Soc. Rev.* **42** 3371-3393.
[7] Gao F P, Cai Y Y, Zhou J, Xie X X, Ouyang W W, Zhang Y H, Wang X F, Zhang X D, Wang X W, Zhao L Y, Tang J T 2010 Pulluan acetate coated magnetite nanoparticles for hyperthermia: preparation, characterization and in vitro experiments *Nano Res.* **3** 23-31.
[8] Almeida T P, Muxworthy A R, Williams W, Kasama T and Dunin-Borkowski R E 2014 Hydrothermal synthesis, off-axis electron holography and magnetic properties of Fe_3O_4 nanoparticle *Journal of Physics: Conference Series* **522** 01206.
[9] Ansar M Z, Atiq S, Alamgir K and Nadeem S 2014 Frequency and temperature dependent dielectric response of Fe_3O_4 nano-crytallites *J. Sci. Res.* **6** **3** 399-406.
[10] Sahoo S K, Agarwal K, Singh A K, Polke B G and Raha K C 2010 Characterization of γ- and α-Fe_2O_3 nano powders synthesized by emulsion precipitation-calcination route and rheological behaviour of α-Fe_2O_3 *International Journal of Engineering, Science and Technology* **2** **8** 118-126.
[11] Hariani P L, Faizal M, Ridwan, Marsi and Setiabudidaya D 2013 Synthesis and properties of Fe_3O_4 nanoparticles by co-precipitation method to removal procion dye *International Journal of Environmental Science and Development* **4** 3.
[12] Wua S, Suna A, Zhaia F, Wanga J, Xua W, Zhanga Q, Volinsky A A 2011 Fe_3O_4 magnetic nanoparticles synthesis from tailings by ultrasonic chemical co-precipitation *Materials Letters* **65** 1882–1884.

IOP Conf. Series: Journal of Physics: Conf. Series **1091** (2018) 012022 doi:10.1088/1742-6596/1091/1/012022

Magnetic properties α-Fe_2O_3Filled MWNT/PolyVinyl Alcohol Composites Film

S. Purwanto, Salim M and Yunasfi
Center for Science and Technology of Advanced Materials, National Nuclear Energy Agency, South Tangerang 15314, Indonesia

Corresponding Author : setyo_p@batan.go.id

Abstract. Magnetic properties of α-Fe2O3 partly filled MWNT/Polyvinyl Alcohol (PVA) composites film have been studied by Raman spectroscopy, Fourier Transform InfraRed (FTIR) spectroscopy, Transmission Electron Microscopy (TEM), Vibrating Sample Magnetometer (VSM), Four Point Probe (FPP). The Composite films were prepared by simply mixing between α-Fe2O3 –filled Multiwalled Carbon Nanotube(MWNT) dispersed in Sodium Dodecyl Sulphate (SDS) with 10% PVA solution for 4 variation concentration with a total volume 5 ml. The result of Raman Spectroscopy on precursor sample showing the existence of α-Fe2O3 phase and MWNT. From FTIR spectroscopy on the film samples observed the attachment of the PVA with iron oxide and MWNT. The paramagnetic properties of the samples film were investigated by means of Vibrating Sample Magnetometer (VSM). The higher value of magnetoresistanceratiowas +40% on the sample with percolation threshold Pc=0.51 wt%.The interaction between α-Fe2O3 filled MWNT in matrix polymer PVA made shrinkage or decreased the localization length inside the grain of composites film.

1. Introduction

In order to realize the vision of printable magnetoelectronic, polymer nanocomposites containing CNT or MWNT have great attention due to the unique nearly one dimensional (1D) electronic properties. The structure of CNT allows electrons to be transported along the tubes without scattering. It is also interesting to filled CNT/MWNT with some magnetic nanoparticle such as Fe,Ni during the synthesisas[1], even with rare earth metal 4-f. Furthermore to compose with conductive polymer such as PANI [2-3], elastomer, [4-5] and studied its some physical properties included magnetotrasport. Since Chakraborty et al.[6]used MWNT-PVA composite film to study the magnetoresistance of the composite film and found MR ratio around +1.2% at RT, with 10% MWNT loading. The aims of this work is to study the effect of α-Fe_2O_3 filled(or trapped) into MWNT in insulator matrix PVA to the magnetic properties and magnetoresistance (MR) of nanocomposite films. It suggesting that to minimize MWNT loading in composite was important,-because high filling in MWNT made intrinsic metallic transport dominant-,therefore the trapped α-Fe_2O_3nanoparticles magnet in MWNT might be improves theproperties of magnetoresistance due to the interaction behavior between iron α-Fe_2O_3-MWNT with matrix insulator polymer PVA.

2. Experimental

2.1. Materialsand Methods
Multi-Walled Carbon Nanotube (MWNT) used in this work were produced by catalytic chemical vapor deposition (CCVD) from Nanostructured & Amorphous Materials, Inc. USA. The MWNT have purity of 95%+, inner diameter of 5- 10 nm and length of 10 – 30 μm. The polymer matrix polyvinyl alcohol (PVA) with ≥ 99+% purity was supplied by Sigma Aldrich. Iron nitrate ($Fe(NO_3)_3.9H_2O$) in crystal form with purity ≥ 98.5%, surfactant material Sodium Dodecyl Sulfate (SDS) with purity ≥ 98.5%.To open the capping of MWNT an amount of MWNT were heated by furnace at 400 °C for 30 minutes. MWNT/SDS solution, was then transferred to a beaker glass containing over saturated $Fe(NO_3)_3.9H_2O$. As prepared solution was then stirred for 27 hours at room temperature until the black slurry was developed. The last step, the sample was purified by annealing of 160°C in ambient condition for five hour to decompose $Fe(NO_3)_3.9H_2O$. The composite films were prepared by simply mixing α-Fe_2O_3-MWNT dispersed in Sodium Dodecyl Sulphate (SDS) with 10% PVA solution for 4 variation of concentration with a total volume of 5 ml. The solution was then poured into a petry dishes to dry naturally over night.

2.2. Characterization
Observations of nanostructure precursor of iron-filled MWNT was performed by using TEM at GadjahMada University. Characterization of the samples were performed at the Advanced Materials Laboratory of PSBTM – BATAN by using Raman spectroscopy (Senterra Raman Spectroscopy) to determine phase in the precursor sample, FTIR spectroscopy to determine bonding in the composite film. Magnetic properties were performed by means vibrating sample magnetometer (VSM-Oxford 1.2Tesla) for obtained M-H curve. Magnetoresistance ratio was measured by Jandel type Four Point probe(FPP) with attach square 4×4 mm^2 film on Si(100) substrate.

3. Results and Discussion
Determination of the precursor phase was performed by using Raman spectroscopy technique with results as shown below. Fig. 1 shows that the peaks at 219, 284, 398, 472, 590 cm^{-1} indicates the presence of α-Fe_2O_3 phase [7]. Two peaks at 219 and 472 cm^{-1} arise from A_{1g} mode of Fe-O bands. The other peaks was attributed to E_g mode of Fe-O band. While the peaks of the D-band and G-band in 1289 and 1586 cm^{-1} are the peak of MWNT [8].The more higher ratio of D/G intensity for sample post treatment compare to original MWNT was indicating the presence of defect in the crystal structuresp^3 carbons due to the acid treated on MWNT.

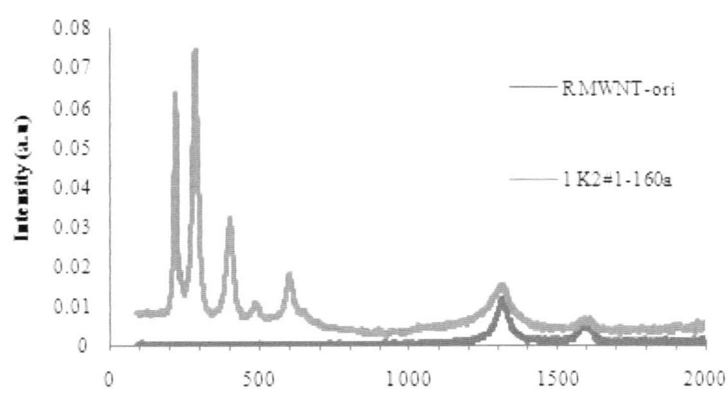

Figure 1. Raman spectrum of sample α-Fe_2O_3-MWNT post heated at 160°C and original MWNT.

The observation of nanostructures on the sample was performed by using Transmission Electron Microscope (TEM) at Gadjah Mada University. The observation condition are 800,000 times for magnification at a voltage of 120kV. Figure 2 shows that the phase of α-Fe_2O_3 (represented by black color) could enter the nanotube with the inner diameter of about 20 nm and the outer diameter of about 50 nm. The phase of α-Fe_2O_3 was impregnated not only through the hollow tube but also through the damaged all of MWNT when interacting with nitric acid [9].

The inclusion of this phase into MWNT through hollow capillary is due to the nature of MWNT during the mixing process taked place. The opened cap of MWNT used in this study allowed the entry of α-Fe_2O_3 at 400°C. This result is slightly different from the techniques developed by Valinejad et al.[10] by heating the MWNT at the temperature of 700°C and 810°C, which generate much less the percent loss of material MWNT.

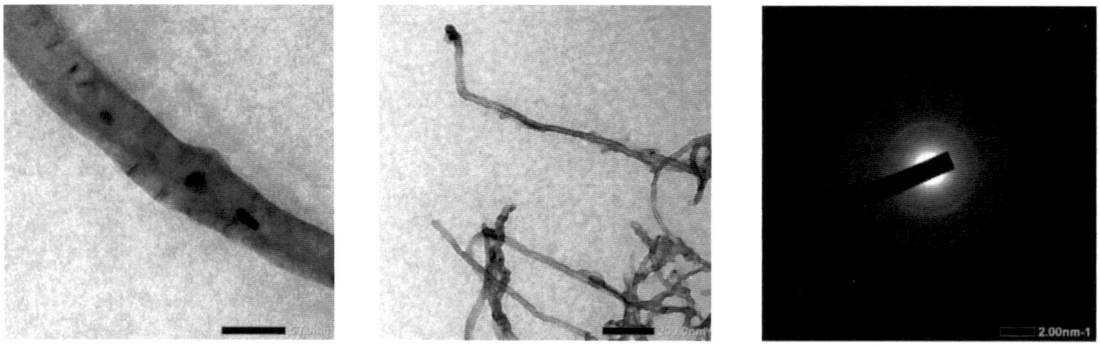

Figure 2. TEM image and SAD from α-Fe_2O_3-MWNT sample at 800.000 x exposure and Voltage = 120kV. Length of bar is 100 nm.

The precursor magnetic properties was performed by using VSM at Advanced Material Laboratory PSTBM. Fig. 3 shows that it precursor sample has ferromagnetic properties with Ms = 0.18 emu/gr, Mr=0.08 emu/g, and Hc=135mT at the external field H=1 Tesla (10kOe).

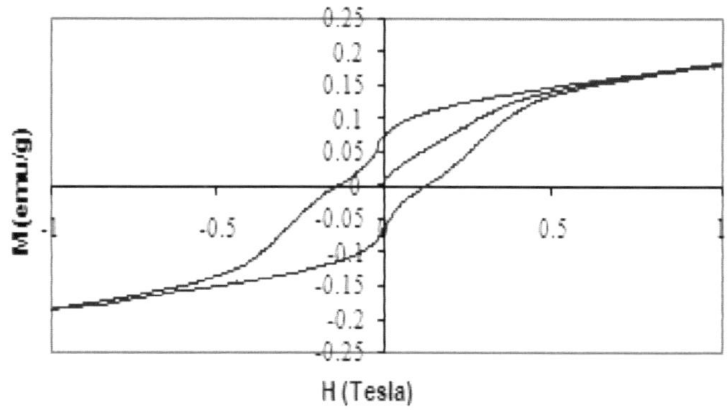

Figure 3. Magnetization curve M-H from powder precursor α-Fe_2O_3-MWNT post heat treatment 160°C

This value was comparable with the report by Zilli et al. [11] for iron nanoparticle trapped in MWNT. However, the curve shows the existence of hysteresis indicating the presence of ferromagnetic spin ordering of α-Fe_2O_3 nanoparticles. Kordas et al.[12] reported the presence of Fe nanoparticles trapped in the CNT will bring hysteresis on the measurement results of the M-H curve.

We have developed four variation samples with fraction filler/matrix PVA as follows;

Table1.Sample code and film Composition

Sample code	Filler/Matrix (vol./vol%)	Weight%
D1	1 : 19	0.24
D2	1 : 9	0.51
D3	1 : 4	1.10
D4	1 : 1	4.60

The quality of composites film was determined by FTIR spectroscopy as described below.

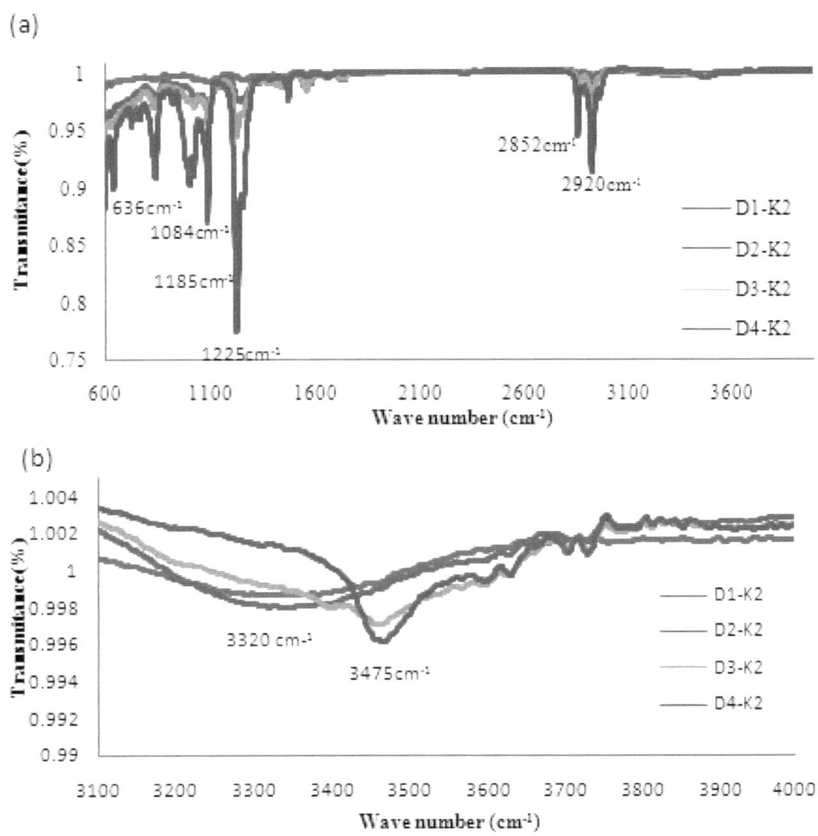

Figure 4. (a) FTIR spectrum of α-Fe2O3/MWNT film at different variation concentration filler/matrix for 600-3100 cm^{-1}; (b) FTIR spectrum for wave number range 3100-4000cm^{-1}

The strong signal at 1185 cm-1 and 1225 cm^{-1} was the proof of C-O bond. These two peaks were clear evidences of successful attachment of PVA on the MWNTs surface. Also, the peak at around 2850-2960 cm^{-1} on IR curve (Fig.4(a)) was indication of CH/CH_2 bonding of PVA. A shown in Fig. 4(a) , the content of iron oxide filled/MWNT increased and the band at 636.5 cm^{-1} and 1084 cm^{-1} increased in intensity. These results suggest a strong interaction between iron oxide filled/MWNT and the polymer matrix with the O-H group while the filler fraction increasing.A broad peak as shown in Fig. 4(b) at around 3320 cm^{-1} verified the presence of a hydroxyl group of –OH, as shown for samples

D1 and D2, also suggesting that broadening the peaks related to existence of uncoated iron oxide/MWNT. Furthermore, these peak become sharper and stonger at around 3400-3500 cm^{-1} for samples D3 and D4 when fraction filler iron oxide-MWNT with matrix PVA increasing.
The magnetic properties measurement was performed by using VSM for the square film with sizes 4x4 mm^2 and magnetic external field perpendicular to film surface up to 10 kOe. All the film sample has paramagnetic properties, except the sample D1 (filler/matrix weight ratio 0.24%) as shown in Figure. Magnetic saturation of D1 sample is around -0.01 emu/g which representing diamagnetic properties from MWNT-PVA film [13]. The other hand, saturated magnetization Ms at H=10 kOe for samples D2, D3 and D4 were 0.04, 0.05 and 0.11 emu/g, respectively. This paramagnetic properties curve shows indicating the presence and degree of random spin interaction between α-Fe_2O_3 nanoparticles trapped inside MWNT where induced by the coated of PVA film matrix.

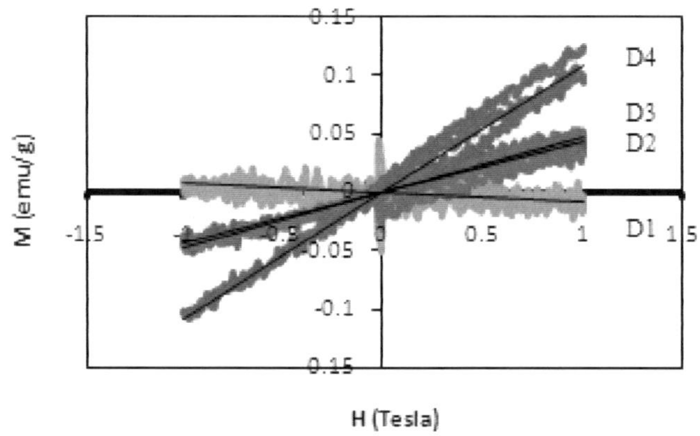

Figure 5. Magnetization curve M-H for composite films with variation volume fraction filler (iron oxide-MWNT)/marix (PVA)

Magnetoresistance (MR) is defined as the difference between the value of resistance of materials in zero magnetic and maximum magnetic field. MR ratio measurement was conducted by using the Four Point probe (FPP) at magnetic external up to H=8kOe. The MR ratio of α-Fe_2O_3-MWNT/PVA thin film on the top of Si (100) were 16%, 40%, 13.6% and 6.2%for positive magnetic field direction with variation volume ratio filler/matrix as shown in Table 2. But this MR behavior were not reversible relatively at low loading filler samples D1 and D2 while the external magnetic field applied in the opposite direction. In the other hand, the MR ratio was reversible for higher volume fraction filler at sample D3 and D4. As shown in Fig.5. The reasonable explanation might be this phenomenon related to the loading volume fraction of α-Fe_2O_3-MWNT in matrix polymer PVA. The filler α-Fe_2O_3-MWNT were well dispersed with more strongly bonding condition with insulator polymer PVA as explained from FTIR spectrum in Fig.4(b).Furthermore, the highest result of MR ratio at 0.51wt% filler/matrix might be related to percolation threshold of carbon nanotube in polymer PVA as predicted by Bauhofer et al.(2009). This result is higher than the phenomenon of magnetoresistance (MR) in MWNT / PVA composite film[6].where MR ratio only +1.2% in the field H = 10kOe with 10 wt % MWNT loading. Thus, the phenomenon of linear positive magnetoresistance (MR) in the above system were coincident with the model for Variable Hopping Transport (VRH) by Nguyen et al.[15]. In this model, a positive MR linear with external field at low magnetic field due to the effect of shrinkage wave function between localization lengths of a grain of filler/matrix. Yosida and Oguro[16] also reported about the phenomenon VRH conduction in bulk samples composed by CNT. Recently, similar phenomenon also evidence in magnetite-Polypyrrole meta composites as reported by Guo et al. [17] and in advanced polyaniline composites [17].

Table 2. The ratio of magneto resistance for α-Fe_2O_3-MWNT/PVA film.

Sample code	Volume ratio Filler/matrix (v/v %)	Weight ratio (wt%)	MR ratio (%) at H=+8 kOe (positive direction)	MR ratio (%) at H= -8 kOe (negative direction)
D1	1 : 19	0.24	+16	+9.8
D2	1 : 9	0.51	+40	+9.8
D3	1 : 4	1.15	+13.6	+8.6
D4	1 : 1	4.61	+6.2	+7.4

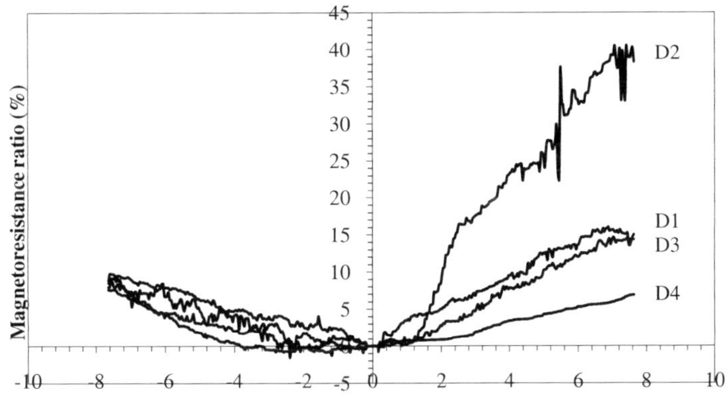

Figure 6. Magnetoresistance ratio in film composite of Fe_2O_3-MWNT/PVA with variation concentration

4. Conclusions

Nanocomposite film α-Fe_2O_3 filled (trapped) into MWNT/Polyvinyl Alcohol (PVA) were successfully synthesized by simple mixing method. TEM images show that iron oxide filled or trapped in MWNT hollows/walls. The filling nanoparticles into MWNT shows α-Fe_2O_3 phase as revealed by Raman spectroscopy. The bonding between filler iron oxide-MWNT with matrix polymer PVA were clearly appear with dependency on the fraction volume filler with matrix as determined by FTIR spectroscopy. This bonding or a coated iron oxide-MWNT by PVA made all film have a paramagnetic properties, except a diamagnetic character at sample D1 with the lowest fraction 0.24wt% filler to matrix PVA. The highest MR ratio obtained in composite film materials are approximately +40% at Ha = 8kOe with a fraction volume filler/matrix 1:9 or 0.51 wt% (sample D2). The linear positive of the MR ratio was interpreted by the wave function shrinkage model between α-Fe_2O_3/MWNT as filler and PVA as the matrix. This behavior strongly related to the trapped or filled some iron oxide filling MWNT in matrix polymer insulator PVA. The interaction between iron-oxide filled MWNT in matrix polymer PVA made shrinkage or decreased the localization length inside the grain of composites.

References

[1]. Borowiak-Palen E, Mendoza E, Bachmatiuk A, Rummeli M H, Gemming T 2006 Iron filled single-wall Carbon nanotubes- A novel ferromagnetic medium *Chem. Phys. Lett* **421** 129-133.
[2]. Chen Y K 1997 Synthesis of carbon nanotubes containing metal oxides and metals of the d-block and f- block transition metals and related studies *J. Mater. Chem.* **7 3** 545-549.
[3]. Suchand Sangeeth C S, Jiminez P, Benito A M, Maser W, Menon R 2010 Charge transport popertiesof water dispersible multiwall carbon nanotube-polyaniline composites *J. of Applied*

Physics **107** 103719.
[4]. Hudziak S, Darfeuille A, Zhang R, Peijs T, Mountjoy G, Bertoni G, Baxendale M 2010. Magnetoresistive phenomena in an Fe-filled carbon nanotube/elastomer composite *Nanotechnology* **21** 25505.
[5]. Karnaushenko D, Makarov D, Yan C, Strubel R, Schmidt O G 2012 Printable Giant Magnetoresistance Devices *Adv. Mater.* **24** 4518- 4522.
[6]. Chakraborty G 2011 Anamalous electrical transport properties of polyvinyl alcohol-multiwall carbon nanotubes composites below room temperature *J. of Appl. Phys.* **109** 033707.
[7]. Chen W, Xiulian P, Bao X 2007 Tuning of Redox Properties of Iron and Iron Oxides via Encapsulation within Carbon Nanotubes *J. Am.Chem. Soc.* **129** 7421-7426.
[8]. Costa S, Borowiak-Palen E, Kruszyñska M, Bachmatiuk A, Kalenczuk J 2008 Characterization of carbon nanotubes by Raman spectroscopy *Materials Science-Poland* **26** **2** 434-441.
[9]. Fu Q 2008 Selective filling of carbon nanotubes with metals by selective washing *New Carbon Materials* **23** **1** 17-20.
[10]. Valinejad M, Shokuhfar A, Arab A 2013 A comparative Study of oxidation and wet chemical methods for Comparative uncapping the multi-walled carbon nanotubes *Journal of Nano and Electronic Physics* **5** **1** 01005-1.
[11]. Zilli D, Chiliotte C, Escobar M M, Bekeris V, Rubiolo G R, Cukierman A L, Goyanes S 2005 Magnetic properties of multiwalledcarbon nanotube-epoxy composites *Polymer* **46** 6090-6095.
[12]. Kordas K 2007 Magnetic-Field Induced Efficient Alignment of Carbon Nanotubes in Aqueous Solutions *Chemistry of Materials* **19** **4** 787-791.
[13]. Chen C S 2009 Preparation and magnetic property of multi-walled carbon nanotube/α-Fe_2O_3 Composites *Trans. Nonferrous Met. Soc. China* **19** 1567-1571.
[14]. Bauhofer W, Kovacs J Z 2009 A review and analysis of electrical percolation in carbon nanotube polymer Composites *Compos. Sci. Tech.* **69** 1486-1498.
[15]. Nguyen V L, Spivak B Z, Shklovskii B I 1985 Tunnel hopping in disordered systems *Sov. Phys. JETP* **62** 1021-1029.
[16]. Yosida Y, Oguro I 1999 Variable range hopping conduction in bulk samples composed of single-walled carbon nanotubes *J. Appl. Phys.* **86** 999.
[17]. Guo J 2013 Magnetite–PolypyrroleMetacomposites: Dielectric Properties and Magnetoresistance Behavior *J. Phys. Chem. Cres.* **117** **19** 10191-10202.
[18]. Gu H 2014 Electrical transport and magnetoresistance in advanced polyanaline nanostructures and Nanocomposites *Polymer* **55** 4405- 4419.

IOP Conf. Series: Journal of Physics: Conf. Series **1091** (2018) 012023 doi:10.1088/1742-6596/1091/1/012023

Characterization and Studies on AC Conductivity of $CaMnO_3$ Material

Sigit Dwi Yudanto[1], Wisnu Ari Adi[2], Agung Imaduddin[1]

[1]Research Center for Metallurgy and Material, Indonesian Institute of Sciences Puspiptek Area, South Tangerang, 15314, Indonesia
[2]Centre for Science and Technology of Advanced Materials, National Nuclear Energy Agency, Puspiptek Area, South Tangerang, 15314, Indonesia

E-mail address: sd.yudanto@gmail.com

Abstract. Synthesis of $CaMnO_3$ has been performed by solid state reaction method using $CaCO_3$ and $MnCO_3$ powder as raw materials. The raw materials were weighed, milled, compacted into a pellet and then sintered at 1250°C. Phase of material, microstructure, and conductivity of the samples were observed. The refinement results of X-ray diffraction pattern shows that $CaMnO_3$ formed as a single phase, which has a structure orthorhombic (P n m a) with lattice parameters, a = 5.277 Å, b = 7.452 Å, and c = 5,261 Å. The atomic density of the refinement result is 4.591 $gr.cm^{-3}$. The morphology of $CaMnO_3$ sample has a good particle homogeneity with the particle size 1- 2 μm. The value of AC conductivity on the $CaMnO_3$ is directly proportional with the increasing ofthe temperature. The highest value of the AC conductivity of the $CaMnO_3$ sample is 2.8 x 10^{-3} S/cm at a temperature of 400°C.

1. Introduction

$CaMnO_3$ material has attracted the attention of researchers because $CaMnO_3$ is an oxide material which can be applied as thermoelectric[1]–[7]. J.W Park, et alhas performed substitution of Bi on the Ca sites and Nb on the Mn sites and caused the thermoelectric properties increased due to electrical conductivity[8]. The indicator of the thermoelectric properties of thematerials istheirfigure of merit (Z) value is relatively high. These Z is defined as follow:

$$Z = \sigma S^2/\kappa \quad (1)$$

where$S$, σ, and κ are respectively the Seebeck coefficient, the electrical conductivity and the thermal conductivity.

$CaMnO_3$ material can be synthesized through a solid-state reaction. J.W. Park, et al carry out synthesis of $CaMnO_3$ using solid-state reaction method[8]. The experiment was begun by mixing raw materials, followed by calcination, compaction and sintering at 1300°C for 12 hours.

Recently, several studies involving successful synthesis of$CaMnO_3$have been reported, especially those performed at high temperatures (≥ 1000°C). The interesting featuresin theCa-Mn-O system are the electrical and magnetic properties that are found in the Ca_2MnO_4, $Ca_3Mn_2O_7$, $Ca_4Mn_3O_{10}$, $CaMnO_3$, and Ca_2MnO_4 compounds[9]. The stable activity of the composition of CaO-MnO occurs in the temperature range of 1100-1300 °C.

Based on the description that $CaMnO_3$ material can be applied as a thermoelectric device, so the aim of this study is to synthesize and characterize $CaMnO_3$ material.

2. Materials and Methods

$CaMnO_3$ was synthesized by solid-state reaction method. In this work, $CaCO_3$(Merck) and $MnCO_3$ (Aldrich) powders were used as the starting materials, by following equation of reaction:

$$CaCO_3(s) + MnCO_3(s) + \tfrac{1}{2}O_2(g) \rightarrow CaMnO_3(s) + 2CO_2(g) \quad (2)$$

This experiment was started by weighing the raw materials in accordance with the stoichiometry. The starting materials were mixed and milled using planetary ball mill. After milling, the mixture was calcined at 800°C in order to eliminate the carbon content in $CaCO_3$ and $MnCO_3$. After calcinations, the mixture was pressed into a pellet with diameter of 25 mm and a thickness of 2 mm at 7 tons. The pellet was sintered at 1250°C for 6 hours in air atmosphere and then cooled to room temperature in furnace.

The phase formation was identified by the PHILIPS PW1710 X-ray diffractometer(XRD) with CoKα radiation (λ = 1.7903 Å), and the quantitative analysis of phases formed in the sample was carried out employing the GSAS software. Then the particle morphology was observed by using the JEOL-JSM 6390AScanning Electron Microscope (SEM), while alternating current (AC) conductivity was measured using the LCR meter instrument in the frequency range of 1to10^5 Hz and the measurements were performed at room temperature, 100°C, 200°C,300°C and 400°C.

3. Results and discussion

3.1. Analysis of X-ray diffraction

Figure 2 shows the results of X-ray diffraction pattern refinementof the sample $CaMnO_3$ using Rietveld method[10].

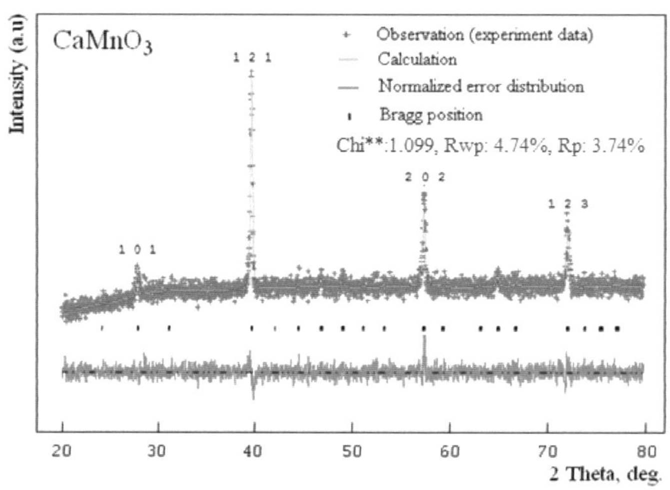

Figure 1. $CaMnO_3$ diffraction pattern refinement using Rietveld method

In the figure 2, the reliability factorsR_{wp}, $R_p$and χ^2 (chi-squared) are relatively good,R_{wp} = 4.74%, R_p = 3.74% and χ^2 = 1.099, because here the curve of normalized error distribution just resembles thebackground intensity and its normal probability plot shows that the values of the observed and the calculated intensity seems to be in good agreement with each other. Qualitative analysis refers to the International Centre for Diffraction Data (ICDD), PDF-2, No. 89-0666[11].The refinement results of x-ray diffraction pattern showsthat the sample consists of a single phase, which has an orthorhombic

(P n m a) structure with lattice parameters a = 5.277 Å, b = 7.452 Å, c = 5.261 Å, α = β = γ = 90°, and V = 206.88 Å^3. The atomic density of the refinement result is 4.591 gr.cm^{-3}.

From X-ray diffraction patterns the crystallite size estimation (L) by using Scherrer equationcouldalso be determinedwith the L value is defined as follow:

$$L = K\lambda / \beta cos\theta \qquad (3)$$

Scherrer equation was developed to calculate the crystallite size of the sample powders by X-ray diffraction method using a specified wavelength λ (nm),by measuring the full width at half maximum of the peaks (β) (rad.) at an angle of 2θ[12]. The calculation results by this approach have yielded the estimatedsample'scrystallite size range of around 31 to44 nm.

3.2. Observation by scanning electron microscope

The surface microstructure of the $CaMnO_3$ sample that was sintered at 1250°C have been observed by using the scanning electron microscope (SEM), as shown in Figure 2. The SEM image in figure 2 shows that the particle morphology of $CaMnO_3$ sample has a good particle homogeneity and particle diffusion is very uniform across the samples surface with the particle size varies from 1 - 2 μm. The SEM image also shows that samples sintered at temperatures of 1250°C is still visibly porous. This is presumably due to the reaction brought about by the release of CO_2 from carbonate raw materials during the sintering process. Besides, it is suspected that air bubbles trapped duringthe compacting process would be the cause of pores formation after the heating process.

Figure 2. The particle morphology of the $CaMnO_3$ sample sintered at 1250°Cfor 6 hours in air atmosphere. The sample was cooled in the furnace.

3.3. Analysis of AC conductivity

The result of conductance measurement (G) can be converted by a factor of sample geometry to obtain alternating current conductivity (σ) which is dependent on the frequency. In other words, if the value of the conductance of a material (G) is 1/R, while the resistivity (ρ) is 1/σ, then the value of the conductivity of the material can be calculated based on the equation as follows:

$$\delta = G\frac{d}{A} \qquad (4)$$

where σ, G, d, and A are respective the conductivity (S/cm), the conductance (Siemens), the thickness of the sample (cm), and the surface area of the sample (cm^2).

IOP Conf. Series: Journal of Physics: Conf. Series **1091** (2018) 012023 doi:10.1088/1742-6596/1091/1/012023

The alternating current conductivity of $CaMnO_3$sample is shown in figure 3.

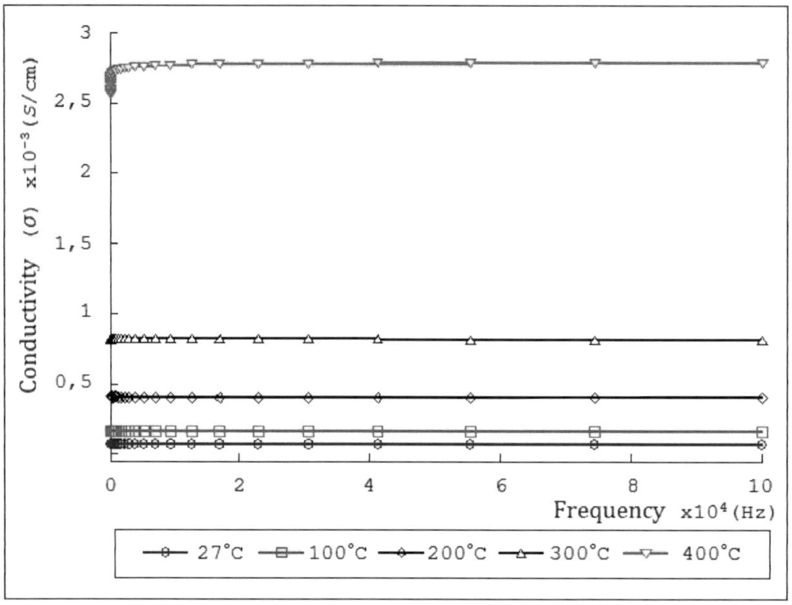

Figure 3. Curveof AC conductivity in the $CaMnO_3$ sample on the frequency as a function of temperature.

In figure 3 it appears that the value of AC conductivity on the $CaMnO_3$ is directly proportional to the increase in temperature. The convergence of conductivity values was seen in all conditions with varying treatment temperature. The convergence value of conductivity at room temperature occurs in the conductivity value of 6.6 x 10-5 S/cm. While the convergence value of conductivity at 100°C, 200°C, 300°C and 400°C respectively are 1.6 x 10^{-4} S/cm; 4,1 x 10^{-4} S/cm; 8.2 x 10^{-4} S/cm and 2.8 x 10^{-3} S/cm.

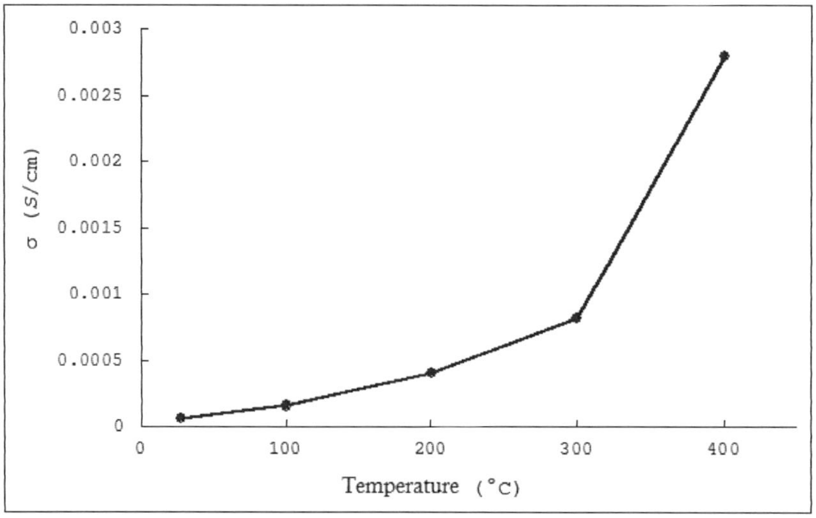

Figure 4. Convergence curve of conductivity values as a function of temperature on the $CaMnO_3$

The conductivity value experience a different trend occuring at 400°C. The conductivity of $CaMnO_3$ increased from 2.5 - 2.7 x 10^{-3} at a frequency of 1 Hz - 3 kHz, and then converged to the conductivity value of 2.8 x 10^{-3} S/cm at a frequency of 5 kHz. This is different from the values of conductivity at room temperature, 100°C, 200°C and 300°C, where the convergence of conductivity value has occurred at a frequency of 1 Hz. Convergence curve of temperature conductivity values is shown in figure 4.

4. Conclusion

Synthesis of $CaMnO_3$ has been successfully carried out using solid state reaction method. The refinement results of X-ray diffraction pattern shows that $CaMnO_3$was formed as a single phase, and has a good particle homogeneity and diffusion of particles is very uniform across the sample surface with the particle size varies from 1-2 μm. The value of AC conductivity on the $CaMnO_3$ is directly proportional to the increasing of the temperature.The highest value of the AC conductivity of the $CaMnO_3$ sample is 2.8 x 10^{-3} S/cm at a temperature of 400°C

References

[1] Phaga P, Vora-Ud A and Seetawan T 2012 *Procedia Eng.* **32** 1050–1053.
[2] Fergus J W 2012 *J. Eur. Ceram. Soc.***32** **3** 525–540.
[3] Seetawan T, Singsoog K, Srichai S, Thanachayanont C, Amornkitbamrung V and Chindaprasirt P 2014 *Energy Procedia* **61** 1067–1070.
[4] Alfaruq D S, Otal E H, Aguirre M H, Populoh S and Weidenkaff A 2012 *J. Mater. Res.* **27** 985–990.
[5] Zhan B, Lan J, Liu Y, Lin Y, Shen Y and Nan C 2014 *J. Mater. Sci. Technol.***30** 821–825.
[6] Paengson S, Pilasuta P, Singsoog K, Namhongsa W, Impho W and Seetawan T 2017 *Mater. Today Proc.* **4** 6289–6295.
[7] Kabir R, Tian R, Zhang T, Donelson R, Tan T T and Li S 2015 *J. Alloys Compd.* **628** 347–351.
[8] Park J W, Kwak D H, Yoon S H and Choi S C 2009 *J. Alloys Compd.* **487** 550–555.
[9] Horowittz H and Longo J 1978 *Mater. Res. Bull.* **13** 1359–1369.
[10] Larson A C and Von Dreele R B 2004 *Los Alamos Natl. Lab. Rep. LAUR* **86** 748.
[11] Santiago-Teodoro M, Hernández-Cruz L, Montiel-Sánchez H, Álvarez-Lucio G, Flores-González M A and Legorreta-García F 2011 *J. Mex. Chem. Soc.***55** 204–207.
[12] Monshi A, Foroughi M R and Monshi M R 2012 *World J. Nano Sci. Eng.* **2** 154–160.

IOP Conf. Series: Journal of Physics: Conf. Series **1091** (2018) 012024 doi:10.1088/1742-6596/1091/1/012024

Synthesis and Characterization of Magnetically Modified Composites (TiNT/CNT/Fe_3O_4)

Slamet[1], Praswasti P. D. K. Wulan[1], Desi Heltina[1], Adel Fisli[2], Davin Philo[1]

[1]Universitas Indonesia, Department of Chemical Engineering, Faculty of Engineering, Depok 16424, Indonesia
[2]Center for Science and Technology of Advanced Materials (PSTBM)-BATAN, Kawasan Pupitek Serpong, Tangerang 15314, Indonesia

Corresponding author: slamet@che.ui.ac.id

Abstract. In this study, Titania nanotube (TiNT) had been modified by combining it with carbon nanotube (CNT) and magnetic material, Fe_3O_4, forming magnetically modified composite (TiNT/CNT/Fe_3O_4). Magnetic properties were added to the composite to overcome photocatalytic application problem for waste water treatment case, especially the catalyst recovery issue. Prior to the modifications, TiNT was synthesized from TiO_2 P25 using hydrothermal process at 130 °C for 6 h. TiNT and CNT were then combined using hetero agglomeration process in acid condition to obtain TiNT/CNT composite. Various amount of Fe_3O_4 nanoparticles were then composed on the surface of TiNT/CNT using ultrasonic assisted in-situ process, producing TiNT/CNT/Fe_3O_4 magnetic composites. The samples were analyzed with various characterizations: zeta potential, FT-IR, FE-SEM/EDX, XRD, and VSM. Experimental results show that TiNT/CNT/Fe_3O_4 composite with good crystallinity and morphology had been successfully synthesized. The optimal amount of Fe_3O_4 in TiNT/CNT/Fe_3O_4 magnetic composite was 0.3 times the amount of TiNT to minimize the photodissolution effect, while the composite was still categorized as superparamagnetic materials (with the saturation magnetization value of 21.1 emu/g and coercivity of 84.4 oe). The magnetic separation test also confirmed that prepared magnetic composites could be effectively separated from the test solution.

1. Introduction

Recently, nanosized semiconductor photocatalyst titania (TiO_2) has attracted the attention of many scientists, due to its wide range of applications (solar panel, super hydrophilic materials, degradation pollutant, hydrogen production, etc). On the further research, titania nanoparticles morphology has been successfully modified into nanotube form [1, 2]. Titania nanotube has been widely developed, because of its good electronic and mechanic characteristic, large surface area,ion-exchangeable ability, remarkable rapid and long-distance electron-transport capability, andenhanced light absorption[3]. Titania nanotube might also be able to interact with other materials relatively better than titania nanoparticles.

In environmental technology sector, photocatalyst material, peculiarlyTiO_2 is generally accepted as one of the most effective photo-induced catalysts and is frequently used to oxidize organic and inorganic compounds in air and water due to its strong oxidative ability and long-term photo-stability [4]. Furthermore, to enhance its performance, TiO_2 is usually modified by combining it with other materials, such as carbon nanotube (CNT) which will increase its adsorption ability and photocatalytic efficiency.However, for the field application that demands repeatable usage,nanosizedphotocatalyst still has a major problem that greatly limits the performance of the catalyst itself: the difficulty to separate catalyst suspension after use for the waste water treatment due to the small dimension of titania that it needs a very complex filtration technology[5] . Many techniques have been applied for solving this problem, for instanceby immobilizing TiO_2 catalysts unto a suitable support [6]. Even though this method can solve the catalyst recovery problem, the activity of the photocatalyst would be lower because the catalyst must be attached on the support, limiting the movement of the catalyst and the contact between the photocatalyst, light, and pollutant. The other disadvantage of this method is the difficulty to maintain the catalyst to strongly attach on certain support. Therefore, a long-term usage of immobilized TiO_2, one of the important parameters for industrial application is difficult to achieve, because a significant decrease of photocatalytic efficiency might occur[5].

Another way to solve the problem is by combining the photocatalyst with magnetic materials, so the photocatalyst can be magnetically separated from waste water [7]. This method is quite effective and simple, where the after-used photocatalyst can be recovered from the solution and used repeatedly [7]. Compared to the immobilization technique, the advantage in using magnetic composite is that the reaction condition is still slurry, so that photocatalyst particles can move freely into all of the liquid body, making the interaction between particles, light, and pollutant more efficient. Related to the usage of magnetic material, it is better to use superparamagnetic iron oxide-based material, like Fe_3O_4 and γ-Fe_2O_3, because these materials are not toxic, and their presences in nature are abundant with relatively low price.

To date, many researchers have conducted studies to produce magnetically recoverable photocatalyst nanoparticles, especially for the application of waste water treatment [8]. However, a study to produce magnetic photocatalyst, using titania with nanotube morphology has not been extensively investigated until now. More than that, many kinds of photocatalyst modifications were done separately. Based on those things, a study to modify photocatalyst TiNT, by combining it with supportive material (CNT) and magnetic materials (Fe_3O_4) simultaneously in an integrated process, producing magnetically modified composite TiNT/CNT/Fe_3O_4 becomes an interesting and strategic topic. This study was aimed to investigate the character of magnetic composite and its potential application to be used for the long term usage in larger scale of waste water treatment application. A magnetic separation test was also conducted to prove the ability of the magnetic composites to be magnetically recovered.

2. Material and Methods

2.1. Material

TiO_2 P25 with a crystallographic mode of 79% anatase and 21% rutile, a 53.6 m^2g^{-1} BET surface area , and an average particle size of 21 nm, used as raw materials, was bought from Evonic Industries. Commercial multi-wall carbon nanotube (MWCNT) was used on this study and purchased from Dong Yang (HK) *Int'l Group Limited*. Chemical such as sodium hydroxide pellet,concentrated nitric acid (HNO_3, 65 %), iron (III) chloride hexahydrate ($FeCl_3.6H_2O$), iron (II) chloride tetrahydrate ($FeCl_2.4H_2O$), and ammonia solution (NH_3, 25%) were purchased from Merck, and fuming hydrochloric acid (HCl, 37%) was bought from Mallinckrodt.

2.2. Preparation of Titania Nanotube (TiNT)

The preparation of titania nanotube (TiNT) was similar to the previous study [1]. TiNT was synthesized by dissolving 6 g of titania P25 powder into 300 mL of 10 M NaOH and put into ultrasonic bed TELSONICfor 30 min prior to hydrothermal treatment at 130 °Cfor 6 h. The sample was then stirred at 600 rpm using a teflon-lined autoclave equipped with adjustable rotating speed mixer. Thereafter, the sample was washed with acid (0.2 NHCl) repeatedly for a certain time until it reached pH 2-3 and then with aquadest until the pH of approximately 5 was reached. TiNT powder was then obtained after drying at 80 °C for several hours and calcined at 500 °C for 2 hours.

2.3. Pre-treatment of CNT

CNT, used in this study was firstly pre-treated through acid treatment to generate negative surface charge of CNT. Firstly, 1 g of MWCNT was mixed with 100 mL HNO_3 65% and put into ultrasonic bath processorfor 1 h. The pre-treatment was continued with reflux process of the mixture for 6 h at temperature 120 °C. After the mixture was cooled enough, it was filtered to obtain the suspension, which was named by pre-treated CNT with negative surface charge. The suspension was dried overnight at 60 °C and the dried product was crushed to obtain modified CNT.

2.4. Synthesis of TiNT/CNT and TiNT/CNT/Fe3O4 Composite

TiNT with weight of 1 g and 0.015 g of pre-treated CNT were mixed in 100 mL aquadest. Then, pH of the mixture was adjusted to 3, using HNO_3 solution (0.1 M). The mixture was subsequently put into ultrasonic bath processor for 5 minutes. After that, evaporation of the mixture was conducted at 80 °C to dry off the water. The dried product was crushed to obtain TiNT/CNT composite in powder form.

Synthesis of TiNT/CNT/Fe_3O_4 was done using ultrasonic assisted in-situ method. Firstly, 2.6 g of $FeCl_3.6H_2O$and 1 g of $FeCl_2.4H_2O$ (mol ratio of Fe^{3+}:Fe^{2+} = 2:1) were dissolved in 5.15 mL of HCl 0.1 N and stirred till all of the solid fully dissolved. Afterwards, the iron salt solution was diluted by adding 7.5 mL of aquadest, followed by stirring for 15 minutes. Then, the TiNT/CNT composite was mixed with 100 mL of aquadest. Various amount of the iron salt solutions (to vary the composition of Fe_3O_4 in composite), prepared beforehand were added to the mixture, followed by stirring for 1 h, in order to disperse the iron ions (Fe^{3+} and Fe^{2+}) in the mixture. Certain amount of NH_4OH, 1.5 N solution (9 mL/mL of iron salt solution) were added drop by drop to the mixture, while put into ultrasonic bath processorat 65 °C. Ultrasonication treatment was continued for 5 minutes. The mixture was then washed using aquadest and magnetically separated until it reached pH 7. The obtained solid suspension was dried at 60 °C overnight and 80 °C for 2 h. The dried solid was crushed to obtain magnetic composite TiNT/CNT/Fe_3O_4.

2.5. Composite Characterization

Prior to the synthesis of the composite, zeta-potential characterization for the raw material (titania and pre-treated CNT) was conducted to investigate the surface charge character of raw materials in various conditions (of pH). Zeta potential of the particles in their aqueous suspensions (with various pH) were measured using Zatsizernano-Zs, Malvern. The magnetic composites were further characterized by several analysis methods.FT-IR characterization was conducted using Fourier Transformation Infra-Red (FT-IR) Spectrophotometer type Shimadzu Prestige 21 to investigate the bond characteristics in samples. The analysis was carried out using IR light with wave number range of 400 – 4000 cm^{-1}. The surface morphology and the material composition of samples were analyzed using a field emission-scanning electron microscope FE-SEMFEI INSPECT F50 equipped with energy dispersive X-ray spectroscopy (EDX) AMETEK which was conducted at 20 kV. The crystalline phase of the composites were determined using powder X-ray diffraction (XRD).The XRD patterns were obtained using a PHILIPS PW1710 with Cu anode tube (l = 0.154184 nm). XRD tuberuns at voltage of 40 kV and current of 30 mA, with scanning rate of 2°/min (with sampling pitch : 0.02°) in the range of 2θ=10 – 80°.Magnetic characteristic of samples were measured using Vibrating Sample Magnetometer (VSM) type 1.2 T Oxford. The samples were given magnetic field from -1 to +1 Tesla, with rate of 0.25 Tesla/minute and the magnetic response or mass magnetization of samples were measured and presented in form of hysteresis curve.

2.6. Magnetic Separation Test

Magnetic separation test was conducted to prove the ability of magnetic composite to be separated magnetically. 0.1 g of magnetic composites were mixed with 100 mL of aquadest and stirred sufficiently. The vessel containing the mixture of composite sol was placed on a rather weak magnetic bar (to make the difference in magnetic strength visually significant). Images of the composite sol condition in the vessel were taken from time to time and the time until all composites particles were fully separated from the liquid phase.

3. Results and Discussion

3.1. Zeta Potential (ζ-potential)Characterization

Figure 1 shows the effect of pH on ζ-potential of TiO_2P25 [9], pure MWCNT [10], and CNT which had been given acid treatment (CNT-AT). In this case, TiO_2 P25 was considered to have same surface charge characteristic with TiNT. It can be seen that TiO_2 and CNT have similar profile of ζ-potential, which means they have similar surface charge characteristic. Associated to that phenomenon, particles with same surface charge (either positive or negative), like TiO_2 (or TiNT) and CNT would repel and couldn't attach to each other. Therefore, certain treatment,such as acid treatment, was needed to modify the surface charge of CNT, making it possible to attach with TiNT. It can be seen at figure 1 that ζ-potential of CNT-AT were negative (-10 up to -29 mV) for pH range of 2-11, indicating the negative surface charge of CNT-AT. This phenomenon is strongly related with the formation of hydroxyl, carbonyl, and carboxylic acid at the wall of CNT, which occurred because of the oxidation process by strong acid [11]. The negative surface charge of CNT-AT would enable the hetero agglomeration process between TiNT and CNT at acid condition, where TiNT would have positive surface charge. Hence, they would attach each other making electrostatic attraction force between them. For further sections, the term CNT would refer to pre-treated CNT (CNT-AT).

3.2. FT-IR Characterization

Figure 2 contains the FT-IR spectra of samples TiNT/CNT and TiNT/CNT/Fe_3O_4 composites. Both samples contain peak at 3400 cm^{-1}, 1630 cm^{-1}, and 400 – 850 cm^{-1}. The peak around 1630 cm^{-1} is due to the bending vibration of the O–H bond of chemisorbed water, and the peak around 3400 cm^{-1} is due to the stretching mode of the O–H bond of free water [12]. The broad band at 400 – 850 cm^{-1} corresponds to the Ti–O–Ti stretching vibrationmode in crystal TiO_2[12]. TiNT/CNT composite specifically has the quite broad peak around 1080 cm^{-1}, indicating the formation of Ti-O-C bond between TiNT and CNT [13]. The strong peak at 1400 cm^{-1} of TiNT/CNT composites signifies the presence of carboxylic acid, C-O-H bond which is deduced to be coming from pre-treated CNT [13]. The last two peaks cannot be found at the FT-IR spectra of TiNT/CNT/Fe_3O_4, because the amount of CNT in the composite was very small which the CNT could be cloistered by Fe_3O_4 nanoparticles, making the bonds which involved CNT could not be detected. Ye, Cai, Zheng, Xiao, Hua and Huang [14]obtain similar result, where almost all strong peaks of SiO_2/TiO_2 sample are covered up after Ag is coated onto the SiO_2/TiO_2 substrate. For the TiNT/CNT/Fe_3O_4 composite, the Fe-O-Febond is supposed to produce peak at 580 cm^{-1}[15]. However, the peaks of Fe-O-Fe bond are overlapping with the broad band of Ti-O-Ti which lies from 400-850 cm^{-1}.This phenomenon makes the FT-IR characterization is not enough to determine the presence of Fe_3O_4 nanoparticles in the magnetic composite (TiNT/CNT/Fe_3O_4).

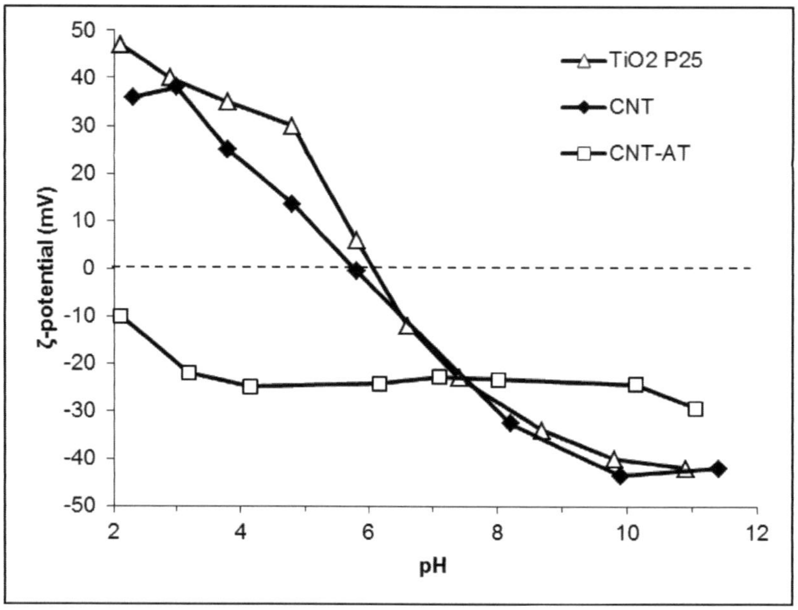

Figure 1. Effect of pH on ζ-potential of TiO_2 P25[9], pure MWCNT [10], and pre-treated CNT (CNT-AT).

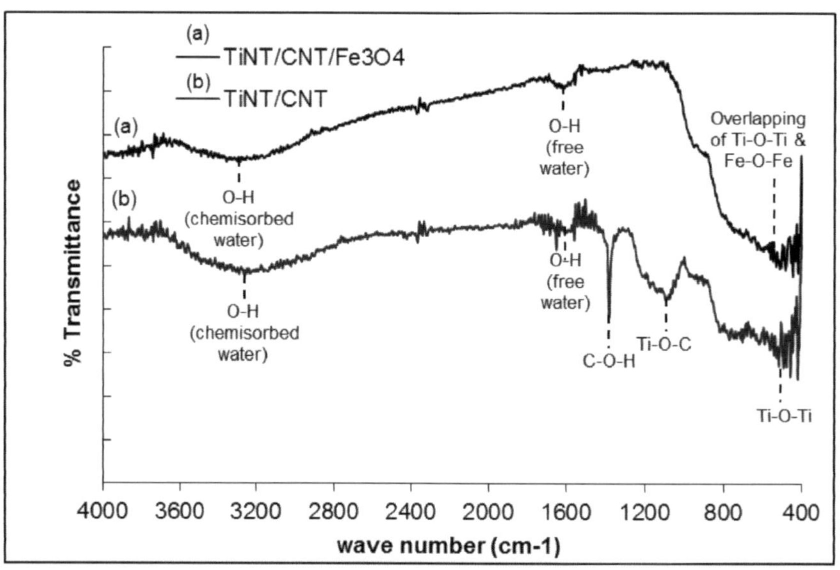

Figure 2. FT-IR spectra of TiNT/CNT and TiNT/CNT/Fe_3O_4 composite

3.3. FE-SEM/EDX Characterization

Figure 3 illustrates the FE-SEM image of TiNT, CNT, Fe_3O_4, TiNT/CNT, and TiNT/CNT/Fe_3O_4. As-prepared TiNT (figure 3 (a)) has morphology of straight tubes with outer diameter of 24-42 nm. The tubes of TiNT are relatively small but long and separated perfectly from each other. CNT (figure 3 (b)) which had been pre-treated beforehand has the form of curly tube, with outer diameter of 45-70 nm.

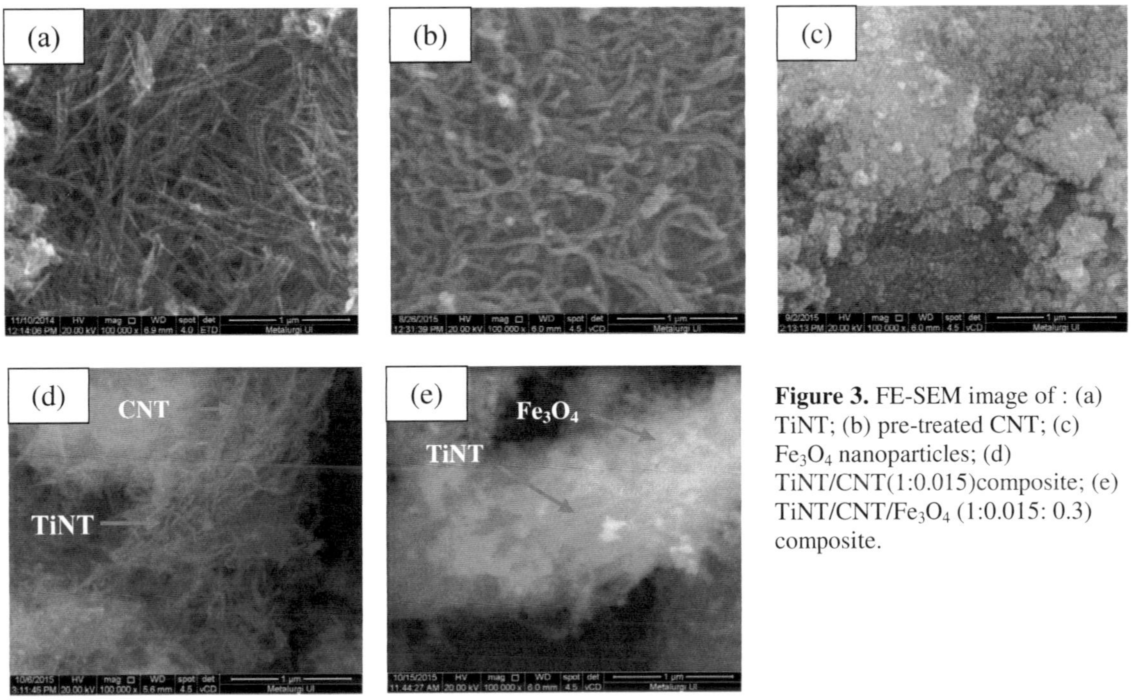

Figure 3. FE-SEM image of : (a) TiNT; (b) pre-treated CNT; (c) Fe_3O_4 nanoparticles; (d) TiNT/CNT(1:0.015)composite; (e) TiNT/CNT/Fe_3O_4 (1:0.015: 0.3) composite.

Table 1. Energy dispersive X-ray spectroscopy analysis of samples.

Elements	% weight				
	TiNT	CNT (pre-treated)	Fe_3O_4	TiNT/CNT (1: 0.015)	TiNT/CNT/Fe_3O_4 (1: 0.015: 0.3)
Ti	47.48	-	-	50.76	37.77
O	52.52	5.79	36.59	44.46	50.82
C	-	94.21	-	4.78	2.96
Fe	-	-	63.41	-	8.45

The magnetic material, Fe_3O_4 (figure 3 (c)) forms the clustered structure, which each particle of Fe_3O_4 sizes of less than 25 nm. The cluster structure would be giving advantage for the separation process of magnetic particles from the solution, because of its relatively big volume, making the magnetic response stronger. Therefore, the superparamagnetic characteristic of each nanoparticle could be maintained [16]. CNT is found to attach on several tubes of TiNT in TiNT/CNT composite (figure 3 (d)). This phenomenon could occur because of electrostatic attraction force between different surface charges of TiNT and CNT as explained before at ζ-potential characterization section. However, CNT is hardly found in TiNT/CNT/Fe_3O_4 (figure 3 (e)), possibly because of the small composition of CNT which make it could be easily covered by Fe_3O_4, similar to FT-IR result of the composite. In the other hand, TiNT can be easily found in form of long straight tubes which are covered by Fe_3O_4 nanoparticles.The cluster structure of Fe_3O_4 is still there, which would make the composite to have quite strong magnetic response so that the magnetic separation process could be done properly [16]. However, related to the closure of TiNT by the particles of Fe_3O_4, there would be negative drawbacks for the photocatalytic process including the strong light absorption by iron oxide or photodissolution effect which could interfere the contact between photocatalyst and the light. As a result, it might decrease the photocatalytic activity . Therefore, it is important to determine the optimal amount or composition of the magnetic materials to minimize the photodissolution effect and maintain the good magnetic characteristic.

Energy dispersive X-ray spectroscopy result of TiNT, CNT, Fe_3O_4, TiNT/CNT, and TiNT/CNT/Fe_3O_4 could be seen in Table 1.TiNT contains only Ti and O elements only, while Fe_3O_4 contains only Fe and O elements, as they should be. For acid treated CNT, besides C element, there is O element which derives from carboxylic acid (as the outcome of the acid treatment). As the result of the combination between TiNT and CNT, TiNT/CNT composite contains the constituent elements of each material, which are Ti, O and C. Table 1 also shows thatTiNT/CNT/Fe_3O_4 composite contains the elements of Ti, O, C, and Fe, which proves the presence of each material: TiNT, CNT, and Fe_3O_4. C element of CNT is found in minor amount which supports the result of FE-SEM and FT-IR analysis.

3.4. XRD Characterization

Figure 4 shows the XRD pattern of TiNT, TiNT/CNT, and TiNT/CNT/Fe_3O_4 with varied compositions of Fe_3O_4. All samples give out the dominant peaks at 2θ of25.6°, 37.6°, 47.9°, 53.7°, and 55.1°, which are conforming tocrystalline structure of anatase TiO_2 (according to JCPDS file no. 21-1272). Each of the magnetic composites (TiNT/CNT/Fe_3O_4) contains magnetite, Fe_3O_4 which produces XRD peaks at 30.2°, 35.5°, 43.3°, 56.8° (JCPDS file no. 03-0863), similar to the magnetite phase for pure Fe_3O_4 (figure 3(f)). As the amount of Fe_3O_4 goes higher in the composite, the peaks of TiO_2 anatase become shorter and the peaks of magnetite grow higher. However, calculation by Scherrer equation shows that the size of anatase crystallite sizes of pure TiNT, TiNT/CNT composite, and TiNT/CNT/Fe_3O_4 composites are not significantly different (see table 2), which means that the addition of Fe_3O_4 and CNT did not affect the crystal of TiNT.

It is noticeable that there is no significant difference between the peaks of TiNT/CNT/Fe_3O_4 (1: 0.015: 0.5) composite and the peaks of TiNT/CNT/Fe_3O_4 (1: 0.015: 0.3) composite, which means their magnetic properties would be similar. This phenomenon might happen due to the instability of Fe_3O_4 in open air, which make it could easily be oxidized to Fe_2O_3along the preparation and storing process . For TiNT/CNT/Fe_3O_4 (1: 0.015: 0.5) composite, the preparation process is relatively longer, especially for the step of adding ammonia (NH_4OH) solution drop by drop, where the composite needs more ammonia due to larger amount of Fe_3O_4. The oxidation process might also be accelerated by the ultrasonication process.

Table 2. Anatase crystallite size, saturation magnetization and coercivity value of samples.

Samples	Anatasecrystalite size (nm)*	Saturation magnetization (emu/g)	Coercivity (oe)
TiNT	11.0	-	-
TiNT/CNT (1 : 0.015)	10.5	-	-
TiNT/CNT/Fe_3O_4 (1 : 0.015 : 0.1)	9.02	4.12	110.4
TiNT/CNT/Fe_3O_4 (1 : 0.015 : 0.3)	9.79	21.1	84.4
TiNT/CNT/Fe_3O_4 (1 : 0.015 : 0.5)	10.4	26.8	72.9
Fe_3O_4	-	79.7	59.5

* calculated by Scherrer equation

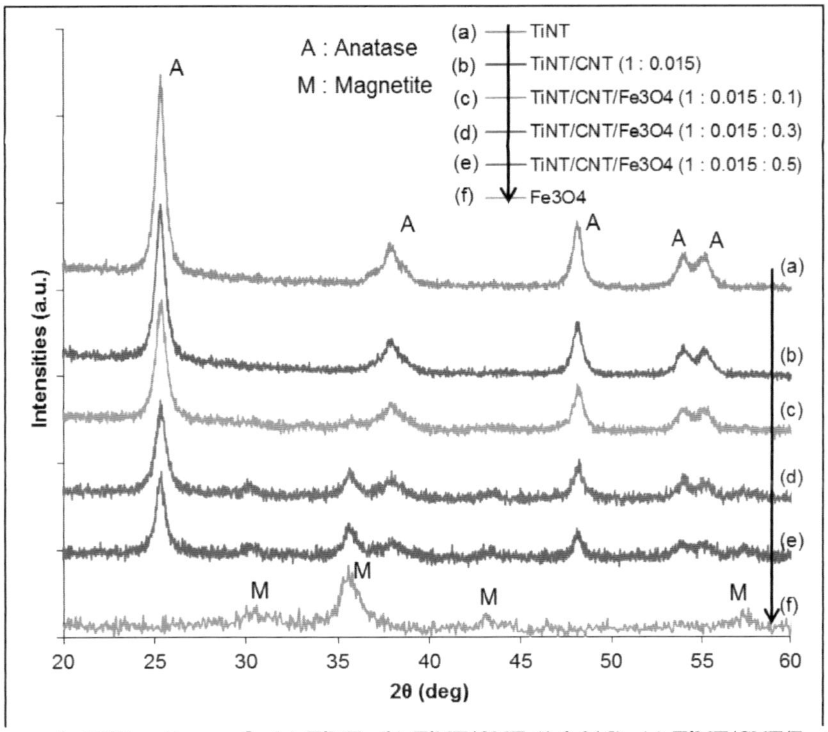

Figure 4. XRD patterns of : (a) TiNT; (b) TiNT/CNT (1:0.015); (c) TiNT/CNT/Fe_3O_4 (1:0.015:0.1); (d)TiNT/CNT/Fe_3O_4 (1:0.015:0.3); (e)TiNT/CNT/ Fe_3O_4 (1:0.015:0.5); (f) pure Fe_3O_4

3.5. Evaluation of Magnetic Properties

The magnetic response of prepared Fe_3O_4 and TiNT/CNT/Fe_3O_4 composites with varied amount of Fe_3O_4were measured at room temperature using VSM analysis, as illustrated in figure 5. The values of saturation magnetization and coercivity are shown in table 2. The observed value of saturation magnetization for the prepared magnetic iron oxide, Fe_3O_4 is 79.7 emu/g. Magnetic properties depend on the amount of magnetic phase (Fe_3O_4) fraction in the materials. The decrease in the saturation magnetization in composite is due to the lower content of magnetic iron oxide in materials and the presence of non-magnetic materials [7].It seems that the reduction of saturation magnetization value is not proportional to the amount of Fe_3O_4. This result is similar to XRD result which shows that the difference between the peak height of magnetite phase of TiNT/CNT/Fe_3O_4 (1: 0.015: 0.05) and TiNT/CNT/Fe_3O_4 (1: 0.015: 0.03) is not significant. As explained before, this phenomenon might happen due to the oxidation process of Fe_3O_4 (see *XRD Characterization* Section). Superparamagnetic behavior of material can be seen from the low value of coercivity. The coercivity value of those two magnetic composites are similar to each other and lower than 100 oe, which indicates that those composites have superparamagnetic behavior [7]. In the other hand, TiNT/CNT/Fe_3O_4 (1: 0.015: 0.01) has low saturation magnetization value and great coercivity value (> 100 oe), which means the composite could not be categorized as superparamagnetic material. Superparamagnetic behavior is needed for the field application to facilitate recovering the magnetic composite efficiently from treated water by the assistance of magnetic field [7].

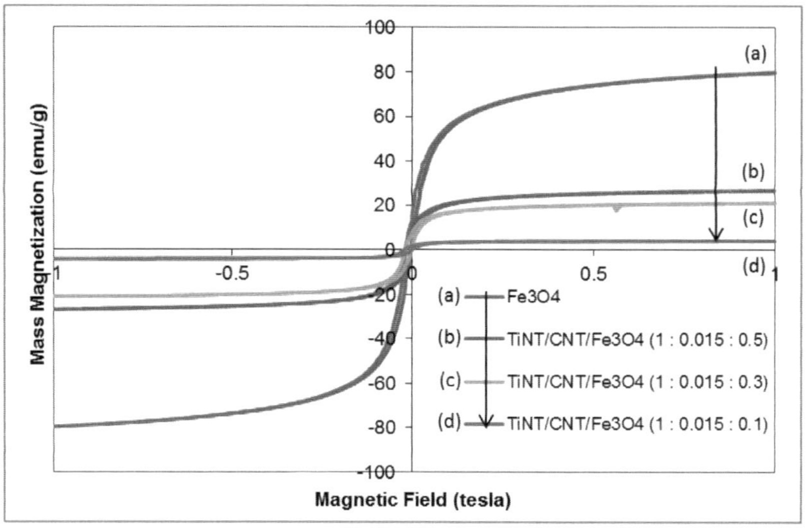

Figure 5. Hysteresis curve of Fe_3O_4 nanoparticles and TiNT/CNT/Fe_3O_4 composites with varied amount of Fe_3O_4.

Magnetic separation test for the samples was conducted and the images of the test can be seen on Figure 6. TiNT/CNT sol was very stable, so that after 60 minutes there were no significant differences. Even though the sol was left for 24 hr, the sol was still stable. Magnetic composites were shown to have magnetic response to the magnetic field, produced by magnetic bar. The separation time until the supernatant became transparent for TiNT/CNT/Fe_3O_4 (1: 0.015: 0.1), TiNT/CNT/Fe_3O_4 (1: 0.015: 0.3), and TiNT/CNT/Fe_3O_4 (1: 0.015: 0.5) composites were 60, 24, and 9 minutes respectively, which means that the separation time of the magnetic composite was inversely proportional to the amount of magnetic phase. However, for TiNT/CNT/Fe_3O_4 (1: 0.015: 0.3) and TiNT/CNT/Fe_3O_4 (1: 0.015: 0.5) composites, it is a bit contrary to the result of VSM analysis which shows that the magnetic strength of those two composites are quite similar. This phenomenon can be understood if we also consider the factor of composite density, which would also increase as the amount of Fe_3O_4 increased. High density of composite would make the composite to sink faster in water. In this case, it can be concluded that the separation time of a composite was determined not only by magnetic response but also the density of the composite.

For field application, it is demandable to have decent rate of composite separation. In this case, using rather small amount of Fe_3O_4, like on TiNT/CNT/Fe_3O_4 (1: 0.015: 0.1) is not recommended, because the separation process would take fairly long time. Using high amount of magnetic phase ($\geq$ 0.5 times the amount of photocatalyst), iron oxide is not recommended too, because it might lead to excessive photo dissolution effect. Therefore, using medium amount of Fe_3O_4 will be sufficient to produce composite which can be efficiently separated from the treated water by the assistance of magnetic field. From these outcomes, it could be concluded that the optimal amount of Fe_3O_4 was 0.3 times the amount of TiNT. The composition was chosen in order to minimize the negative drawbacks to photocatalytic activity because of the presence of magnetic phase. Yet, the composition still can make magnetic composite owned supermagnetic behavior to enable the magnetic separation process for composite recovery. Photodegradation test of the samples to investigate the performance of the composites for eliminating organic waste and the photo dissolution effect is still in progress.

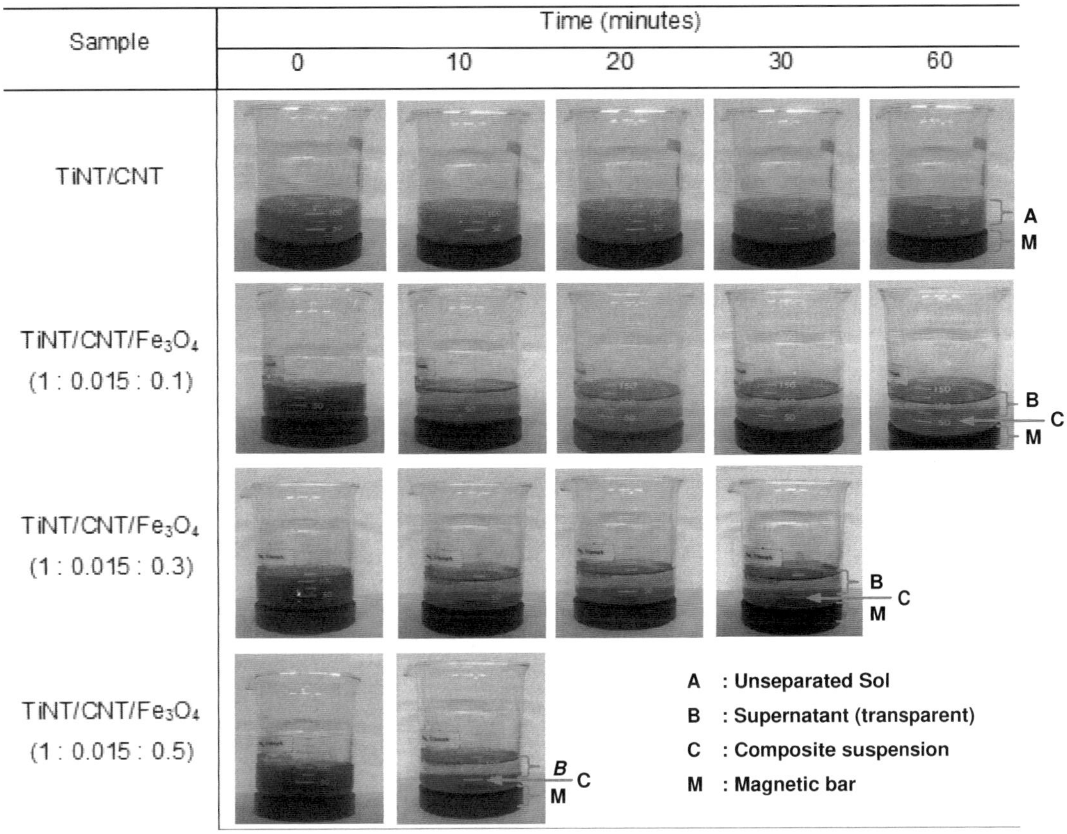

Figure 6. Photos of magnetic separation test for the samples.

4. Conclusion

The TiNT/CNT/Fe_3O_4composite had been successfully prepared by ultrasonic assisted in-situ process, which the composite had good magnetic properties. In the magnetic composites, TiNT and CNT were attached to each other due to electrostatic attraction force and Fe_3O_4 nanoparticles were formed on the surface of TiNT/CNT in cluster structure. Based on the experimental results, the optimum amount of Fe_3O_4 in the magnetic composite was 0.3 times the amount of TiNT to minimize the photo dissolution effect. Yet, with the chosen composition, the superparamagnetic behavior could be maintained, with the saturation magnetization and coercivity values were 21.1 emu/g and 84.4 Oe respectively. It would not be effective to add more Fe_3O_4, because the magnetic characteristic would insignificantly increase. On the contrary, it could increase the photo dissolution effect. Photo degradation test of the samples is still in progress.

Acknowledgments

The authors would like to sincerely acknowledge the Directorate of Research and Community Service in Universitas Indonesia for the financial support on this research project, grant no. 1829/UN2.R12/HKP.05.00/2015.

References

[1] Slamet, Tristantini D, Valentina and Ibadurrohman M 2013 Photocatalytic hydrogen production from glycerol–water mixture over Pt-N-TiO_2 nanotube photocatalyst *International Journal of Energy Research* **37** 1372-81

[2] Slamet, Ratnawati, Gunlazuardi J and Dewi E L 2017 Enhanced photocatalytic activity of Pt deposited on titania nanotube arrays for the hydrogen production with glycerol as a sacrificial agent *International Journal of Hydrogen Energy* **42** 24014-25
[3] Liu N, Chen X, Zhang J and Schwank J W 2014 A review on TiO_2-based nanotubes synthesized via hydrothermal method: Formation mechanism, structure modification, and photocatalytic applications *Catalysis Today* **225** 34-51
[4] Binas V, Venieri D, Kotzias D and Kiriakidis G 2017 Modified TiO2 based photocatalysts for improved air and health quality *Journal of Materiomics* **3** 3-16
[5] Delnavaz M, Ayati B, Ganjidoust H and Sanjabi S 2015 Application of concrete surfaces as novel substrate for immobilization of TiO 2 nano powder in photocatalytic treatment of phenolic water *Journal of Environmental Health Science and Engineering* **13** 58
[6] Shavisi Y, Sharifnia S, Zendehzaban M, Mirghavami M L and Kakehazar S 2014 Application of solar light for degradation of ammonia in petrochemical wastewater by a floating TiO_2/LECA photocatalyst *Journal of Industrial and Engineering Chemistry* **20** 2806-13
[7] Fisli A, Saridewi R, Dewi S H and Gunlazuardi J 2013 Preparation and Characterization of Fe_3O_4/TiO_2 Composites by Heteroagglomeration *Advanced Materials Research* **626** 131-7
[8] Joo J, Ye Y, Kim D, Lee J and Jeon S 2013 Magnetically recoverable hybrid TiO_2 nanocrystal clusters with enhanced photocatalytic activity *Materials Letters* **93** 141-4
[9] Sentein C, Guizard B, Giraud S, Yé C and Ténégal F 2009 Dispersion and stability of TiO_2 nanoparticles synthesized by laser pyrolysis in aqueous suspensions *Journal of Physics* **170**
[10] Sun J and Gao L 2003 Development of A Dispersion Process for Carbon Nanotubes in Ceramic Matrix by Heterocoagulation *Journal of Carbon* **41** 1063-8
[11] Saleh T A and Gupta V K 2012 Photo-catalyzed Degradation of Hazardous Dye Methyl Orange by Use of A Composite Catalyst Consisting of Multi-walled Carbon Nanotubes and Titanium Dioxide *Journal of Colloid and Interface Science* **371** 101-6
[12] Cai J-H, Huang J-W, Yu H-C and Ji L-N 2012 Synthesis, Characterization, and Photocatalytic Activity of TiO_2 Microspheres Functionalized with Porphyrin *International Journal of Photoenergy* **10**
[13] Urlaub R, Posset U and Thull R 2000 FT-IR spectroscopic investigations on sol–gel-derived coatings from acid-modified titanium alkoxides *Journal of Non-Crystalline Solids* **265** 276-84
[14] Ye X, Cai S, Zheng C, Xiao X, Hua N and Huang Y 2015 SiO2/TiO2/Ag multilayered microspheres: Preparation, characterization, and enhanced infrared radiation property *Applied Surface Science* **345** 279-85
[15] He Z, Liu D, Li R, Zhou Z and Wang P 2012 Magnetic solid-phase extraction of sulfonylurea herbicides in environmental water samples by Fe3O4@dioctadecyl dimethyl ammonium chloride@silica magnetic particles *Analytica Chimica Acta* **747** 29-35
[16] Makovec D, Sajko M, Selišnik A and Drofenik M 2011 Magnetically recoverable photocatalytic nanocomposite particles for water treatment *Materials Chemistry and Physics* **129** 83-9

Influence Composition Fe_2O_3 of Isotropic Magnet $BaFe_{12}O_{19}$ on Microstructure and Magnetic Properties.

Suprapedi [1*], Priyo Sardjono[1], Muljadi[1*], Kerista Sebayang[2]

[1)]Research Center for Physics- Indonesia Institute of Sciences Kawasan PUSPIPTEK Serpong, Tangerang Selatan, Prov. Banten, Indonesia
[2)]Physics-MIPA University of North Sumatera, Indonesia

[*)]email : suprapu@gmail.com, muljadi2002@yahoo.com

Abstract. The isotropic magnets Barium Hexsaferit has been made by using a sintering process and the composition of Fe_2O_3 on barium hexaferrite was carried by the addition of Fe_2O_3 (0, 0:25, and 0.5, % wt). The raw materials were used such as: commercial powder $BaFe_{12}O_{19}$ and hematite Fe_2O_3 from e-Merck. Both the raw materials were weighed according to the composition, and then refined for 48 hrs by using ball mill. After milling processes, the mixed powder was measured particle size distribution by using Laser Particle Size Analyzer, and it was obtained average particle size about 17,15 μm. After that, the powders were mixed with 2 % white. Celuona WE 518 as a binder, then the sample powders were formed by using cold pressing with a force about 5 tons becomes pellets with a diameter of 13.10 mm and thickness of 7.12 mm. Then samples pellet were sintered by using Electrical Thermolyne Furnace at temperature: 1150 ° C and holding time for 2 hours. The sintered samples were measured microstructure by using XRD, magnetic properties by using Permeagraph and Gausmeter. The variation of composition Fe_2O_3 on barium ferrite can influence significantly on crystal structure and magnetic properties. It was found two phases such as: $BaFe_{12}O_{19}$ phase and Fe_2O_3 phase on the samples with the addition of 0.5 % Fe_2O_3, but for sample without the addition of Fe_2O_3 and with 0.24% Fe_2O_3 have only single phase $BaFe_{12}O_{19}$. The results of measurement magnetic properties show that Magnet BaFe12O19, with addition 0.25% wt. Fe_2O_3 has high value for flux magnetic density and Br, but the Hc value of magnet BaFe12O19 becomes 2 times higher with the addition of 0.5% wt. Fe_2O_3.
Keywords: Barium Ferrite, hematite Fe_2O_3 , isotropic magnet, sintering, Flux magnetic density, Permeagraph

1. Introduction

Barium hexa ferrite is a ferromagnetic oxides with a hexagonal crystal structure [1]. This magnetic material has been developed at the Philips laboratory in 1955. But it is still used until now in widely application, for example: as a component of motor, sensors, component speakers, magnetic recording media and so on [2,3]. The hexa ferrite is called permanent magnetic ferrites with basic formula $MO.6Fe_2O_3$ (M: Ba, Sr, Pb) [4]. Barium ferrite is scientifically and technologically attractive because of its relatively high Curie temperature, high coercive force and high magnetic anisotropy field, as well as its excellent chemical stability and corrosion resistivity [5]. Magnetic properties of ferrite magnets are not very high, but their coercivity being higher than the coercivity of AlNiCo magnets.

The Curie temperature of ferrite magnets is 450°C. Table 1 shows the properties of magnetic materials as a permanent magnet.

Table 1. Magnetic Properties of Permanen Magnet Materials[1].

Materials	**Magnetic Properties**				
	Br, (T)	**Hcj, (kA/m)**	**Hcb, (kA/m)**	**(BH)max, (kJ/m)**	**Temperature Curie, °C**
Ba-Ferrite (sintered)	0.38	180	155	27,0	450
Sr-Ferrite (sintered)	0,40	275	265	30,0	450
NdFeB (sintered)	1,43	950	915	398	310
Sm2Co17	0,96	1600	700	175	825
AlNiCo	0,85-0,90	115-127	121-136	35,8-63-3	860

The behaviour of ferrite magnets in different temperatures is different than that of other kinds of permanent magnets. With elevated temperature coercivity increases, but remanence decreases [6,7]. The disadvantages of ferrite magnets are their brittleness and high shrinkage after sintering [8]. They need machining after sintering process. Ferrite magnets can be prepared as sintered or bonded magnets and isotropic or anisotropic in character. Permanent magnets of this type are the most popular, mainly due to their low price. Ferrite magnets are nevertheless still produced and applied up to this point, because raw materials cost, the production costs and product prices are cheaper than raw materials based on rare earth metal. The production of permanent magnet based on barium ferrite consists of several stages, the first stage is powder preparation to produce powder barium ferrite with a hexagonal structure ($BaFe_{12}O_{19}$), the second stage is forming process and third stage is a densification process at high temperature or the sintering process, and the final stages are a finishing process and the process of magnetization [1]. The milling process of Fe_2O_3 and $BaCO_3$ mixture leads to increase the content of Fe_2O_3 phase and decrease the content of $BaCO_3$, and milling process causes enriching of mixed powder particles by Fe_2O_3 [5,9]. The effect of Fe_2O_3 composition on magnetic properties and microstructure have been studied [5]. The increasing of Fe_2O_3 content in material Ba-Ferrite can influence magnetic properties, the remanence (Br) will decrease and coercivity (Hc) will increase [5]. The manufacturing of Ba-ferrite by milling and heat treatment process or solid-solid reaction should pay attention to factors of increased impurity materials (Fe_2O_3) during the process of milling [10]. The increasing of Fe_2O_3 content during milling process can affect the characteristics of the permanent magnet [7,10,11]. In this paper, the effect of variation of Fe_2O_3 composition on manufacturing Ba-Ferrrite to the microstructure, physical properties and magnetic properties were investigated.

2. Experimental Method

The sample preparation was used a commercial powder of barium ferrite and hematite (Fe_2O_3) powder. Both the raw materials were weighed according to the composition, and then refined for 48 hrs by using ball mill. After milling processes, the mixed powder was measured particle size distribution by using Laser Particle Size Analyzer, and it was obtained average particle size about 17,15 µm. After that, the powders were mixed with 2 % white. Clone WE 518 as a binder, then the sample powders were formed by using cold pressing with a force about 5 tons becomes pellets with a diameter of 13.10 mm and thickness of 7.12 mm. Then samples pellet were sintered by using Electrical Thermolyne Furnace at temperature: 1150 ° C and holding time for 2 hours. The sintered samples were measured microstructure by using XRD, magnetic properties by using Permeagraph and Gausmeter.

IOP Conf. Series: Journal of Physics: Conf. Series **1091** (2018) 012025 doi:10.1088/1742-6596/1091/1/012025

3. Results and Discussion

Figure 1 shows curves of particle size distribution for 48 hrs milled powder, and figure1 shows that its obtained average particle size about 17.55 μm after milling process of commercial $BaFe_{12}O_{19}$ powder for 48 hrs.

.

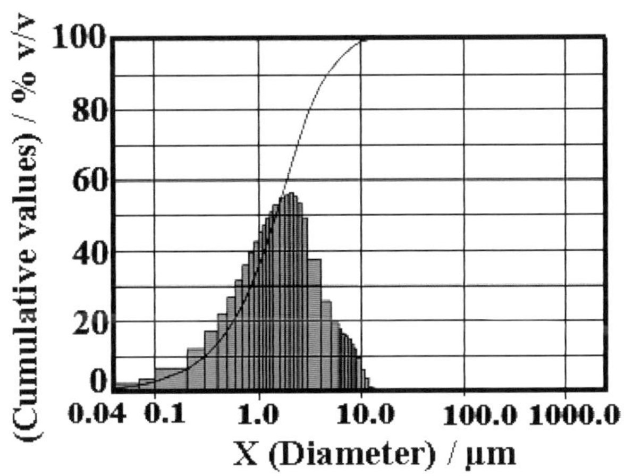

Figure 1. Particle Size Distribution Curve for Powder After Milling 48 h

The results of XRD analysis for sample milled powder with a variation of Fe_2O_3 composition are shown in figure 2. Figure 2 indicates that sample without additive Fe_2O_3 (original sample) and after sintering 1150°C has one phase of BaFe12O19. The XRD patterns of sample with 0.25% Fe_2O_3 is similar with the original sample, but for sample with 0.5% Fe_2O_3 is different and it has 2 phases (BaFe12O19 and Fe_2O_3). This experiment shows that the additive of Fe_2O_3, with content 0.5 % gives effect on the crystal structure

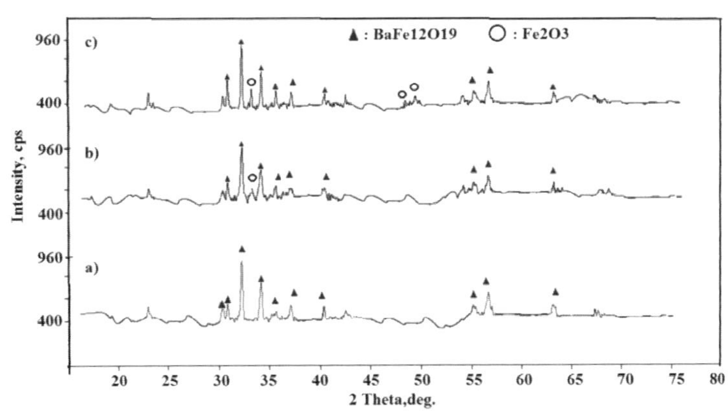

Figure 2 XRD patterns of sample after sintering 1150 °C for (a) sampel with 0% Fe_2O_3, (b) sample with 0.25 % Fe_2O_3 and (c) sample with 0.5% Fe_2O_3

Flux magnetic density is a value of magnetic field on the surface of the sample and it was measured by using Gaussmeter. Table 2 shows the value of flux magnetic density for all samples, and value of flux magnetic density decrease with increasing of percentage of Fe_2O_3, but value of flux magnetic density of samples with 0% and 0.25% Fe_2O_3 are not much different.

Table 2. Value of flux magnetic density

% wt. Fe_2O_3	**Flux magnetic density, Gauss**
0	595.3
0.25	589.7
0.5	495.5

Decreasing of flux magnetic density is due to changing of crystal structure or the existing of Fe_2O_3 phase. This is because the presence of Fe_2O_3 phase which is paramagnetic in the sample will reduce the value of flux magnetic density. Samples with 0.25% and 0.5%wt. Fe_2O_3 were measured hysteresis curve by using Permeagraph to know the value of Remanence (Br), Coercivity (Hc), and Energy Product (BHmax). The results of measurement are shown in figure 3.

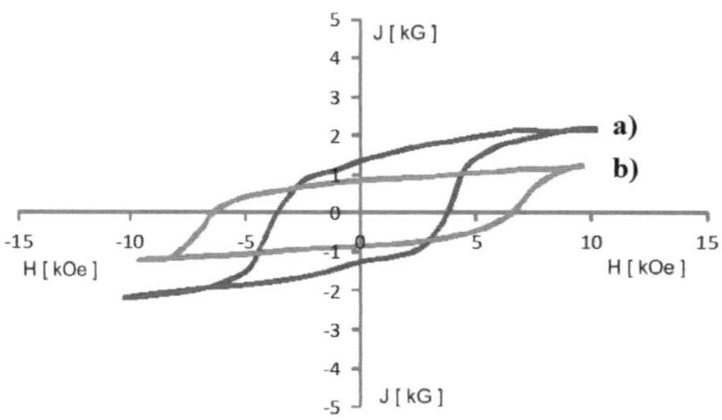

Figure 3. Hysterisis Curve of Sample With a) 0.25 %wt. Fe_2O_3 and b) 0.5% wt. Fe_2O_3

The addition of Fe_2O_3 can influence magnetic properties, where the value of Br decreases and the value of Hc increases in sample with higher Fe_2O_3 composition. Table 3 shows value of Br, Hc and BHmax. Increasing of Fe_2O_3 composition from 0.25 % to 0.5 % can increase value of Hc two times and decrease value of Br from 1.51 kGauss to 0.95 kGauss. The decline in the value of Br with the existing of Fe_2O_3 phase due to material properties of Fe_2O_3, that is paramagnetic Fe_2O_3.

Table 3 shows value of Br, Hc and BHmax.

% wt. Fe_2O_3	Br , kGauss	Hc , kOe	BHmax, MGOe
0.25	1.51	3.7	0.52
0.50	0.95	6.49	0.21

A novelty of the results of this research is to add 0.25% of Fe_2O_3 being 0.5% can increase the value of Hc reach almost twice. Surely it is the resilience of the demagnetization $BaFe_{12}O_{19}$ samples with the addition of 0.5% Fe_2O_3 might increase anyway.

IOP Conf. Series: Journal of Physics: Conf. Series **1091** (2018) 012025 doi:10.1088/1742-6596/1091/1/012025

4. Conclussion

The percentage of addition Fe_2O_3 can give significantly influence to crystal structure (microstructure) and magnetic properties. Magnet $BaFe_{12}O_{19}$, with addition 0.25% wt. Fe_2O_3 has high value for flux magnetic density and Br, but the Hc value of magnet $BaFe_{12}O_{19}$ becomes 2 times higher with the addition of 0.5% wt. Fe_2O_3

References

[1]. Buschow K H J 2009 *Handbook of Magnetic Materials* **18** 1 15-28.
[2]. Bahadur R D, Kumar S R and Kumar A 2006 *J.Chem.Sci.* **118** 1 15-21.
[3]. Moosa I S 2013 *International Journal of Advanced Research in Engineering and Technology [IJARET]* **4** 6 127-141.
[4]. Tang X, Yang Y G 2009 *Materials Science-Poland* **27** *2.*
[5] Nowosielski R, Babilas R, Wrona J 2007 *Achievements in Materials and Manufacturing Engineering* **20** 1-2 307-309.
[6]. Mallick K K, Shepherd P, Green R J 2007 *Journal of Magnetism and Magnetic Materials* **312**, 418–429.
[7] Hshiang H I, Yao R Q 2007 *Materials Chemistry and Physics* **104** 1-4.
[8]. Jančárik V, Grusková A, Sláma J, Dosoudil R 2006 *Journal Of Electrical Engineering* **57** 8/S 163-166.
[9]. Shah I, Abbas T, Ullah Z, Hassan N, Rauf A, Ullah K and Naseem S 2015 *Armenian Journal of Physics* **8** 4 185-190.
[10]. Doni W R, Manaf A, Sardjono P 2014 *IJBAS-IJENS* **13** 4 63-68
[11]. Molaei M J, Ataie A, Raygan S, Picken S J, Mendes E, Tichelaar F D 2012 *Powder Technology* **221** 292–295.

IOP Conf. Series: Journal of Physics: Conf. Series **1091** (2018) 012026 doi:10.1088/1742-6596/1091/1/012026

Morphology Modification of Barium Hexaferrite to Nanorod Particle Using Chitosan and Starch as Template

Syahwin[1], Nasruddin MN[1], Muhammad Zarlis[1], Bambang Sunendar[2]

[1]University of North Sumatera, Jln. Bioteknologi - USU, Medan, 20155, Indonesia
[2]Bandung Institute of Technology, Jln. Ganesha No.10 - Bandung 40132, Indonesia

Abstract. Barium hexaferrite is known as a permanent magnet that has hexagonal or granules morphologies. Magnetic field of barium hexaferrite can be increased by changing its morphology. A rod-shaped barium hexaferrite is expected to result in higher magnetic field compared to granules even though the volumes of both are the same. This study aims to synthesize a rod-shaped barium hexaferrite and to evaluate its magnetic properties. Barium hexaferrite was synthesized using sol-gel method by mixing barium nitrate and iron (III) nitrate nonahydrate with molar ratio of 1:12. Chitosan 1% was added as dispersant and starch 0.5% (w/v) was used as template with variation in volume (5%, 10% and 15% (v/v)). The solution was dried at 100°C for 48 hours and resulting xerogel was then calcined at 1000°C for 2 hours to obtain barium hexaferrite powder. X-Ray D result demonstrated the presence of barium hexaferrite with small fraction of hematite and pure barium hexaferrite. Scanning electron microscopy images showed different concentrations of starch generated different size and morphologies of resulting barium hexaferrite. A rod shaped barium hexaferrite was formed for all samples that prepared using starch template, however hexagonal shape was still found in all samples. The size of resulting barium hexaferrite is decrease by increasing the concentration of starch. Theparticle size of barium hexaferrite measured was 125nm thickness with of500nmlength for 5% starch, 109 nm thickness and 406 nm length for 10% starch and 62.5 nm thickness and 406 nm length for 15% starch, Barium hexaferrite synthesized without using starch and chitosan produced hexagonal morphology with the thickness of 240-574 nm and length of 574-957 nm. Vibrating Sample Magnetometer showed hysteresis curve as the percentage variation of starch was used, higher amount of starch used tends to increase the values of saturation and remanence. Avery significant increase in coercivity value amounted to 457.1% occurred in the use of 10% starch. It can be concluded that starch can be used as rod-shaped template in sol-gel method to produce barium hexaferrite. By using chitosan and starch in the process of synthesis of barium hexaferrite using sol-gel method can increase the value of the magnetization and the coercivity barium hexaferrite.

1. Introduction

Barium hexaferrite ($BaFe_{12}O_{19}$) is one of high-coercive magnetic material that is among the most commonly used and it is still promising because of its superior properties such a slow production costs, high curie temperature, high coercivity, chemical stability and corrosion resistance as reported by Burak Kaynar M. et al. , Dursun Set al.,Samira Mandizadeh et al.,Mahgooba A.and Hudeish A. Y.,

Jun Young Kwak et al[1-5]. Barium hexaferrite is one of the M-type hexagonal ferrite that belong to magnetoplumbite group. The molecular structure of M-type ferrite is made up of one R and one S block with an overlap of hexagonally and cubically packed layersby Chauhan et al [6]. Hexagonal ferrites, also called ashexaferrites, are magnetic materials that are important and useful for various industrial applications such as magnetic recording, data storage devices, microwave devices, and permanent magnets. The most famous type of hexaferrite arethe M-type, Y-type, Z-type, W-type, X-type and N-type as reported by Pullar R.C. [7]. Barium hexaferriteis also known as an organic-based magnet because it constitutes its morphology oxygen tends to form hexagonal. Organic molecule-based magnet has all the traits commonly associated with metal-based (or atom based) magnets, for example, hysteresis and remanence, saturation magnetization and coercivity it as reported by Joel S. Miller[8]. Some work reported that the most common morphologies of barium hexaferrite were hexagonal shapes [9-17]. Based on the morphology of the barium ferrite, a magnetic field can be increased by forming the rod-shaped[18].This study aims to modify the morphology of the barium hexaferrite to a nanorod shape. In this study barium hexaferritewas prepared using wet-chemical Sol-Gel method.The sol–gel method can realize molecular level mixing and that the dispersant can improve the activity of the reactants, which helps to form the single-phase spinel barium ferrite [19], by using thesol-gel method it is easy to add the dispersant and nanorod template. In this research-weused chitosan 1% as a dispersant and 0.5% starch with different volume ratio as a template to obtain a rod-shaped morphology which will increase the value of magnetization.

Chitosan is a biosourced polysaccharide obtained by alkaline deacetylation of chitin [20,21].Chitosan acts as a structural dispersant as it contain amino and hydroxyl functional group that are the able to bind the metal ions. Chitosan is agood dispersant for iron oxide [22].

Starch is an insoluble glucan composed of two polymers of glucose, amylopectin and amylose[23]. The amylopectin has branchedstructure while amylase has linear structure. The amylose content in the starch is expected to be able to act as nanorod template.

2. Materials and Method

Barium nitrate (Ba $(NO_3)_2$, 99.9%, Merck), Iron(III) Nitrate nanohydrate (Fe $(NO_3)_3.9H_2O$, 99.9, Merck) were used as a precursor with a molar ratio of 1:12 according to the stoichiometric ratio. Acetate acid (CH_3COOH, 99.8%, Merck) was used as solvents chitosan, low molecular weight chitosan, starch commercial products, Demineralized water (DM water) as the solvent. Before hand prepared solution of 1% chitosan (a ratio of 1g chitosan was dissolved in100 ml of 2% acetate acid) here after starch solution 0.5% was prepared by dissolving 0.5 g of starch in 100ml DM water at a temparature of 70^0C.

Precursor solution was made by inserting the $Ba(NO_3)_2$ into DM water while stirring using a magnetic stirrer for 10 minutes then $Fe(NO_3)_3.9H_2O$ was added into the precursor solution and stirred constantly for 10 minutes. To the precursor solution was added a solution of chitosan 1% of 10% v/v solution was continuously stirred for 15 minutes then added a 0.5% starch solution: 5% v/v, 10% v/v and 15% v/v of the solution keep stirring for 30 minutes, then the solution was stirred using homogenizer at 1000 rpm rotational speed in order to form a homogeneous solution.After the solution was dried in an oven with a temperature of 100^0C for 48 hours, resulting xerogel wastaken and crushed using a mortar to obtain a powder. Powder was then calcined in a furnace at a temperature of 1000^0C with holding time for 2 hours to obtain a compound $BaFe_{12}O_{19}$. The powder was tested contentcrystal using a X-Ray Difraction (XRD) Philips Analytical type: PW1710 basedthen photographed morphology using a Scanning Electron Microscope (SEM) JEOL-JSM-6510 LV and last tested properties of magnetic using a Vibration Sample Magnetometer (VSM) Oxford type: VSM 1.2H while the flow chart of the experiment can be seen in Figure 1.

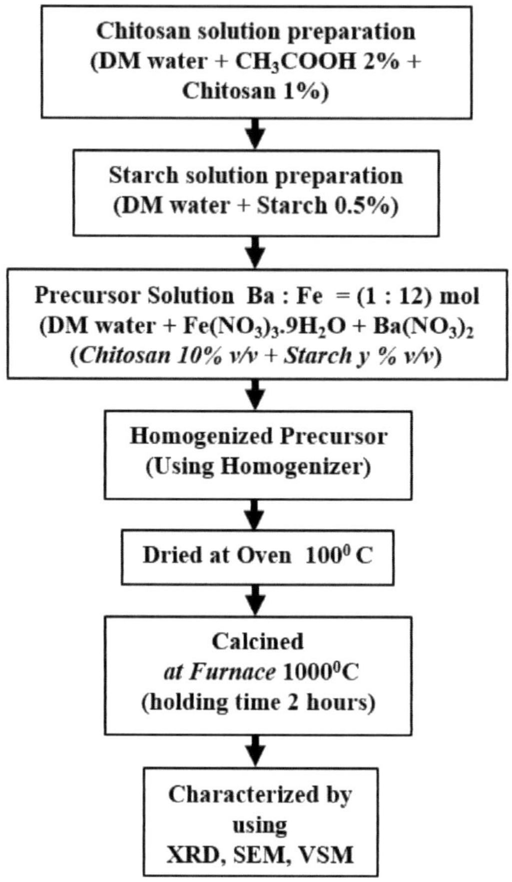

Figure 1. The flow chart of barium hexaferrite powders synthesis.

3. Results and Discussion

3.1. XRD-analysis.

hjThe XRD diffraction patternshows that barium hexaferrite is successfully formed as can be seen in Figure 2, as also indicated by reflection peaks at:32.60°; 34.46°; 35.37°; 37.29°; 40.62° (2θ).The sample prepared using 5% (v/v) starch still contains hematite while in the oneusing 10% (v/v) starch no hematite has been found indicating it is a pure $BaFe_{12}O_{19}$.

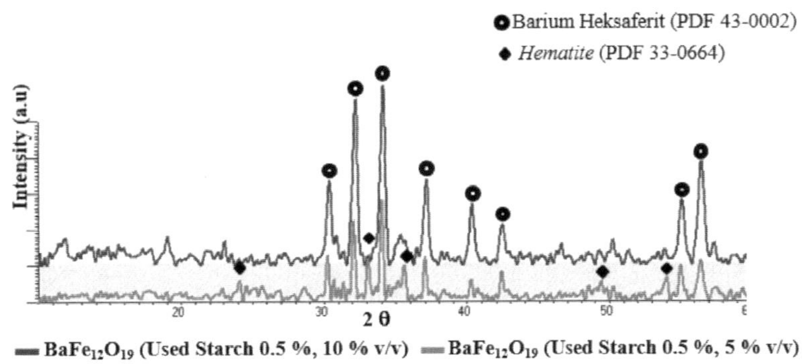

Figure 2. Diffraction pattern of barium hexaferrite.

3.2. SEM-analysis

The SEM microgram images of barium hexaferrite morphologies can be seen in Figure3. The addition of starch would cause the formed morphology tends to decompose is due to the dispersant, and the obtained rod-shaped morphology is caused by the template while the size of its morphology is likely to first get the thick size and then dwindling to reach a size of 62.5nm, the amount of the morphology of rod-shaped is still less than the morphology of hexagonal shape, and it is likely due to the lower amount of amylase than amylopectin. The chemical structure is linear amylase it binds morphological precursor that has been separated by 1% Chitosan solution. Thus Starch can be used as a template to obtain nanorod morphology.

Figure 3. The shape and size $BaFe_{12}O_{19}$ morphology of various samples. (a) $BaFe_{12}O_{19}$ (unused chitosan and starch); (b) $BaFe_{12}O_{19}$(starch 5%v/v); (c) $BaFe_{12}O_{19}$(starch 5%v/v); (d) $BaFe_{12}O_{19}$(starch 15%v/v).

Table 1. SizeMorphologyof $BaFe_{12}O_{19}$.

No	Starch Template (v/v)	Morphology	Thick (nm)	Length (nm)
a	0 %	Hexagonal & Granular	240 – 574	574 - 957
b	5 %	Hexagonal &Rod	125 – 775	500 – 625
c	10 %	Hexagonal &Rod	109 – 312,5	406 – 531,5
d	15 %	Hexagonal &Rod	62.5 – 140.6	406 – 1250

From Table 1 shows the result obtained by adding starch rod-shaped morphology are getting smaller.

3.3. VSM-analysis.
Results of characterization using Vibration Sample Magnetometer (VSM) can be seen in Table 2 and Figure 4.

Table 2. Magnetization of $BaFe_{12}O_{19}$

No	Starch Template (v/v)	Saturation *Ms* (emu/g)	Remanence *Mr* (emu/g)	Coersivity *Hc* (tesla)
A	0 %	58.6	33	0.07
B	5 %	54	31.6	0.21
C	10 %	59.8	36.4	0.32
D	15 %	62.2	37.5	0.28

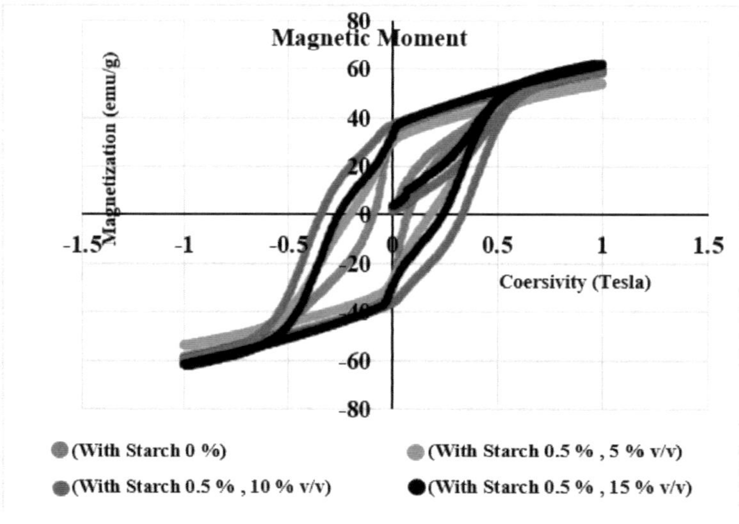

Figure 4. Hysterisis curve of $BaFe_{12}O_{19}$ at different used strach

Based onTable 2, and Figure 4 an increasingly high percentage of starchis used which tends to increase the value of saturation, remanence and coercivity of barium hexaferritebut the optimal value of coercivity is obtained on the use of 10% (v/v)starch,to demonstrate thatthe use of starchwouldenhance the magnetic properties of barium hexaferrite. Magnetization values obtained in this study is still lowerthan the theoretical value of about 72 emu/gas reported by the Haneda K. And Kojima[24], this is likely due to the morphology of the formed stem, in addition the value of the magnetization is also influenced by many factors,such asthe chemical composition, particle size, degree of crystallinity, magnetic anisotropy, and so on. Nevertheless, the novelty introduced through the use organic template derived from natural polymer starch has improved the value of the magnetization of barium hexaferrite.

4. Conclussion

Materials which have been synthesized in this study are barium hexaferrite as was verified or confirmed by the experimental inpection employing the X-Ray Difraction technique. Barium hexaferrite morphology can be modified into a nanorod shape through the wet chemical process using Sol-gel method by adding a 1% solution of chitosan as a dispersant and 0.5% starch solution as a template. Magnetization values barium hexaferrite obtained increased after use starch as a template that is characterized by the formation of nanorod morphology. Ther results also obtained a significant increase in coercivity value amounted to 457.1% the use of starch 10% (v/v) in the precursor solution. For further research it is expected to increase the amount of amylose which is used in a manner that reduces the amount amylopectin exist in starch that can multiply the number of nanorod morphology formed it is expected to further increase the value of the magnetization of barium hexaferrite.

References

[1] Kaynar M B, Özcan S, Shah S I 2015 Synthesis and magnetic properties of nanocrystalline $BaFe_{12}O_{19}$ *Ceramic International*. (http://.dx doi.org/10.1016/ j.ceramint. 2015.05078)

[2] Dursun S, Topkaya R, Akdogan N, Alkoy S 2012 Comparison of the structuraland magnetic properties of submicron barium hexaferrite powders preparedby molten salt and solid state calcination routes *Ceram Int* **38** 3801–6.

[3] Mandizadeh S, Soofivand F, Salavati-Niasari M 2015 Sol–gel auto combustion synthesis of $BaFe_{12}O_{19}$ nanoceramics by using carbohydrate sugars as a novel reducing agent *Advanced Powder Technology* xxx

[4] Mahgooba A and Hudeish A Y 2012 Thermal annealing effect on the structural and magnetic properties of barium hexaferrite powders *American Academic & Scholarly Research Journal* **4** 5.

[5] Kwak J Y, Lee C S, Kim D and Kim Y I 2012 Characteristics of Barium Hexaferrite NanoparticlesPrepared by Temperature-Controlled Chemical Coprecipitation. *Journal of the Korean Chemical Society* 2012, **56** 5.

[6] Chauhan C C, Jotania R B, Jotania K R 2013 Structural Properties Of CobaltSubstituted Barium HexaferriteNanoparticles Prepared By A ThermalTreatment Method *Nanosystems: Physics, Chemistry, Mathematics* **4 3** 363–36.

[7] Pullar R C 2012 Hexagonal ferrites: A review of the synthesis, properties and applications of hexaferrite *Ceramics* **57 1** 191

[8] Miller J S 2014 Oragnic and molecule-based magnet.Review *Materials Today* **17** 5

[9] Jie L, Wu Z H, Xun L Y, Li L Y, Bing M Y 2012 The structural and magnetic properties of barium ferrite powders prepared by the sol gel method *Chin. Phys. B* **21 1** 017501.

[10] Tan G, Chen X 2013 Structure and multiferroic properties of barium hexaferrite ceramics. *Journal of Magnetism and Magnetic Materials* **327** 87–90.

[11] Zhao L, Lv X, Wei Y, Ma C, Zhao L 2013 Hydrothermal synthesisof pure $BaFe_{12}O_{19}$ hexaferrite nanoplatelets under high alkaline system *Journal of Magnetism and Magnetic Materials* **332** 44–47.

[12] An G H, Hwang T Y, Kim J, Kim J B, Kang N, Jeon K W, Kang M, Cho Y H 2014 Novel Method For Low Temperature Sintering Of Barium Hexaferrite Withmagnetic Easy-Axis Alignment. *Journal of the European Ceramic Society* **34** 1227–1233.

[13] Ahmed N M A, Helmy S I 2013 Innovative methodology for the synthesis of ba-m hexaferrite $BaFe_{12}o_{19}$ nanoparticles *Materials Research Bulletin* **48** 3394–3398.

[14] Chen D, Liu Y, Li Y, Zhong W, Zhang H 2012 Low-temperature sintering of m-type barium ferrite with $BaCu(B_2O_5)$ additive *Journal of Magnetism and Magnetic Materials* **324** 449–452

[15] Khorrami S A, Islampour R., Bakhtiari H., Naeini Q S M 2013 The Effect Of Molar Ratio on Structural and Magnetic Properties of $BaFe_{12}O_{19}$ nanoparticles prepared by sol-gel auto-combustion method *Int. J.Nano Dimens*. **3** **3** 191-197.

[16] Mosleh Z, Kamelin P, Ranjbar M,Salamati H 2014 Effect of annealing temperature on structural and magnetic properties of $BaFe_{12}O_{19}$ hexaferrite nanoparticles *Ceramics International* **40** 7279–7284.

[17] Xu X, Park J, Hong Y K, Lane A M 2015 Synthesis and characterization of hollow mesoporous $BaFe_{12}O_{19}$ spheres *Journal of SolidState Chemistry* **222** 84–89.

[18] Mohapatra J, Mitra A, Tyagi H, Bahadur D, Aslam M 2015 Iron oxide nanorods as high-performancemagnetic resonance imaging contrast agents *The Royal Society of Chemistry*.

[19] Jie L, Wu Z H, Xun L Y, Li L Y and Bing M Y 2012 The structural and magnetic properties of barium ferrite powders prepared by the sol gel method *Chin. Phys. B* **21** **1** 017501

[20] Elsabee M Z and Abdou E S 2013 Chitosan based edible films and coatings: *A Review. Mater Sci Eng.* C **33** 1819–41.

[21] Baouche N M, Elchinger P H, de Baynast H, Pierre G, Delattre C, Michaud P 2014 Chitosan as an adhesive *European Polymer Journal* **60** 198–213

[22] Tsai Z.-T 2010 In situpreparation of high relaxivity iron oxide nanoparticles by coating with chitosan: A potential MRI contrast agent useful for cell tracking *J. Magn. Magn. Mater.*, **322** 208–213.

[23] Corre D L, JBras J and Dufresne A 2010 Starch Nanoparticles: A Review *Biomacromolecules* **11** 1139–1153.

[24] Haneda K and Kojima J H 1973 *Appl. Phys* 44, 3760

Synthesis of $REBa_2Cu_3O_{7-\partial}$ (RE=Nd, Eu, Gd) Nanoparticles by Wet-Mixing Method

W.G. Suharta[1], P. Suardana[1], Windaryoto[1], S. Pratapa[2], S. Suasmoro[2], D. Darminto[2], Y. Ishii[3], W. Isao[3]
[1] Department of Physics, Faculty of Mathematics and Natural Sciences, Udayana University
[2] Department of Physics, Faculty of Mathematics and Natural Sciences, Institutes of Technology Sepuluh Nopember, Kampus ITS Sukolilo, Surabaya 60111, Indonesia
[3] Advanced MESON Science, Laboratory, RIKEN, 2-1 Hirosawa, Wako, Saitama, 351-0198, Japan

E-mail: wgsuharta@gmail.com

Abstract. The $Nd_{0.33}Eu_{0.33}Gd_{0.33}Ba_2Cu_3O_{7-\partial}$ (NEG-123) nanoparticles have been successfully synthesized by wet-mixing method using HNO_3 as dissolving agent. The NEG-123 samples are divided into two which were treated differently. The first and second samples were mixed respectively, for one and 24 hours. All the samples were then calcined at 600°C for 3 hours, and sintered at relatively low temperature (750 to 900°C), each for 30 minutes. Samples were characterized using X-ray diffraction (XRD), transmission electron microscopy (TEM), and vibrating sample magnetometer (VSM). Rietveld analysis shows the lattice parameter of *a*-axis decreases with the increasing sintering temperature from 750 to 900°C, while the lattice parameter of *b* and *c*-axes increases. Further, it was known based on the XRD analysis that the higher the sintering temperature, the greater the peak intensities and average crystal size. The NEG-123 samples with the average crystal size of hundreds nanometer exhibit paramagnetism, while those with average crystal size of ten nanometers demonstrate superparamagnetism at room temperature.

Keywords : nanoparticle; wet-mixing method; paramagnetism; superparamagnetism

1. Introduction

Since the discovery high-temperature ferromagnetism that reported in alkaline-earth hexaboride, research about nano-particle more intensively conducted in several countries. Practically, magnetic nano-particle used as active components of ferrofluids, in recording tapes, and in biomedical applications [1, 2].

Ferromagnetism properties at room temperature in nano-particle of non-magnetic and magnetic oxides such as CeO_2, TiO_2, MgO and Al_2O_3 have been widely studied. The origin of ferromagnetism in these nano-particle suggested due to magnetic moments arising from oxygen vacancies at the surfaces of the nanoparticle, as has been reported by Sundaresan et.al [3]. Ghosh et.al has synthesized zinc sulphide nano-particles to be used as devices of field emission [4]. Also, oxygen vacancies at the surfaces of the nanoparticle has been studied by Gang Niu et.al, on Pr doped CeO_2 [5].

Research on nanoparticle has not been only limited to the synthesis of composite materials, but also on superconductors. In the case of superconducting materials, the $Ba_2Cu_3O_{7-\delta}$ (YBCO, Tc = 91 K)

IOP Conf. Series: Journal of Physics: Conf. Series **1091** (2018) 012027 doi:10.1088/1742-6596/1091/1/012027

nanoparticle was firstly reported to exhibit ferromagnetism at room temperature, while its YBCO crystalline bulk, obtained by heating the nanoparticle at 940°C, showed a linear magnetization curve. Across the superconducting transition temperature, the magnetization curve changes from that of a soft ferromagnetic to a superconductor, as has been reported by Shipra et.al [6]. Ferromagnetism in 2212 phases of Bi-Sr-Ca-Cu-O nano-superconductors were also obtained by Baqiya et.al [7]. Furthermore, M. Muralidhar et.al [8] fabricated nano-structured high Tc superconductor of NEG-123 with 30 mol % Gd-211 for application in medicine. V.S. Vinila et.al has also done research on making ceramic nanocrystalline superconductor of GBCO at variation of temperature [9].

In this study, we fabricated nano-particles of $Nd_{0.33}Eu_{0.33}Gd_{0.33}Ba_2Cu_3O_{7-\delta}$ (NEG-123) by wet-mixing method. In this case, the Y element on the YBCO superconductor is replaced by the rare earth element. Substitution and replacement of the Y element with the rare earth element in YBCO superconductor have been done by many researchers, such as by Suharta et.al. He has succeeded in synthesizing RE-123 superconductor with variations of RE elements [10]. Addition or replacement element with the rare earth element is not only done on the YBCO superconductor, but also done on the BSCCO superconductor, as has been done by Suharta et.al [11].

2. Experimental Method

$Nd_{0.33}Eu_{0.33}Gd_{0.33}Ba_2Cu_3O_{7-\delta}$ samples were prepared by wet-mixing method with HNO_3 as dissolving agent. Powders of Nd_2O_3, (99,9 %), Eu_2O_3 (99,9 %), Gd_2O_3 (99,9 %), $BaCO_3$ (99,9 %), and CuO (99,9 %) were used as starting materials. Synthesis process was as follows : the first, the raw material of Nd-123 was mixed and stirred in HNO_3 for 30 minutes, calcined at 600°C for 3 hours and sintered with various temperature (750-900°C), each for 30 minutes. The second, the raw materials of NEG-123 were mixed and stirred in HNO_3 for 24 hours, calcined at 600°C for 3 hours and sintered with variations of temperature (750-900°C), each for 30 minutes. The structure of the samples was examined using X-ray diffraction (Cu$K\alpha$ radiation, 10-90° 2θ-range with a step size of 0,02°). Rietveld refinements to acquire the crystal data of phases for each synthesized material performed using Rietica software by employing Thompson-Cox-Hasting pseudo-Voigt peak shape function. TEM images of the samples were collected to investigate the morphology and distribution of the grains and determine their size. The magnetic hysteresis was measured using a vibrating sample magnetization (VSM Oxford 1,2 T).

3. Results and Discussion

The XRD patterns of the NEG-123 samples prepared by mixing for 30 minutes and sintering at varying temperature for 30 minutes are shown in Fig. 1. While those with mixing for 24 hours and sintering with various temperature for 30 minutes are given in Fig. 2. In general, the diffraction pattern shows sharp peaks, indicating that the crystalline phase has grown well with dominant NEG-123 phase having orthorhombic structure. The $BaCO_3$, $BaNd_2O_4$, $BaCuO_2$, Nd_2BaCuO_5 compounds as minor impurities observed at around 27-32°. The additional sintering temperature has been performed to eliminate impurities and to achieve a single NEG-123 phase. It can also be seen in Fig. 1 and 2 (insert) demonstrating the degradation of $BaCO_3$ compound. A further additional sintering temperature from 750 to 900°C has increased the volume fraction of the NEG-123 phase from 84.3 to 95.9 %.

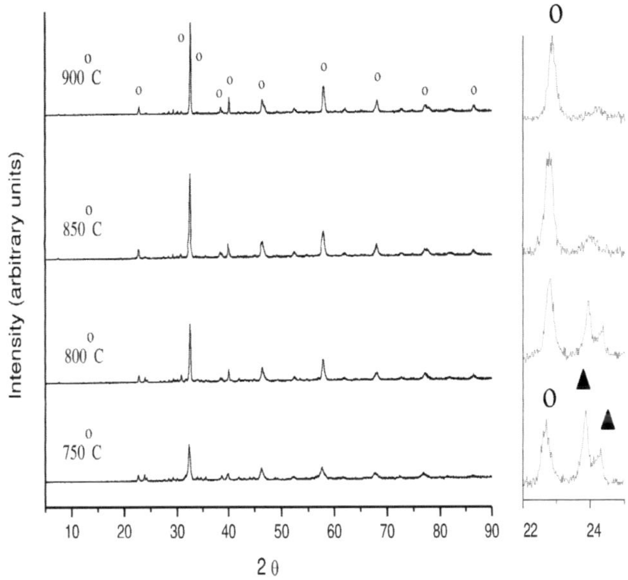

Figure 1. XRD patterns (Cu*Kα* radiation) of the NEG-123 samples prepared with mixing for 30 minutes and sintering for 30 minutes. Insert : NEG-123 samples at 2θ=23-31°.
Symbol O NEG-123, ▲ $BaCO_3$

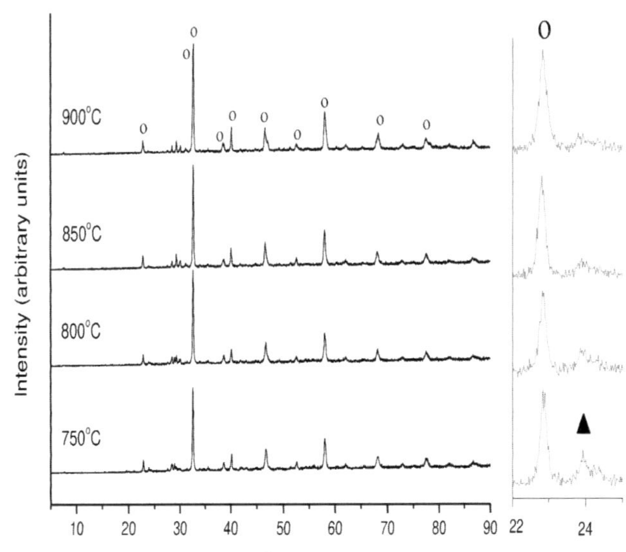

Figure 2. XRD patterns (Cu*Kα* radiation) of the NEG-123 samples prepared with mixing for 24 hours and sintering for 30 minutes. Insert : NEG-123 samples at 2θ=23-31°.
Symbol O NEG-123, ▲ $BaCO_3$

The Rietveld analysis on the XRD data in Fig. 1 has produced lattice parameter values (a, b, c) for each sample as shown in Fig. 3(a) for NEG-123 samples. The curve shows the lattice parameter of *a* decreasing with the additional sintering temperature, while *b* and *c* increases. The value of *a* is smaller than that of *b* on sintering temperature below 800°C. It would be resulting negative values of orthorhombic strain. The *a* is greater than *b* on sintering temperature above 800°C, producing positive orthorhombic strain. Furthermore, Fig. 3(b) shows the value of the lattice parameter of NEG-123 samples analyzed from data in Fig. 2. The results are consistent with the former group of NEG-123 samples, where the lattice parameter of *b* and *c* increases with additional sintering temperatures, while *a* decreases.

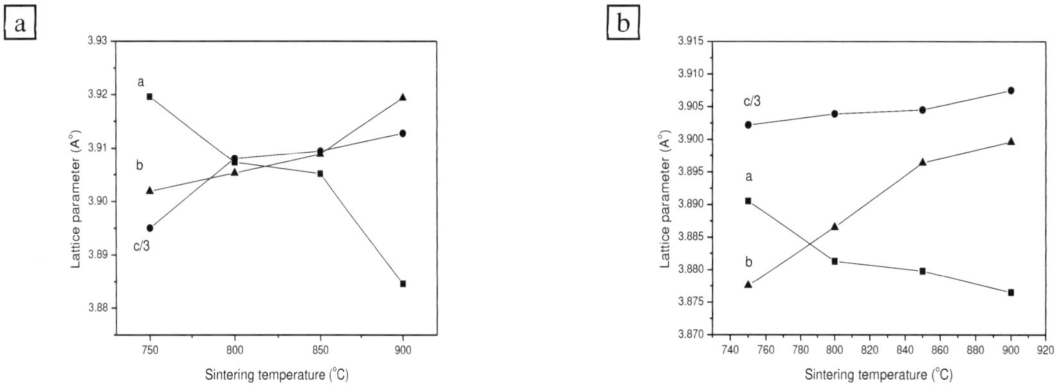

Figure 3. The lattice parameter of NEG-123 samples prepared with various sintering temperature (from 750 to 900°C) and mixing for (a) 30 minutes, and (b) 24 hours.

The solid particle size of the crystal can be determined using Scherrer equation and Rietica software. The effect of sintering temperature on the mean particle size for NEG-123 samples that has been mixed for 30 minutes and sintered for 30 minutes showed the average particle size was greater than 150 nm (Fig. 4a). This indicates that the sintering temperature increasing from 750 to 900°C has already led to agglomeration. So, it can said that is, the higher the sintering temperature, the greater the average crystal size. Similar to the results of the first sample, the average particle size of the second sample increases with increasing sintering temperature, as shown in Fig. 4(b).

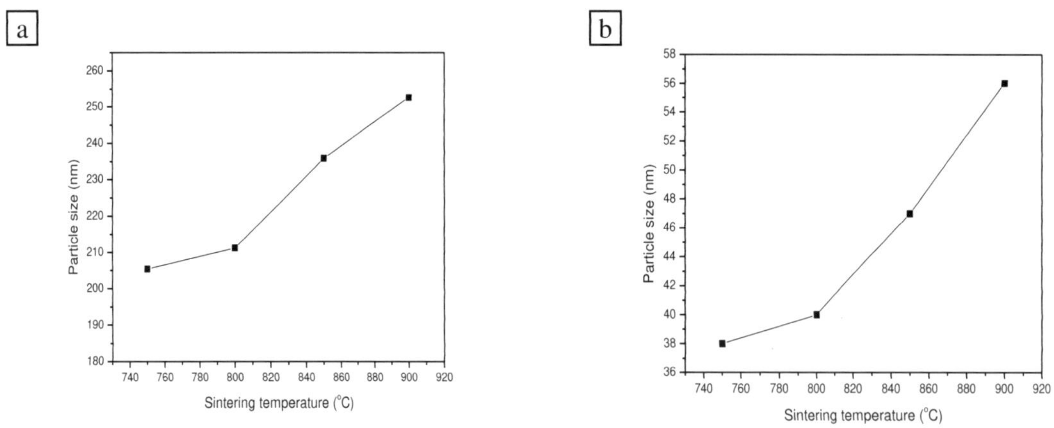

Figure 4. Particle size with increasing sintering temperature of the NEG-123 samples prepared using a set of mixing/sintering time : (a) 30 minutes/30 minutes, (b) 24 hours/30 minutes.

TEM morphology of NEG-123 samples that mixed for 30 minutes and sintered at a temperature of 750, 800 and 900°C for 30 minutes showed in Fig. 5. It showed the particle size of the sample with a sintering temperature of 750 and 800°C in a scale of 100 nm, while the particle size of the sample with sintering temperature of 900°C in the scale of 500 nm. In general shape of NEG-123 particles at a temperature of the 750-900°C showed rod-shape.

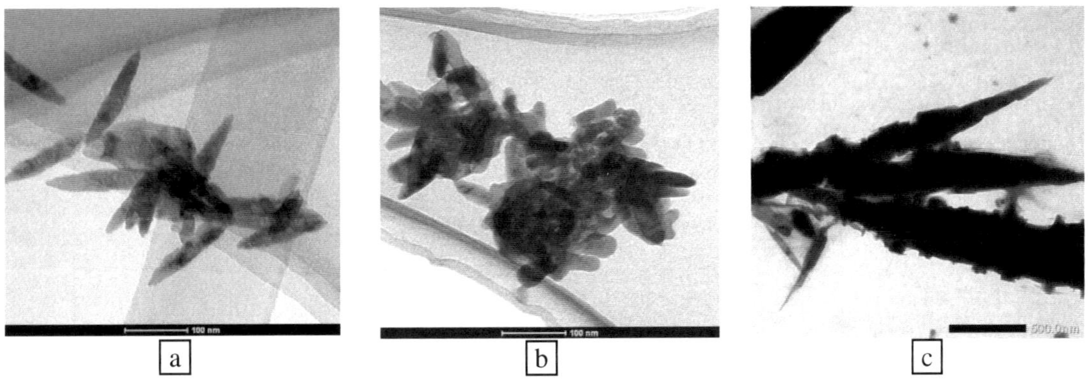

Figure 5. TEM images of NEG-123 samples which were mixed for 30 minutes and sintered for 1 hour at : (a) 750°C, (b) 800°C, (c) 900°C

The morphology of NEG-123 samples that mixed for 24 hours and sintered at 750, 800 and 900°C for 30 minutes observed using TEM are shown in Fig. 6. The morphology is seen to be more homogeneous distribution of particles with average particle size between 20-30 nm as shown in Fig. 6(a) and Fig. 6(b). Meanwhile, the sample that was sintered at 900°C has a bigger average particle size between 20-250 nm and shows a less homogeneous distribution, indicating that agglomeration has already begun because of sintering as shown in Fig. 6(c). In general shape of NEG-123 particles at a temperature of 750-900°C showed spherical-shaped.

The difference in treatment resulted in differences in the shape and size of the particles. There are two possibilities, the first, because longer mixing produces the ionic bond. The second, because of the agglomeration process, in which a short time sintering produces crystal size smaller than a longer sintering time.

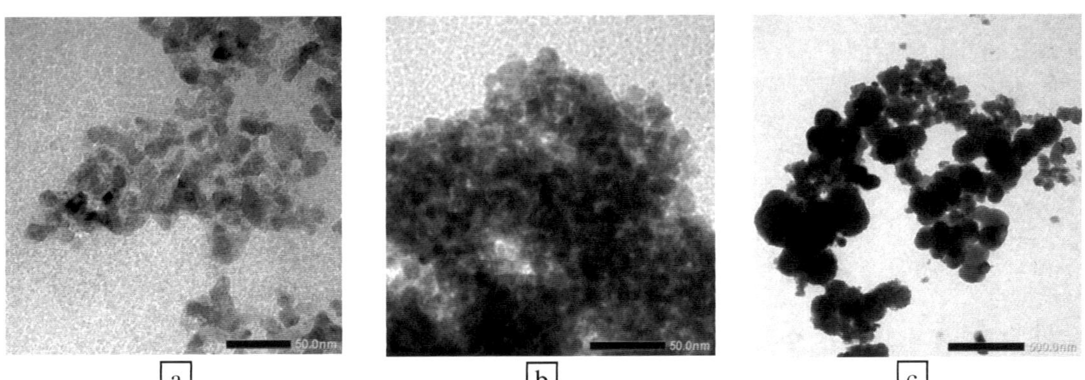

Figure 6. TEM images of NEG-123 samples that were mixed for 24 hours and sintered for 30 minutes at : 750°C, (b) 800°C and (c) 900°C

Going further to the magnetic properties of NEG-123 samples, measured using VSM at room temperature. The NEG-123 samples that mixed for 30 minutes and sintered at 750, 800 and 900°C did not show any hysteresis, but only exhibited paramagnetic properties as shown in Fig. 7(a). Contrarily, for the samples that mixed for 24 hours and sintered at 800, 850, 900°C and 750°C (insert) showed a superparamagnetic properties at room temperature, as show in Fig. 7(b). This effect may probably come from the oxygen vacancies on the sample surface because of having a much larger surface area in the lowering particle size, as suggested previously.

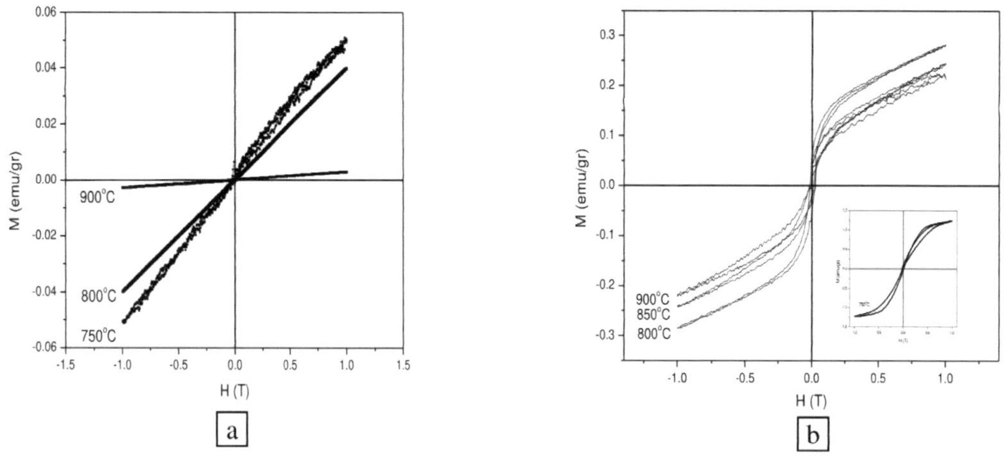

Figure 7. VSM images of (a) NEG-123 samples mixed for 30 minutes and sintered at 750, 800 and 900°C for 30 minutes, (b) NEG-123 samples mixed for 24 hours and sintered at 750, 800, 850 and 900°C for 30 minutes

The magnetization curve showed that the addition of the sintering temperature of 750°C to 900°C resulted in the value of the saturation magnetization (M_S) gets smaller. These results are consistent with the results of TEM characterization, where the addition of the sintering temperature causes the particle size increases and then result in coercive and saturation decreases because the particles form a multidomain. Similar results were obtained by Sundaresan et al., in which the magnetic moment saturation of nano particles is reduced by the addition of particle size.

The nanoparticles have a large surface area, whereas the formation of the surface energy, lower than in the samples in bulk. Low energy on nano particles cause defects on the surface in an amount sufficient. The magnetization can explained by the presence of magnetic cluster which has superparamagnetic properties. However, the sintering temperature increase resulted the larger of energy, so that the particle size increases and the concentration of defects decreases and then magnetic properties disappear.

Conclusions

The NEG-123 nanocrystals have successfully been synthesized by using wet-mixing method. The introduction of additional sintering temperature from 750°C to 900°C can improve the crystallization of the sample marked with the made-up diffraction peak intensity. The increasing sintering temperature has led to the decreasing lattice parameter of a-axis, as well as increasing the lattice parameter of b and c-axis. And then, the higher of sintering temperature, the greater of average crystal size. NEG-123 samples with average crystal size of hundreds nano-scale showed paramagnetic

IOP Conf. Series: Journal of Physics: Conf. Series **1091** (2018) 012027 doi:10.1088/1742-6596/1091/1/012027

properties, while NEG-123 with the average crystal size of tens of nano-scale showed superparamagnetic properties at room temperature.

References

[1] Akbarzadeh A, Samiei M and Davaran S 2012 *Nanoscale Res Lett* **7** **1** 144.

[2] Issa B, Obaidat I M, Albiss B A and Haik Y 2013 *Int. J. Mol. Sci.* **14** **11** 21266-21305.

[3] Sundaresan, Bhargavi R, Rangarajan N, Siddesh U, Rao CNR 2007 *Phys. Rev. B* **74** 161306.

[4] Ghosh S C, Thanachayanont C and Dutta J 2014 *The 1st ECTI Annual Conference (ECTI-CON2004) Pattaya* 145-148.

[5] Niu G, Hildebrandt E, Schubert M A, Boscherini F, Zoellner M H, Alff L, Walczyk D, Zaumseil P, Costina I, Wilkens H and Schroeder T 2014 *ACS Appl. Mater. Interfaces* **6** **20** 17496-17505.

[6] Shipra, Gomathi A, Sundaresan A, Rao CNR 2007 *Solid State Communications* **142** 685-688.

[7] Baqiya M A, Widodo H, Rochmawati L, Darminto, Adachi T and Koike Y 2012 *AIP Conf. Proc.* **1454** 260.

[8] Muralidhar M, Koblischka MR and Tomita M 2012 *Current Microscopy Contrbution to Advances in Science and Technology* 1468-1479.

[9] Vinila V S, Jacob R, Mony A, Nair H G, Issac S, Rajan S, Nair A S, Satheesh D J, Isac J 2014 *Journal of Crystallization Process and Technology* **4** 168-176

[10] Suharta W G, Suasmoro S, Pratapa S, Darminto 2013 *AIP Conference Proceedings* **1555** 62-66.

[11] Suharta W G, Widagda I G A, Putra K, Suyanto H 2016 *Journal of Physics: Conference Series* 012006.

IOP Conf. Series: Journal of Physics: Conf. Series **1091** (2018) 012028 doi:10.1088/1742-6596/1091/1/012028

Effects of the geometry factor on the reflection loss characteristics of the modified lanthanum manganite

Wisnu Ari Adi[1], Yosef Sarwanto[1], Yana Taryana[2], and Bambang Soegijono[3]

[1]Centre for Science and Technology of Advanced Materials – National Nuclear Energy Agency, Kawasan Puspiptek Serpong, Banten, Indonesia
[2]Center for Research of electronic and telecommunication - Indonesia Institute of Sciences, Bandung, Indonesia
[3]Dept of Physics, Faculty of Mathematics and Natural Sciences – University of Indonesia, Jakarta, Indonesia

Corresponding author : dwisnuaa@batan.go.id

Abstract. The synthesis and characterization of modified lanthanum manganite materials by using mechanical milling technique have been performed. This magnetic material is prepared by oxides, namely lanthanum oxide, barium carbonate, iron oxide, titanium dioxide, and manganese carbonate. The mixture was milled for 10h, compacted at 5000 psi into pellets with three kinds of different thickness (d = 1.5, 2.0 and 4.0 mm, respectively) and then sintered at temperature of 1000 °C for 10h. The refinement results of x-ray diffraction pattern showed that the sample is single phase. The sample had composition in accordance to stochiometry composition. The geometry factor consists of the particle morphology observed using scanning electron microscope and thickness measured millimeter device. The microstructure analyses shows that the particle shapes was aggregates with the particle sizes distributed homogeneously on the surface of the sample. The results of the microwave absorption indicated that there were three of absorption peak frequency at 9.9 GHz, 12.0 GHz, and 14.1 GHz. The microwave absorption of the sample increases with increasing thickness of absorption area. We concluded that the increasing thickness of absorption region resulted new resonance frequencies, and the new resonance frequencies are joined to each other to form a wider resonance frequency and it is known as a broadband absorber. The first section in your paper.

1. Introduction

The demand for the reduction of electromagnetic wave interference is a topic of interest at the moment. Electromagnetic wave interference is a emission of high frequency noise that poses a problem for electronic devices which are working at high frequency due to its ability to reduce the performance of the electronic equipments. Therefore, an absorber material is needed which is capable to absorb electromagnetic waves in such manner so that the electromagnetic wave interference could be eliminated. Recently some modified materials such as hexaferrite-based, perovskite-based, and carbon-based have been found to exhibit the microwave absorption characteristics [1-5]. Development of magnetic material as absorber materials is required by a structural-engineering technique in which it is expected that the end product should be a superior material suitable for microwave absorber applications [6-9].

There are at least two factors that affect the high performance of absorbing material, namely intrinsic and extrinsic factors. The prerequisite of intrinsic factor that is required of an electromagnetic wave absorber material is that a material must have both the permeability and the permittivity, meanwhile the extrinsic factor is the geometry factor such as particle shape and thickness of absorption area [10-12] have reported that the authors have succeeded in the modification of a manganite-based material in which they performed a material modification of Fe substituted lanthanum manganite system in order to increase the material's permeability. The nano-crystalline powders, is a type of an excellent microwave absorbing material, which could be synthesized by the sol–gel method. The authors determined the relationship between reflection loss and frequency for their samples with three different thicknesses (d = 1.80, 2.00 and 2.21 mm, respectively). The absorption peaks increase with the thickness [12] reported that they have carried out a convenient graphical method for the determination of the optimal thickness for single-layer absorbers under normal wave incidence. Matching condition demands coincidence of thicknesses of equal amplitudes and phases for reflection and transmission coefficients of the wave reflected from the first-boundary air-absorber and the wave transmitted and reflected from second-boundary absorber-metal, respectively. Reflection loss maps are constructed as functions of layer thickness and one of the material parameters $tg\varepsilon$ or $tg\mu$ for constant ratio ε_r/μ_r and another parameter $tg\mu$ or $tg\varepsilon$ ((ε_r and μ_r are the real parts of complex relative permittivity and permeability and $tg\varepsilon$ and $tg\mu$ are the dielectric and magnetic loss tangents of the material) [13]

In this work, the authors performed a further studyon this particular topic based in part on the results of previous studies. More specifically, in this work synthesis of modified lanthanum manganite nanoparticles is carried out, and in addition,the effect of absorption thickness on the absorption properties of the materialhas also been investigated. The aim of this study is to synthesize nanoparticles of modified lanthanum manganite and to understand the changes in the extrinsic factors of microwave absorption performance of this material caused by the particle shape and thickness of absorption area.

2. Materials and Methods

Modified lanthanum manganite referred is a perovskite-based magnetic material and its molecular formula is $La_{0.8}Ba_{0.2}Fe_{0.3}Mn_{0.35}Ti_{0.35}O_3$.The material is synthesized by a solid-state reaction method through mechanical milling technique using a high-energy milling (HEM) of the type Spex 8000. The synthesis process of the material was carried out according to research result of Azwar et al [14]. However, the sample was re-milled for 30 hours to obtain nanocrystallites powders. The aim of the 30 hours re-milling was to obtain a more perfect quality improvement over the previous research results. The re-milled material was compacted at 5000 psi into pellets with three different thickness (d = 1.5, 2.0 and 4.0 mm, respectively).

The characterization of the product materials consist of phase analysis using Miniflex Rigaku diffractometer (XRD) equipped with a tube of cooper anode (CuKα). An additional Rietveld analysis was performed applying the GSAS program [15]. Elemental composition of the material was measured by using energy dispersive spectroscopy (EDS) method, meanwhile the particle morphology of the sample is observed by using the scanning electron microscope (SEM). The microwaves absorption characteristic investigation was carried out by using vector network analyzer (VNA) in order to obtain the effects of reflection and transmission of electromagnetic waves from signal source with a certain frequency. The reflection loss (RL) property of the samples was measured by using a Vector Network Analyzer ADVANTEST type R3770 in the frequency range of 300 kHz - 20 GHz (Figure 1(a)). For the analysis of microwave reflection and transmission, testing was conducted in the frequency range of 5 GHz to 18 GHz with a prototype form of bulk samples with a diameter of 1.5 mm.

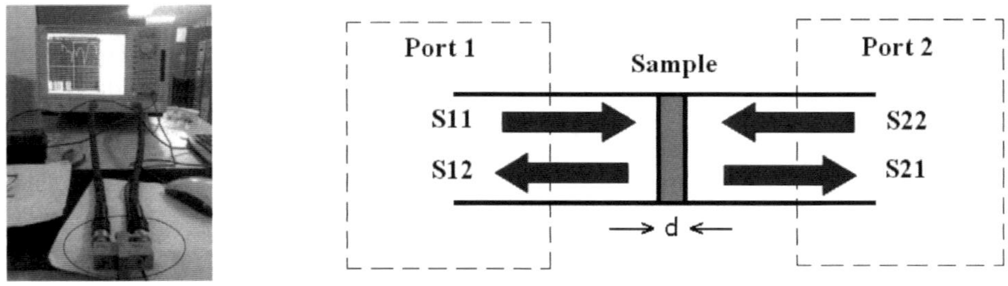

(a) Instrument and adapter

(b) Electromagnetic wave propagation scheme in adapter

Figure 1. Vector Network Analyzer – Advantest type R3770 [8]

The electromagnetic wave propagation scheme in adapter of vector network analyzer (VNA) is shown in Figure 1(b).Wave coming from the Port 1 and the reflected wave (S11) are also received by the Port 1 and wave transmission (S21) was received by the Port 2. Polarization of electromagnetic waves in the waveguide is very useful for analyzing the electrical properties of the material thickness d. To obtain accurate data on the measurement of reflection and transmission coefficients, the measurement needs to be performed using TRL calibration (reflection transmission line).

3. Results and Discussion

Figure 2(a) shows the refinement results, which indicate that a very good quality of fitting with a small R_{wp} factor of 7.63 and a goodness of fit value χ^2 (chi-squared) of 1.302 have been achieved; and which also are in good agreement with the established rietveld refinement standards [15]. The refinement results of x-ray diffraction pattern show that the sample is a single phase of lanthanum manganite.

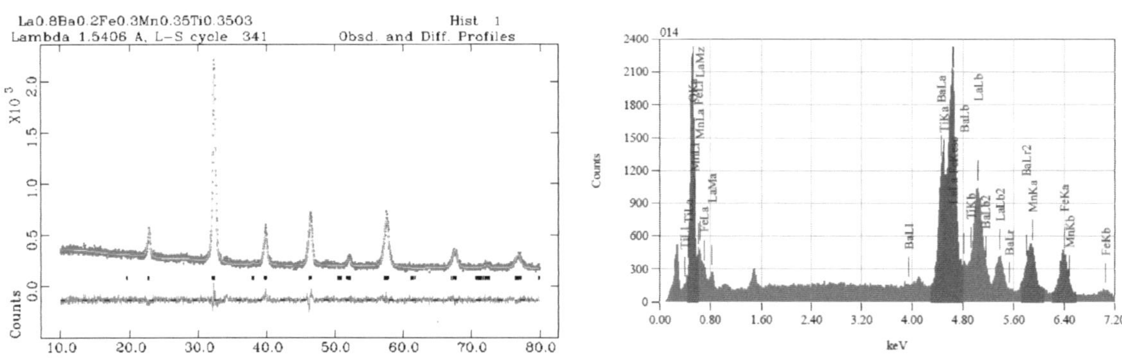

(a) XRD profile of material (b) EDS spectra of the material

Figure 2. Phase and element analysis of the modified lanthanum manganite

Table 1. Structure parameter of the sample compared with previous research [14]

• Sample	• Previous research [11]
Monoclinic,	Monoclinic,
Space group : I 1 2 / a 1,	Space group : I 1 2 / a 1,
a = 5.5386(9) Å, b = 5.577(1) Å, c = 7.8465(1) Å,	a = 5.5182(8) Å, b = 5.5442(8) Å and c = 7.822(1) Å
$\alpha = \gamma = 90^{o}$, $\beta = 89.80(2)^{o}$	Å$\alpha = \gamma = 90^{o}$ and $\beta = 89.63(1)^{o}$
V = 242.88(7) Å^3, and ρ = 6.407 gr.cm^{-3}	V = 239.68(3) Å^3, ρ = 6.559 gr.cm^{-3}
Single phase	Single phase
Crystallite size = 41 nm	Crystallite size = 42 nm

In Table 1 the lattice parameter values, the volume unit cell, and atomic density values (which are approximately similar to the results of a previous work) are tabulated so that required support data to exact elemental content in the sample. The elemental content of the $La_{0.8}Ba_{0.2}Fe_{0.3}Mn_{0.35}Ti_{0.35}O_3$ is shown in Figure 2(b) and Table 2.

Tabel 2. The results of element analysis by using energy dispersive spectroscopy

No.	Unsure	Content (wt.%)	Content (wt.%) [14]
1.	Lanthanum (La)	48.64 ± 0.49	16.64
2.	Barium (Ba)	12.44 ± 0.40	4.30
3.	Titanium (Ti)	4.10 ± 0.19	4.07
4.	Manganese (Mn)	6.27 ± 0.29	5.42
5.	Iron (Fe)	7.17 ± 0.29	6.10
6.	Oxygen (O)	21.38 ± 0.14	63.47

EDS spectra shows that the elements compositionconsist oflanthanum (La), barium (Ba), titanium (Ti), manganese (Mn), iron (Fe) and oxygen (O). This means that the sample had composition in accordance with the stochiometry composition.
The Williamson-Hall method was used to calculate the mean crystallite size by X-ray broadening peaks analysis of the sample [14].

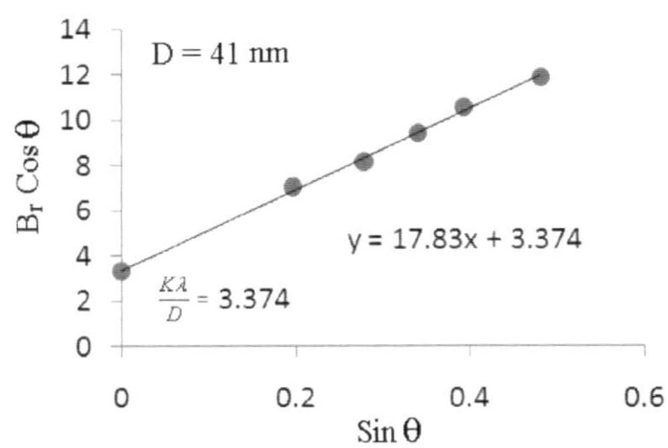

(a) Williamson hall curve

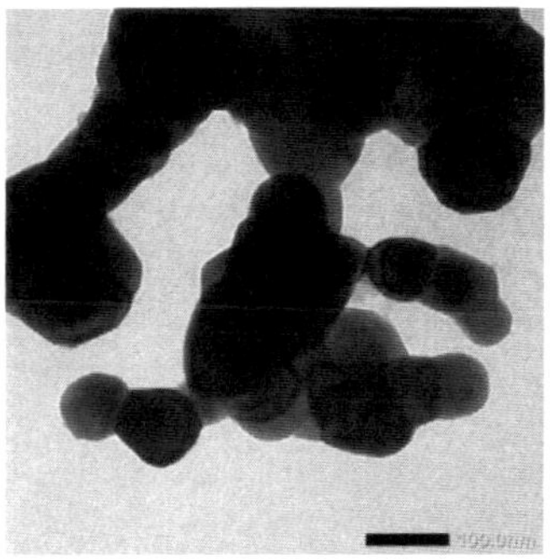

(b) TEM image

Figure 3. The crystallite size and particle size of the samples

Figure 3(a) shows a curve of B_r Cos θ vs. Sin θ plot for the sample according to the results based on the Williamson-Hall calculation method. The linear regresion graph from the curve can be used to determine the mean crystallite size of the sample. The mean crystallite size of the sample is around 41 nm. Figure 3(b) is the particle size of the sample observed by using transmission electron microscope (TEM). TEM image shows that the sample has particles with a varying diameter of around 100 – 200 nm and a spherical shape. Thus, the analysis result of X-ray difraction profile shows that the synthesis of modified lanthanum manganite have been successfully accomplished with a single phase and the mean crystallite size of 41 nm. This result signifies that the method is reproducible and has a similarity with the previous research result [14].

For further analysis, the authors measured the performance of the sample's absorbing properties with three kinds of different thickness (d = 1.5, 2.0 and 4.0 mm, respectively). The measurement result of the total absorption of the sample in the frequency range 9-15 GHz is shown in Figure 4. The sample has a relatively significant absorption rate, which occurs in a relatively wide range. There is an indication that an absorption of microwaves in the frequency range of 9 to 15 GHz has been found.

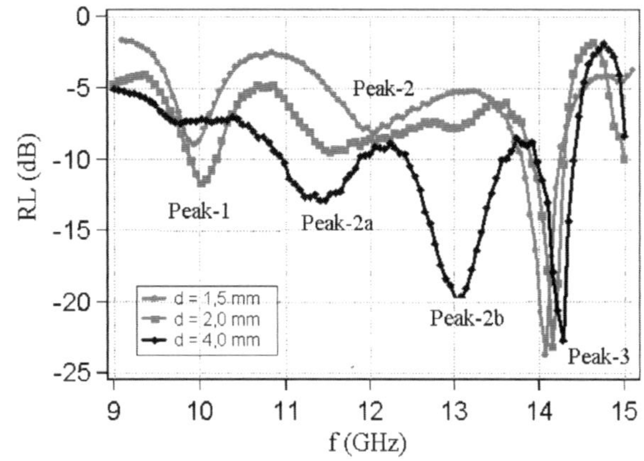

(a) Reflection loss of the sample

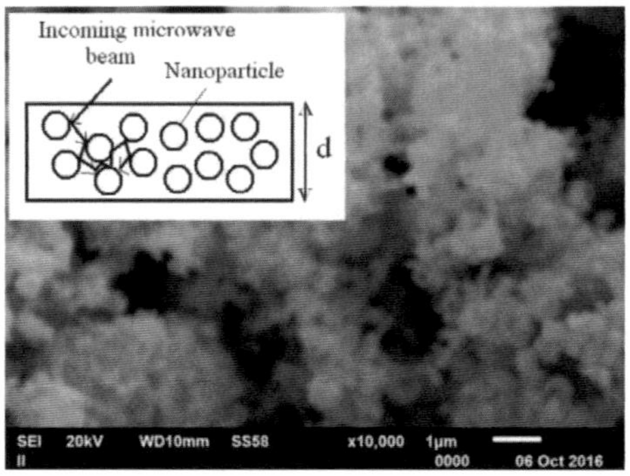

(b) SEM image of the sample

Figure 4. Reflection loss curve and geometry factor of sample.

Microwave absorption in the samples increases with increasing thickness of the absorption field. For the thickness of the sample d = 1.5 mm it seems that there is an absorption of microwaves in the frequency range between 9-15 GHz with three peaks of frequency absorption at 9.9 GHz (peak – 1), 12.0 GHz (peak – 2), and 14.,1 GHz (peak – 3), with reflection loss of ~ 9 dB, ~ 8 dB, and ~ 23.5 dB, respectively. For the thickness of d = 2.0 mmit seems that there are four absorption peaks, namely at the frequencies of 10.1 GHz (peak – 1), 11.5 GHz (peak – 2a), 13.0 GHz (peak – 2b), and 14.2 GHz (peak – 3) with reflection loss of ~ 13 dB, ~ 10 dB, ~ 7 dB and ~ 23.0 dB, respectively. Then when the thickness d is equal to 4.0 mm, it seems that there are also four absorption peaks, appearing at the frequencies of 9.8 GHz (peak – 1), 11.3 GHz (peak – 2a), 13.1 GHz (peak – 2b), and 14.3 GHz (peak – 3) with reflection loss of ~ 7 dB, ~ 13 dB, ~ 20 dB and ~ 22.5 dB, respectively.

The three sharp peaks (peak – 1, peak – 2 and peak – 3) are known as the resonance absorber when the thickness d is equal to 1.5 mm. Resonance absorber is the absorbing characteristics of microwaves which occur at certain frequencies. It is suspected that these peaks are part of the microwaves absorption characteristics in this material. Then the thickness increased up to d = 2.0 mm, absorbing energy (peak – 1) also increaseswhich is characterized by the reflection loss (RL) value increase from ~ 9 dB up to ~ 13 dB (the absorption percentage increases from 70 % to 80 %). And then the absorbing energy (peak –2) also increasesfrom ~ 9 dB up to ~ 10 dB but the resonance frequency shifts to the left because it displays a new peak of resonance frequency at 13 GHz (peak – 2b) with RL of ~ 9 dB. Absorbing energy (peak – 3) appears to be relatively fixed, but it is slightly shifted to the right due to the presence of the resonance frequency on the peak – 2b increases. Then, when the thickness of the sampleis increased to 4.0 mm, the absorbing energy (peak – 1) decreases but this peak is broad.It shows that this peak occurs at the resonance frequency more evenly and it is known as a broadband absorber. Absorbing energy (peak – 2a) increases from ~ 10 dB up to ~ 13 dB (the absorption percentage up to 80 %). The absorbing energy (peak – 2b) increases significantly from ~ 7 dB up to ~ 20 dB (the absorption percentage up to 90 %). And the absorption energy (peak – 3)appears relatively fixed and is shifted slightly to the right due to the presence of the resonance frequency on the peak – 2b increase. Increasing reflection loss allegedly came from the influence of the geometry (shape of the particles) in the material as shown in Figure 4 (b). The microstructure analysis shows that the particles' shape was spherical aggregates with the particle sizes distributed homogeneously on the surface of the sample. If there is a microwave radiation incident upon the material, the microwave beam will be scattered in all directions so that the beam is not reflected back, as though it was being absorbed due to the scattering process. This scattering factor increases with increasing sample thickness (geometrical factors). In the end, the geometry factor (extrinsic) can increase the absorption of microwave in addition to the intrinsic factors such as the materials permeability and permittivity.

4. Conclussion

Sample of modified lanthanum manganite have been prepared by mechanical milling technique to study the effect of the geometrical factor on the reflection loss characteristic. The sample is single phase with a monoclinic structure, having a crystalline phase of I12/a1, the particle size of 100-200 nm, and the particles have a spherical shape. The microstructure analyses showed that the particles morphology consists of aggregates structures with the particle sizes distributed homogeneously on the surface of the sample. The microwave absorption of the sample increases with increasing thickness of absorption area. We concluded that the increasing thickness of absorption region have resulted in the appearance of new resonance frequencies, and the new resonance frequencies are joined to each other to form a wider resonance frequency which is known as a broadband absorber.

Acknowledgements

The authors gratefully acknowledge the support of the Center for Science and Technology of Advanced Materials for the research facilities. We are also thankful to Asep Yudhi Hercuadi, S.Si. for his help to measure the absorption of microwaves using a vector network analyzer (VNA) in PPET - LIPI Bandung.

References

[1] Adi W A and Manaf A 2012 Structural and Absorption Characteristics of Mn-Ti Substituted Ba-Sr Hexaferrite Synthesized by Mechanical Alloying Route J. Basic. Appl. Sci. Res. 2 8 7826-7834.

[2] Zhang S, Cao Q 2012 Electromagnetic and microwave absorption performance of some transition metal doped $La_{0.7}Sr_{0.3}Mn_{1-x}TM_xO_{3\pm1}$ (TM = Fe, Co or Ni) Materials Science and Engineering B 177 678–684.

[3] Mashadi, Yunasfi and Adi W A 2016 Structural, Magnetic Properties and Microwave Absorbing Behavior of Iron Based Carbon Nanocomposites Malaysian Journal of Fundamental and Applied Sciences 12 1 6-11.

[4] Gunanto Y E, Cahyadi L and Adi W A 2016 Effect of Mn and Ti substitution on the reflection loss characteristic of $Ba_{0.6}Sr_{0.4}Fe_{11-z}MnTi_zO_{19}$ (z = 0, 1, 2 and 3) AIP Conference Proceedings 1725 020023.

[5] Mandal A, Ghosh D, Malas A, Pal P and Das C K 2013 Synthesis and Microwave Absorbing Properties of Cu-Doped Nickel Zinc Ferrite/$Pb(Zr_{0.52}Ti_{0.48})O_3$ Nanocomposites 2013 ID 391083.

[6] Adi W A, Manaf A 2013 Microstructure and Phase Analysis of $La_{0.8}Ba_{0.2}Ti_xMn_{(1-x)}O_3$ system for Microwave Absorber Material (x = 0 – 0.7) Advanced Materials Research 789 97-100.

[7] Gunanto Y E, Jobiliong E and Adi W A 2016 Microwave Absorbing Properties of $Ba_{0.6}Sr_{0.4}Fe_{12-z}Mn_zO_{19}$ (z = 0 – 3) Materials in X Band Frequencies J. Math. Fund. Sci 48 155-65.

[8] Hercuadi A Y, Adi W A and Purwanto S 2012 A Study On The Behaviour Of M-Type Strontium Hexagonal Ferrite Based Microwave Absorbing, Proceeding of The 6th National Radar Seminar And The first International Conference On Radar, Antenna, Microwave, Electronics And Telecommunications (ICRAMET) Bali.

[9] Kiani E, Rozatian A S H, Yousefi M H 2014 Structural, magnetic and microwave absorption properties of $SrFe_{12-2x}(Mn_{0.5}Cd_{0.5}Zr)_xO_{19}$ ferrite J. Magn. Magn. Mater. 361 25-29.

[10] Mohit K, Gupta V R and Rout S K 2014 Microwave Dielectric Properties of $Ni_{0.2}Cu_xZn_{0.8-x}Fe_2O_4$ for Application in Antenna Progress In Electromagnetics Research B 57 157-175.

[11] Duggal S, Aul G D 2014 Review on Effect of Electric Permittivity and Magnetic Permeability over Microwave Absorbing Materials at Low Frequencies International Journal of Engineering and Advanced Technology 3 5 12-19.

[12] Kaur H, Aul G D 2014 A Review Based on Effects of Change in Thickness and Number of Layers on Microwave Absorbing Materials, Intern. Journal of Science and Research 3 5.

[13] Giannakopoulou T, Kontogeorgakos A, Kordas G 2003 Single-layer microwave absorbers: influence of dielectric and magnetic losses on the layer thickness, Journal of Magnetism and Magnetic Materials 263 173–181.
[14] Manaf A and Adi W A 2014 Characterization of single phase of $La_{0.8}Ba_{0.2}Fe_{0.3}Mn_{0.35}Ti_{0.35}O_3$ nanoparticles as microwave absorbers, Advanced Materials Research 896 428-433.
[15] Idris M S, Osman R AM 2013 Structure Refinement Strategy of Li-Based Complex Oxides Using GSAS-EXPGUI Software Package Advanced Materials Research 795 479-482.

Structural, Band Gap Energy, and Magnetic Characters of $Fe_{2.9}Cr_{0.1}O_4$ Nanoparticles for Preparing Ferrofluids

A Taufiq*, N Wahyuni, W Elyani, S Sunaryono, N Hidayat, N Mufti and A Hidayat

Department of Physics, Faculty of Mathematics and Natural Sciences, Universitas Negeri Malang, Jl. Semarang 5, Malang 65145, Indonesia

*Email: ahmad.taufiq.mipa@um.ac.id

Abstract: Ferrite nanoparticles have become interesting materials originating from their performances in many applications. In this paper, the preparation metal-doped ferrites in the system of $Fe_{2.9}Cr_{0.1}O_4$ nanopowders is reported. The $Cr_{0.1}Fe_{2.9}O_4$ nanopowders were utilized to produce ferrofluids. The Indonesian sand was used as a raw material to prepare the samples using co-precipitation route. The $Fe_{2.9}Cr_{0.1}O_4$ particles structured in cubic spinel with the particle size and lattice parameter of about 9.4 nm and 8.36 Å, respectively. The band gap energy of the $Cr_{0.1}Fe_{2.9}O_4$ nanopowders was 2.26 eV. Furthermore, the saturation magnetization of the powder was higher than that of the fluid as the effect of particle size and aggregation.

Keywords: $Fe_{2.9}Cr_{0.1}O_4$, ferrofluid, nanopowder, aggregation, iron sand.

1. Introduction

In nanoscience and nanotechnology fields, one of the materials that intensively investigated is magnetic nanoparticles, especially associated with their biomedical applications. Recent article based on the Elsevier database reported that number of articles of the magnetic particles associated with their biomedical applications increase significantly in an exponential growth for the last decade [1]. Furthermore, in the form of Fe_3O_4 ferrofluids, the magnetic nanoparticles have a specific excellent performance regarding their physical, chemical, and biological characters. Ferrofluids can be defined as a stable colloidal suspension containing superparamagnetic magnetic particles with particle size in superparamagnetic dispersed in organic or inorganic liquid carriers [2].

Commonly, organic or inorganic materials can be used as coated or layered agents to prevent aggregations of the Fe_3O_4 particles in ferrofluids. Based on the structure, Fe_3O_4 becomes one of the spinel groups with a general chemical formula of AB_2O_4, where A is the divalent metal (Fe^{2+}) and B is the trivalent metal (Fe^{3+}). The invers spinel structure can be formed when all divalent metallic ions place B site, a half of trivalent metallic ions places B site and a remained half of the trivalent metallic ions places A site [3]. Therefore, the invers spinel of the Fe_3O_4 particles can be expressed in the formula of $Fe_A^{3+}O^{2-}[Fe^{2+}Fe^{3+}]_BO_3^{2-}$. Such formula guides us to rearrange easily the structure by incorporating other metallic ions such as Mn, Co, Zn, Cu, Cr, and other to enhance the performances of the particles.

In this experiment, we introduce chromium (Cr) as one of the metallic transitions in the group VI-B having special character due to its stable oxidation. Cr doped Fe_3O_4 in the system of $Fe_{2.9}Cr_{0.1}O_4$ nanoparticles have three cation formations such as Fe^{2+}, Fe^{3+}, and Cr^{3+} that would change the ionic arrangement for a better specific behavior. In general, the three metallic ions will place randomly in B and A sites. Quadro and co-workers identified that the Cr and Fe-based catalysts have been utilized in a long time of period for commercial processes due to their high stability [4]. To be more effective and efficient in mass production, despite the important in introducing Cr ions to incorporate Fe_3O_4 particles, it is also important to exploit an iron sand from local place in Indonesia as a raw material. Moreover, in this paper, the detailed nanostructure, functional group, band gap energy, and magnetic behaviors of the prepared $Fe_{2.9}Cr_{0.1}O_4$ in nanopowders and in ferrofluids are also reported.

2. Experimental Method

The main precursor (magnetite powder) was firstly purified from Indonesian sand through mechanical process using a magnetic bar. The magnetite powder was mixed with HCl at room temperature using a magnetic sitter to produce iron chloric solution. The solution was then mixed with $CrCl_3$ solution and following by dropping NH_4OH to obtain precipitate. To obtain the powder sample, the precipitate was rinsed by using H_2O and following by a calcination process at 100 ^0C. Meanwhile, to obtain the ferrofluid sample, the precipitate was directly coated by TMAH and dispersed in H_2O. The obtained samples was characterized by means of XRF, XRD, VSM, FTIR spectrometer, and UV-Vis spectrometer. The data from all characterizations were finally analysed based on the qualitative and quantitative approaches.

3. Results and Discussion

Before investigating the structure and phase purity of the $Fe_{2.9}Cr_{0.1}O_4$ particles, the elemental composition was investigated using an XRF machine at room temperature. Based on the data analysis, it was shown that the sample had the elemental content with the Cr: /(Cr + Fe) ratio of 4.1 %. Therefore, to study the detailed crystal structure and phase purity of the $Fe_{2.9}Cr_{0.1}O_4$ particles, we characterized the sample using an XRD machine at room temperature and its result is shown in Figure 1.

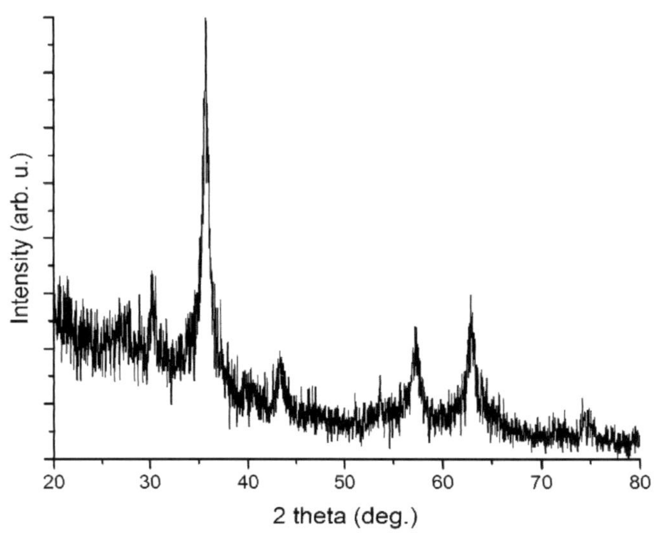

Figure 1. XRD pattern of the prepared $Fe_{2.9}Cr_{0.1}O_4$ particles

Based on the qualitative data analysis, the XRD pattern has similar pattern to the patterns of the Fe_3O_4 and Mn-doped Fe_3O_4 nanoparticles as reported in the literature [5]. It means that the prepared

$Fe_{2.9}Cr_{0.1}O_4$ particles from iron sand crystallize in a mono phase. Further quantitative data analysis of the XRD pattern presented that the $Fe_{2.9}Cr_{0.1}O_4$ particles has a spinel structure with the particle size of about 9.4 nm. The XRD pattern presents the crystal structure in the space group of *F d -3 m Z* in the absence of other phases. Using a Rietveld analysis by performing ICSD No. 29129, it was resulted that the lattice parameters of $a = b = c$ were of approximately 8.396 Å. Nguyen *et al.* reported that the Cr-doped Fe_3O_4 particles increase the lattice parameters of the system by increasing Cr composition [6]. Additionally, their work presented that the shape and morphology of the magnetic nanoparticles are still in spherical shapes with agglomeration. In order to investigate the structural behavior, the functional group of the sample was investigated using a FTIR spectrometer and its result is presented in Figure 2.

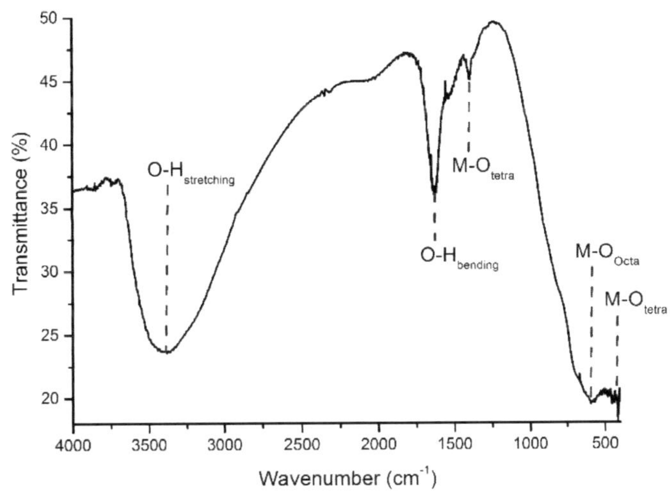

Figure 2. FTIR spectrum of the prepared $Fe_{2.9}Cr_{0.1}O_4$ nanoparticles

The FTIR spectrum in the wavenumber range of 4000 - 400 cm^{-1} exhibits several transmission peaks at ~3400, ~1620, ~1400, ~580, and ~410 cm^{-1}. The transmission peaks at wavenumber of 3400 and 1620 cm^{-1} are resulted from stretching and bending functional groups of O-H [7,8]. Furthermore, the peaks at wavenumber of ~1400, ~580, and ~410 cm^{-1} are resulted from the functional groups of M-O (metal-oxygen). It means that the divalent and trivalent ions of Fe and Cr were successfully occupied the tetrahedral and octahedral positions.

After investigating the structure of the $Fe_{2.9}Cr_{0.1}O_4$ nanoparticles, it is followed by investigation of the magnetic behaviors of the samples. The magnetic characters were performed using *M-H* experiment at ambient temperature as presented in Figure 3-4. In the *M-H* experiment, the external magnetic field was varied from -1 T to 1 T. From Figure 3-4, it is easy to be known that all hysteresis curves have *S* patterns indicating all samples exhibit as superparamagnetic materials. Theoretically, the superparamagnetic state of the samples can be calculated or fitted using a Langevin function [3], as written in Equation 1.

$$\frac{M}{M_S} = \coth\left(\mu H / k_B T\right) - \frac{1}{\left(\mu H / k_B T\right)} \qquad (1)$$

IOP Conf. Series: Journal of Physics: Conf. Series **1091** (2018) 012029 doi:10.1088/1742-6596/1091/1/012029

Where respective M and M_S are the magnetization and saturation magnetization, μ and H are the magnetic moment and magnetic field, T is the temperature, and k_B is the Boltzmann constant.

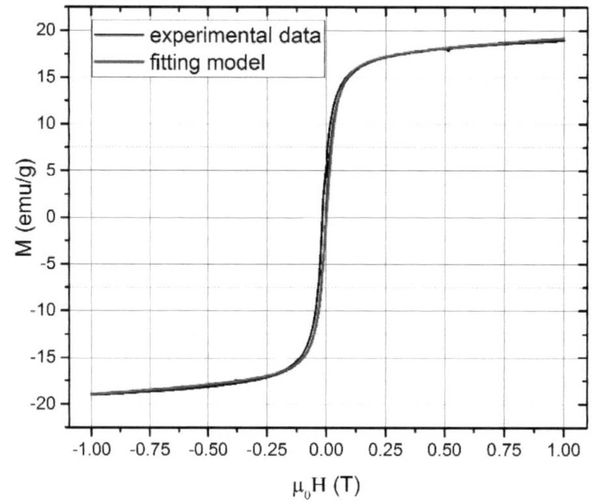

Figure 3. M-H curve of the prepared $Fe_{2.9}Cr_{0.1}O_4$ nanopowders

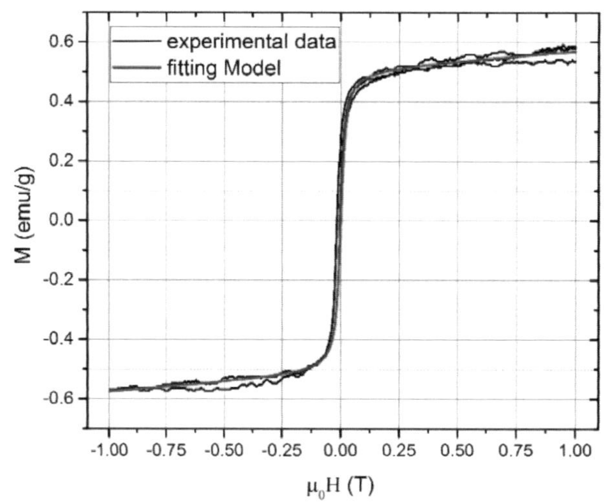

Figure 4. M-H curve of the prepared $Fe_{2.9}Cr_{0.1}O_4$ ferrofluids

Regarding Figure 3-4, the Langevin function is fitted well the experimental data both for the samples in fluids and in powders. The M_S of the $Fe_{2.9}Cr_{0.1}O_4$ nanopowders has a value, which is higher than that of the ferrofluids. The phenomenon is predicted as the effect of the particles size and its clustering. In general, the secondary particle size and clustering of the magnetic nanoparticles in powders has higher than that of the ferrofluids. Physically, in the fluids, the secondary and clustering are affected by the presence of the coated magnetic particles using TMAH dispersing in H_2O [9]. In this situation, the building block of the magnetic nanopowders subjected to their primary particles is similar to the building block of the ferrofluids. As superparamagnetic material, the particle size of the $Fe_{2.9}Cr_{0.1}O_4$ both in the powders and in fluids should be smaller than a Weiss domain of the corresponding particles in bulk forms [10].

In this work, another fundamental character of the prepared $Fe_{2.9}Cr_{0.1}O_4$ nanoparticles was investigated using UV-Vis spectrometer and its result is presented in Figure 5. Such experiment was performed to study the optical band gap energy (*Eg*) of the sample. The *Eg* of the prepared $Fe_{2.9}Cr_{0.1}O_4$ nanoparticles was determined by plotting the absorption coefficient $(\alpha hv)^2$ and photon energy (hv) using Tauc's equation as shown in Equation 2 [11]. Based on the fitting analysis, the *Eg* of the prepared $Fe_{2.9}Cr_{0.1}O_4$ nanoparticles was about 2.26 eV indicating that the value is in the range of the semiconductor character [12].

$$(\alpha hv)^2 = B(hv - E_g) \quad (2)$$

Where *B* is the constant.

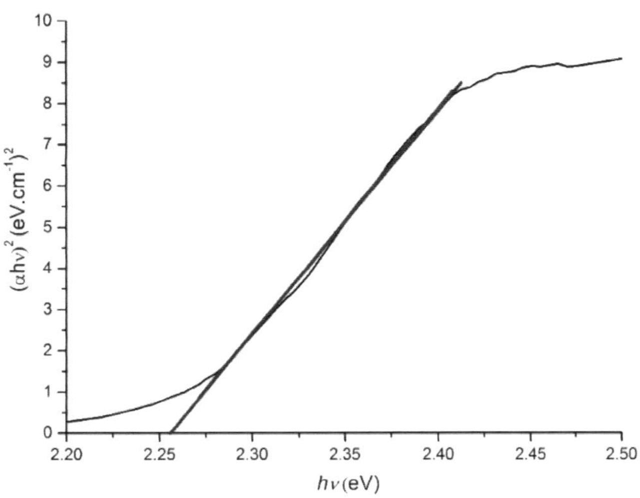

Figure 5. Band gap energy of the prepared $Fe_{2.9}Cr_{0.1}O_4$ nanoparticles

Based on the above discussion regarding band gap energy, the $Fe_{2.9}Cr_{0.1}O_4$ ferrofluids from Indonesian sand open potential applications such as for magnetic sensor based on magneto-optics. In our previous work, the Fe_3O_4 ferrofluids from Indonesian sand prepared by coprecipitation method had excellent performance related to optical character polarization angle of the fluids under external magnetic field [13]. Moreover, the ferrofluids performed a good response in a linear relation originating from the relation of the intensity obtained by photodetector and the external magnetic field varying from zero to 140 mT. Therefore, it is necessary to develop further research by employing the the $Fe_{2.9}Cr_{0.1}O_4$ ferrofluids resulted in this work for magnetic sensor.

4. Conclusion

The $Fe_{2.9}Cr_{0.1}O_4$ nanoparticles formed in spinel structure in nanometric scale of about 9.4 nm. The Fe-O, Cr-O, and H-O bonds dominated the functional groups of the $Fe_{2.9}Cr_{0.1}O_4$ nanoparticles. The samples had a superparamagnetic characteristics associating with *S* shape of the *M-H* curves. The $Fe_{2.9}Cr_{0.1}O_4$ nanopowders has M_S value higher than that of the ferrofluids predicted as the effect of particles size and clustering of the magnetic particles. Moreover, the *Eg* of the $Fe_{2.9}Cr_{0.1}O_4$ nanoparticles was about 2.26 eV presenting a semiconductor material.

Acknowledgement
This work was partially supported by "Hibah PPT/STRANAS" for AT (2016-2017) from DRPM KEMENRISTEKDIKTI RI.

5. References

[1] Ghazanfari M R, Kashefi M, Shams S F and Jaafari M R 2016 Perspective of Fe_3O_4 Nanoparticles Role in Biomedical Applications *Biochem. Res. Int.* **2016** 1–32
[2] Scherer C and Figueiredo Neto A M 2005 Ferrofluids: properties and applications *Braz. J. Phys.* **35** 718–27
[3] Cullity B D and Graham C D 2008 *Introduction to Magnetic Materials* (Hoboken, N.J: Wiley-IEEE Press)
[4] Quadro E B, Dias M de L R, Amorim A M M and Rangel M do C 1999 Chromium and Copper-Doped Magnetite Catalysts for the High Temperature Shift Reaction *J. Braz. Chem. Soc.* **10** 51–9
[5] Taufiq A, Sunaryono, Rachman Putra E G, Okazawa A, Watanabe I, Kojima N, Pratapa S and Darminto 2015 Nanoscale Clustering and Magnetic Properties of $Mn_xFe_{3-x}O_4$ Particles Prepared from Natural Magnetite *J. Supercond. Nov. Magn.* **28** 2855–63
[6] Hoang Duy Nguyen, Trung Dzung Nguyen and Phuong Tung Nguyen 2014 Magnetic properties of Cr doped Fe_3O_4 porous nanoparticles prepared through a co-precipitation method using surfactant *Adv. Nat. Sci. Nanosci. Nanotechnol.* **5** 035017
[7] Hamad H, El-Latif M A, Kashyout A E-H, Sadik W and Feteha M 2015 Synthesis and characterization of core–shell–shell magnetic ($CoFe_2O_4$–SiO_2–TiO_2) nanocomposites and TiO_2 nanoparticles for the evaluation of photocatalytic activity under UV and visible irradiation *New J. Chem.* **39** 3116–28
[8] Mandal S and Natarajan S 2015 Adsorption and catalytic degradation of organic dyes in water using $ZnO/Zn_xFe_{3-x}O_4$ mixed oxides *J. Environ. Chem. Eng.* **3** 1185–93
[9] Taufiq A, Sunaryono, Hidayat N, Hidayat A, Putra E G R, Okazawa A, Watanabe I, Kojima N, Pratapa S and Darminto 2017 Studies on Nanostructure and Magnetic Behaviors of Mn-Doped Black Iron Oxide Magnetic Fluids Synthesized from Iron Sand *Nano* **12** 1750110
[10] Laurent S, Henoumont C, Stanicki D, Boutry S, Lipani E, Belaid S, Muller R N and Vander Elst L 2017 Magnetic Properties *MRI Contrast Agents* (Singapore: Springer Singapore) pp 5–11
[11] Tauc J, Grigorovici R and Vancu A 1966 Optical Properties and Electronic Structure of Amorphous Germanium *Phys. Status Solidi B* **15** 627–37
[12] El Ghandoor H, Zidan H M, Khalil M M and Ismail M I M 2012 Synthesis and some physical properties of magnetite (Fe3O4) nanoparticles *Int J Electrochem Sci* **7** 5734–45
[13] Puspitaningrum A, Taufiq A, Hidayat A, Sunaryono, Hidayat N and Samian 2017 Optical Properties of Fe_3O_4 Magnetic Fluid from Iron Sand *IOP Conf. Ser. Mater. Sci. Eng.* **202** 012054

Nanopowder and Magnetic Fluid Synthesis of $Zn_{0.2}Fe_{2.8}O_4$ Particles and Their Structural and Magnetic Behaviors

A Taufiq[1,*], D Yuliantika[1], Sunaryono[1], N Hidayat[1], M Diantoro[1] and Mujamilah[2]

[1]Department of Physics, Faculty of Mathematics and Natural Sciences, Universitas Negeri Malang, Jl. Semarang 5, Malang 65145, Indonesia
[2]Centre for Science and Technology of Advanced Materials, National Nuclear Energy Agency of Indonesia, Tangerang 15314, Indonesia

*Corresponding author's email: ahmad.taufiq.mipa@um.ac.id

Abstract: In this paper, the authors report the fabrication of $Zn_{0.2}Fe_{2.8}O_4$ particles from Indonesian sand in the forms of nanopowder and magnetic fluid. The fabrication was conducted by employing a combined coprecipitation-sonochemical method. The analysis of the XRD data presented that the $Zn_{0.2}Fe_{2.8}O_4$ was crystallized in a spinel structure with a lattice parameter of about 8.378 Å. The data analysis of the TEM image showed the $Zn_{0.2}Fe_{2.8}O_4$ particles distributed in nanometric size with a spherical shape. The magnetic investigation showed that the magnetic nanopowder and fluid both had superparamagnetic character. Moreover, the band gap of the $Zn_{0.2}Fe_{2.8}O_4$ particles was about 2.28 eV. These characters open a significant opportunity for the prepared samples to be applied for sensory and antibacterial applications.

Keywords: nanopowder, magnetic fluid, $Zn_{0.2}Fe_{2.8}O_4$, spinel, superparamagnetic.

1. Introduction

Nanotechnology research using the material of iron oxide has frequently been performed for the past few years to produce a material with an optimum magnetic response. One example material of iron oxide is magnetite (Fe_3O_4) nanoparticle that has been developed for many applications due to its excellent magnetic response showing superparamagnetic properties [1]. However, unfortunately, nanometric Fe_3O_4 particles have a lower saturation magnetization compared to bulk Fe_3O_4 [2]. Therefore, the magnetization of Fe_3O_4 particles could be improved by performing a transitional metal substitution forming a $(M_xFe_{1-x})_A(Fe_{2-x})_BO_4$ system where M could be Cu [3], Mn-Co [4], Ni-Zn [5], and Mg [6] with x as substitutional fraction notation.

Zink (Zn) as one of the metal substitutions is superior in improving the performance of the Fe_3O_4 particles. In the form of Zn-ferrites ($Zn_xFe_{3-x}O_4$), the material is biocompatible with a low level of toxicity [7], having relatively high capacitance [8], transmission and band gap [9], and showing a paramagnetic property at room temperature [10]. The previous report showed that the substantial fraction of $x = 0$ - 0.4 for the $Zn_xFe_{3-x}O_4$ nanoparticle could maintain the inverse spinel distribution [11] and its superparamagnetic properties [3]. Other reports presented that the highest saturation magnetization was obtained at $x = 0.2$ forming a $Zn_{0.2}Fe_{2.8}O_4$ system in the forms of nanopowder and ferrofluid.

IOP Conf. Series: Journal of Physics: Conf. Series **1091** (2018) 012030 doi:10.1088/1742-6596/1091/1/012030

Its excellent magnetic property makes ferrofluid become one of the critical materials to be developed, especially to vary its applications. The most recent report shows that ferrofluid could be applied as a hyperthermia treatment agent [7], solar cell photovoltaic system [12], energy harvester [13] and even as an electromagnetic hybrid generator [14]. In this research, the $Zn_{0.2}Fe_{2.8}O_4$ ferrofluid was fabricated from $Zn_{0.2}Fe_{2.8}O_4$ nanoparticles dispersed in water (H_2O) as a carrier medium and TMAH ($C_4H_{13}NO$) as a surfactant. Interestingly, the natural Fe_3O_4 from Indonesian iron sand was employed as a main source. Such nanomagnetic particle is layered by the surfactant to result in an active surface that could efficiently bind the surface of the dispersant medium in producing the homogeneous colloidal suspension with good superparamagnetic characteristics.

2. Experimental Method

The purified iron sand was dissolved in HCl using a magnetic stirrer at ambient temperature to prepare the $Zn_{0.2}Fe_{2.8}O_4$ particles. In order to separate the non-dissolved solid after the chemical reaction, the chloride iron solution was filtered and then added with $ZnCl_2$ that has been diluted by the stoichiometry to produce the $Zn_{0.2}Fe_{2.8}O_4$ particles. Such mixture was then reacted with NH_4OH via a coprecipitation process under the effect of ultrasonic radiation. The reaction of the $Zn_{0.2}Fe_{2.8}O_4$ particles fabrication is shown in Equation 1.

$$2FeCl_3 + FeCl_2 + 0.21ZnCl_2 + 8.42NH_4OH \rightarrow 1.07Zn_{0.2}Fe_{2.8}O_4 + 8.43NH_4Cl + 8.21H_2O \quad (1)$$

The sediment of the $Zn_{0.2}Fe_{2.8}O_4$ particles was then neutralized through washing technique. The sediment was then calcined to produce the sample in nanopowders. In general, the synthesis procedure was maintained by following our previous work [15]. The fabrication of sample in the form of $Zn_{0.2}Fe_{2.8}O_4$ ferrofluid was performed by dispersing in the water the $Zn_{0.2}Fe_{2.8}O_4$ particle sediment in $C_4H_{13}NO$. Subsequently, the sample was characterized using XRD, SEM, FTIR, and VSM.

3. Results and Discussion

In Fig. 1, the TEM image exhibits that the $Zn_{0.2}Fe_{2.8}O_4$ particle has a spherical shape with a mean particle size of 12.6 ± 0.4 nm with an inhomogeneity level of 3.4 %. Another research group reported that the magnetic nanoparticles prepared by a chemical combustion method and using Tween 80 as a surfactant produced the particle size of 36.5 nm with an inhomogeneity level of 52 % [16]. It means that this work has successfully provided the magnetic nanoparticles in powders. It also indicates that without a surfactant, the ultrasonic irradiation during sonochemical process successfully formed magnetic particles with a narrower size and high homogeneity. The smaller particle size of the magnetic particles was resulted by the micro-emission process of the ultrasonic field during the sonication process that produced local heat and bubble cavitation in the material. Therefore, the obtained smaller particles were originated from the cavitation process and the local heat that increases the crystal growth of the magnetic particles [17]. Interestingly, the prepared $Zn_{0.2}Fe_{2.8}O_4$ particles in this work, produced by a combination sonochemical-coprecipitation method have a particle size that was smaller than that of the $Zn_{0.2}Fe_{2.8}O_4$ produced by a coprecipitation method [18].

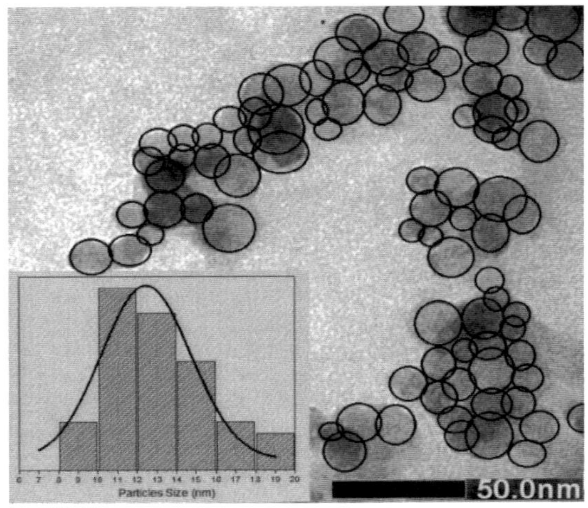

FIGURE 1. TEM image of $Zn_{0.2}Fe_{2.8}O_4$ nanoparticles

The formation of the $Zn_{0.2}Fe_{2.8}O_4$ particles was analyzed based on the FT-IR graphic pattern shown in Figure 2. The FT-IR spectrum at wavenumbers of ~1635 cm^{-1} and 3441 cm^{-1} representing the vibration of the hydroxyl functional group for symmetrical hydroxyl bond [19] and asymmetrical hydroxyl bond [20] while the weak ~2350 cm^{-1} spectrum was Co_2 binding group [21]. The transmittance peak at a wavenumber of ~402 cm^{-1} was the vibration of the metal-oxygen (M-O) bond at an octahedral position [22]. Further, the Zn-O-Fe bond at the tetrahedral position was broken into two transmittance peaks, namely ~655 cm^{-1} and ~1340 cm^{-1} that indicated a successful substitution of Fe ion by Zn ion occupying a tetrahedral position. Therefore, it shows that the Zn ion substitution in Fe_3O_4 that formed the $Zn_{0.2}Fe_{2.8}O_4$ system still maintain its inverse spinel crystal structure.

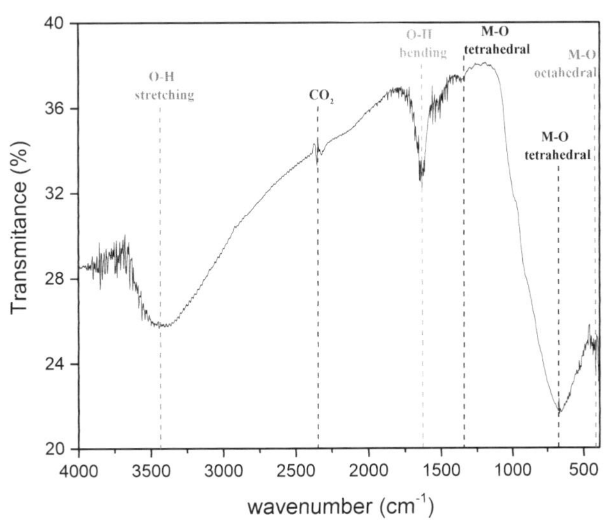

Figure 2. FTIR spectrum of $Zn_{0.2}Fe_{2.8}O_4$ nanoparticles

In this work, a further structural characterization was conducted via XRD using Cu-Kα (l = 0.15406 nm). The diffraction pattern of the $Zn_{0.2}Fe_{2.8}O_4$ nanoparticles is shown in Figure 3a, and its crystal structure is shown in Figure 3b. The figure presents the *hkl* planes of (220), (311), (400), (422), (511), and (440) showing a cubic structure with the space group *F d -3 m* [23]. The refinement analysis via

Rietveld method using AMCSD 7766 as a model data presented that the $Zn_xFe_{3-x}O_4$ nanoparticles have an inverse cubic spinel structure with an ionic distribution that is expressed in the system of $[Zn_{1-x}^{2+}Fe_x^{3+}]_A[Fe_{2-x}^{3+}]_BO_4$. The A symbol represents the tetrahedral position, and the B symbol represents the octahedral position. The XRD data analysis coincides with the analysis of the FT-IR data as discussed above. Another investigation to analyze the elemental composition of the sample was done using XRF. The data presented that the sample has elemental contents of Zn: Fe of about 0.6:1. A further calculation of the mass fraction of Zn and Fe in the sample was 1/3 for [Zn]/[Zn+Fe] and 2/3 for [Fe]/[Zn+Fe]. Thus, the distribution of the inverse spinel can be expressed by $[Zn_{1/3}Fe_{2/3}]_A[Fe_{5/3}]_BO_4$.

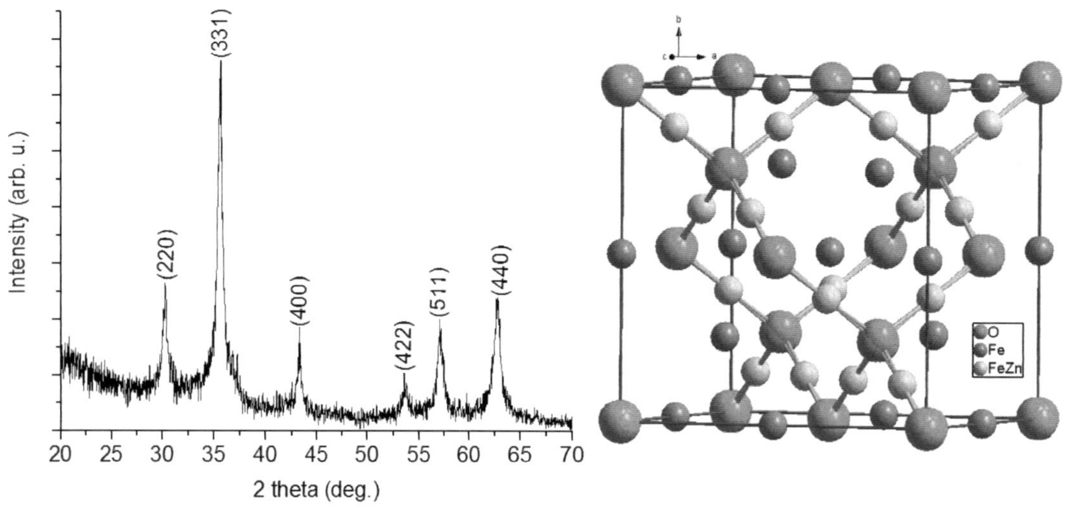

(a) (b)

FIGURE 3. XRD pattern of (a) the $Zn_{0.2}Fe_{2.8}O_4$ nanoparticles and (b) its crystal structure

The quantitative analysis of the XRD data via Rietveld analysis presented information of structural parameters such as lattice constants ($a = b = c$) and unit cell volume (V). The crystallite size (D) of the sample was analyzed for crystal plane (311) by using Scherer's formula as shown in Equation (2). The result as shown in Table 1 exhibited that the crystallite size was 11.95 nm that was slightly different with the particle size from TEM analysis that was 12.65 nm.

$$D = \frac{k\lambda}{\beta\cos\theta} \quad (2)$$

Where k = lattice constant (0.98 for cubic cell), λ = wavelength (0.5406 nm), β = full width half at the maximum intensity/FWHM (rad.), and θ = Bragg's peak angle [24].

TABLE 1. Lattice parameters and particle size of the $Zn_{0.2}Fe_{2.8}O_4$ nanoparticle

Data	$a = b = c$ (Å)	Particle size (nm)		Unit sell volume (Å^3)
		TEM	Scherer's equation	
$Zn_{0.2}Fe_{2.8}O_4$	8.378	12.65	11.95	588.13

The UV-Vis data of the $Zn_{0.2}Fe_{2.8}O_4$ particle that were analyzed by using a Tauc Plot are shown in Figure 4. The Tauc Plot exhibits information of optical parameter of the sample associating with optical band gap. The value of energy band gap depends on crystal structure, thermal effects, and also defects

[25]. Zn doping substitution could cause defects in the crystal due to the radius difference between Zn and Fe. A Zn substitution in Fe_3O_4 formed $Zn_{0.2}Fe_{2.8}O_4$ nanoparticles that produced an energy band of 2.28 eV of which value was higher than the Fe_3O_4 nanoparticle, namely 2.2 eV [26]. This phenomenon is caused by the Zn substitution that has electron affinity (15 Kj/mol) lower than that of Fe (0 Kj/mol), which lead to locating the energy level in the bottom of the conduction band [9]. Based on the quantum mechanics, the extra energy is required by valence electrons leading to an increase in the energy band gap to excite to higher energy states in the conduction band. Another group reported that the magnetic nanoparticles with the energy of 2.2 ± 0.2 eV could be used as photocatalytic agent producing photocatalytic activity of up to 96.7 % [27]. Therefore, the sample in this work that has energy an band gap of 2.28 eV opens a significant potential for the application as a photocatalytic agent in the visible light spectrum.

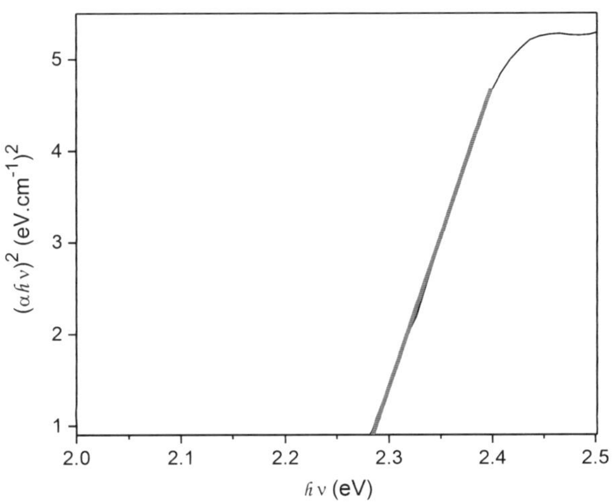

FIGURE 4. Band gap energy of $Zn_{0.2}Fe_{2.8}O_4$

The hysteresis (M-H) curves are presented in Figure 5. The figures show the respective circles and solid lines representing the experimental data and fitting model using Langevin's equation as shown in Equation 3 [28]. The data analysis presents the saturation magnetization (*Ms*) values of the $Zn_{0.2}Fe_{2.8}O_4$ powder and ferrofluid are 33.530 ± 0.040 and 1.263 ± 0.003 emu/g, respectively. Figure 5 also presents the *S* shape of the $Zn_{0.2}Fe_{2.8}O_4$ powder and ferrofluid that represent the superparamagnetic characters of the samples [29].

$$L(x) = M_S\left(\coth(Bx) - \frac{1}{Bx}\right) \quad (3)$$

Where $B = \mu/k_BT$, $x = H$, and Ms is saturation magnetization.

Theoretically, the Zn^{2+} substitution changes the ionic distribution in the spinel structure of Fe_3O_4 and leads to a change in the magnetic behaviors of the samples. In the system of $Zn_{0.2}Fe_{2.8}O_4$, the Zn^{2+} ions may replace the Fe^{3+} ions in tetrahedral position. As a consequence, the Fe^{2+} ions in octahedral changed to Fe^{3+} ions. The magnetic moments in the octahedral site are antiparallel with the tetrahedral site. The imbalance metal ions between the octahedral and tetrahedral sites resulted in some unpaired magnetic spins producing net magnetic moments that can be used to calculate the saturation magnetization of the spinel system [30]. Our previous works presented that the magnetic behaviors of the magnetic nanoparticles in the powders and ferrofluids are also affected by the metal ion composition, primary and secondary particle sizes, fractal structures, and nanoclusters of the samples [29,31].

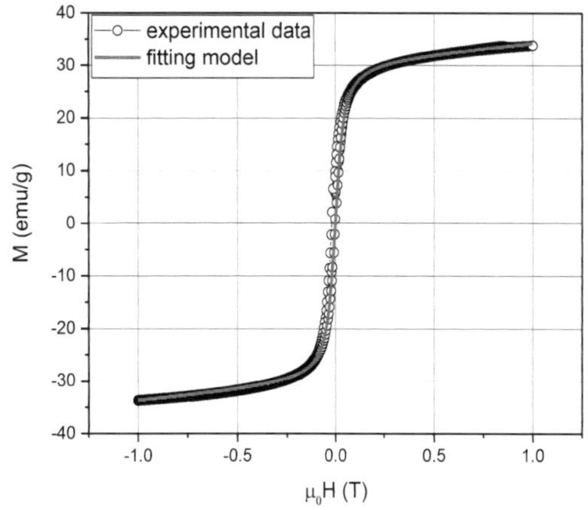

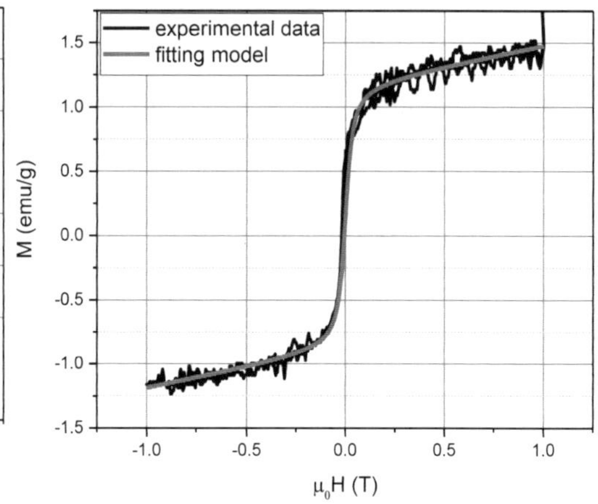

FIGURE 5. Hysteresis curve of superparamagnetic $Zn_{0.2}Fe_{2.8}O_4$ in (a) nanopowder (b) ferrofluid

The characteristic of ferrofluids is a combination of liquid character and magnetic nanoparticles that are easy to be controlled by an external magnetic field. In the magnetic fluids, the saturation magnetization is lower than that of nanopowders originating from the effect of viscosity of the fluids. On powders, the magnetic particles did not have any liquid carriers that differ from the ferrofluids coated by $C_4H_{13}NO$ dispersed in H_2O. The effect of magneto-viscous explains that the high viscosity of the ferrofluids leads to a reduction in the thermal effect [32]. Under external magnetic field, the low thermal energy decreases the Brownian motion of the particles in the ferrofluids and reduces the magnetic response. The low saturation magnetization of the ferrofluids is also affected by the chemical reaction between the surfactant and magnetic particles. Moreover, the spin canting in the surface and pinned spins that were close to the surface also provided a further explanation of why the saturation magnetization of the ferrofluids was lower than that of the nanopowders [33].

The higher magnetic moment of ferrofluids leads to an increase in the aggregation as the effect of van der Waals forces and dipoles interaction [34]. Using a small-angle neutron scattering, the ferrofluids form an aggregation in a one dimension chain-like structure as reported in the previous work [35]. Under an external magnetic field, the chain-like structure builds pre-existing chains. Despite resulting from a single domain particle, the saturation magnetization of ferrofluids also results from the magnetization of the pre-existing chain particles [36]. Magnetic moments of the ferrofluids build a symmetric fan arrangement along the chain axis. Therefore, the magnetizing and demagnetizing processes in the M-H curve of the ferrofluids showed the quasi-magnetic hysteresis that differs from the M-H curve of the nanopowders as shown in Figure 5.

4. Conclusion

The $Zn_{0.2}Fe_{2.8}O_4$ particles form an inverse cubic spinel structure in a nanometric size and a spherical shape. The combined sonochemical-coprecipitation method has succeeded in producing the $Zn_{0.2}Fe_{2.8}O_4$ particles with high homogeneity. Both of $Zn_{0.2}Fe_{2.8}O_4$ nanopowder and ferrofluid exhibit superparamagnetic phenomena with a saturation magnetization of 33.530 ± 0.040 emu/g, and 1.263 ± 0.003 emu/g, respectively. All samples exhibited superparamagnetic behaviors at room temperature. Furthermore, the prepared samples with energy band gap of 2.28 eV open a significant potential application as a photocatalytic agent.

Acknowledgement
This reserach was partly supported by DRPM KEMENRISTEKDIKTI RI for AT (2016-2017) under "Hibah PDUPT" research grant.

5. References

[1] Ramimoghadam D, Bagheri S and Abd Hamid S B 2014 Stable monodisperse nanomagnetic colloidal suspensions: An overview *Colloids Surf. B Biointerfaces* **133** 388–411

[2] Yuan H, Liu E, Yin Y, Zhang W, Wong P J, Zheng J-G, Huang Z, Ou H, Zhai Y and Xu Q 2016 Enhancement of magnetic moment in $Zn_xFe_{3-x}O_4$ thin films with dilute Zn substitution *Appl. Phys. Lett.* **108** 232403

[3] Tehrani F S, Daadmehr V, Rezakhani A T, Akbarnejad R H and Gholipour S 2012 Structural, magnetic, and optical properties of zinc-and copper-substituted nickel ferrite nanocrystals *J. Supercond. Nov. Magn.* **25** 2443–55

[4] Giri J, Pradhan P, Somani V, Chelawat H, Chhatre S, Banerjee R and Bahadur D 2008 Synthesis and characterizations of water-based ferrofluids of substituted ferrites [$Fe_{1-x}B_xFe_2O_4$, B = Mn, Co (x = 0–1)] for biomedical applications *J. Magn. Magn. Mater.* **320** 724–30

[5] Li O A, Lin C-R, Chen H-Y, Hsu H-S, Shih K-Y, Edelman I S, Wu K-W, Tseng Y-T, Ovchinnikov S G and Lee J-S 2016 Size dependent magnetic and magneto-optical properties of $Ni_{0.2}Zn_{0.8}Fe_2O_4$ nanoparticles *J. Magn. Magn. Mater.* **408** 206–12

[6] Akhtar M N, Rahman A, Sulong A B and Khan M A 2017 Structural, spectral, dielectric and magnetic properties of $Ni_{0.5}Mg_xZn_{0.5-x}Fe_2O_4$ nanosized ferrites for microwave absorption and high frequency applications *Ceram. Int.* **43** 4357–65

[7] Zargar T and Kermanpur A 2017 Effects of hydrothermal process parameters on the physical, magnetic and thermal properties of $Zn_{0.3}Fe_{2.7}O_4$ nanoparticles for magnetic hyperthermia applications *Ceram. Int.* **43** 5794–804

[8] Lv Z, Wang Q, Bin Y, Huang L, Zhang R, Zhang P and Matsuo M 2015 Magnetic behaviors of Mg-and Zn-Doped Fe_3O_4 nanoparticles estimated in terms of crystal domain size, dielectric response, and application of Fe_3O_4/Carbon nanotube composites to anodes for lithium ion batteries *J. Phys. Chem. C* **119** 26128–42

[9] Rezay Marand Z, Helmi Rashid Farimani M and Shahtahmasebi N 2014 Study of magnetic and structural and optical properties of Zn doped Fe_3O_4 nanoparticles synthesized by co-precipitation method for biomedical application *Nanomedicine J.* **1** 238–47

[10] Güner S, Esir S, Baykal A, Demir A and Bakis Y 2014 Magneto-optical properties of $Cu_{1-x}Zn_xFe_2O_4$ nanoparticles *Superlattices Microstruct.* **74** 184–97

[11] Liu X, Liu J, Zhang S, Nan Z and Shi Q 2016 Structural, magnetic, and thermodynamic evolutions of Zn-doped Fe_3O_4 nanoparticles synthesized using a one-step solvothermal method *J. Phys. Chem. C* **120** 1328–41

[12] Sardarabadi M, Passandideh-Fard M, Maghrebi M-J and Ghazikhani M 2017 Experimental study of using both ZnO/water nanofluid and phase change material (PCM) in photovoltaic thermal systems *Sol. Energy Mater. Sol. Cells* **161** 62–9

[13] Khairul M A, Doroodchi E, Azizian R and Moghtaderi B 2017 Advanced applications of tunable ferrofluids in energy systems and energy harvesters: A critical review *Energy Convers. Manag.* **149** 660–74

[14] Seol M-L, Jeon S-B, Han J-W and Choi Y-K 2017 Ferrofluid-based triboelectric-electromagnetic hybrid generator for sensitive and sustainable vibration energy harvesting *Nano Energy* **31** 233–8

[15] Taufiq A, Bahtiar S, Sunaryono, Hidayat N, Hidayat A, Mufti N, Diantoro M, Fuad A, Munasir, Rahmawati R, Adi W A, Pratapa S and Darminto 2017 Preparation of Superparamagnetic $Zn_{0.5}Mn_{0.5}Fe_2O_4$ Particle by Coprecipitation-Sonochemical Method for Radar Absorbing Material IOP Conference Series: Materials Science and Engineering vol 202

[16] Prabhu Y T, Rao K V, Kumari B S, Kumar V S S and Pavani T 2015 Synthesis of Fe_3O_4 nanoparticles and its antibacterial application *Int. Nano Lett.* **5** 85–92

[17] Saffari J, Mir N, Ghanbari D, Khandan-Barani K, Hassanabadi A and Hosseini-Tabatabaei M R 2015 Sonochemical synthesis of Fe_3O_4/ZnO magnetic nanocomposites and their application in photo-catalytic degradation of various organic dyes *J. Mater. Sci. Mater. Electron.* **26** 9591–9

[18] Taufiq A, Bahtiar S, Sunaryono, Hidayat N, Hidayat A, Fuad A, Diantoro M, Pratapa S and Darminto 2012 Ahmad Taufiq, S. Bahtiar, Sunaryono, Nurul Hidayat, A. Fuad, M. Diantoro, Arif Hidayat, Suminar Pratapa, Darminto. 2012. Analysis of Crystal Structure and Dielectric of Zn^{2+} Ion Doped Nanopartilces Magnetic Based on Iron Sand Synthesized by Coprecipitation Method *Indonesian Journal of Materials Science* **13** 153–6

[19] Hamad H, El-Latif M A, Kashyout A E-H, Sadik W and Feteha M 2015 Synthesis and characterization of core–shell–shell magnetic ($CoFe_2O_4$–SiO_2–TiO_2) nanocomposites and TiO_2 nanoparticles for the evaluation of photocatalytic activity under UV and visible irradiation *New J. Chem.* **39** 3116–28

[20] Mandal S and Natarajan S 2015 Adsorption and catalytic degradation of organic dyes in water using ZnO/$Zn_xFe_{3-x}O_4$ mixed oxides *J. Environ. Chem. Eng.* **3** 1185–93

[21] Gatelytė A, Jasaitis D, Beganskienė A and Kareiva A 2011 Sol-Gel Synthesis and Characterization of Selected Transition Metal Nano-Ferrites **17**

[22] Nawle A C, Humbe A V., Babrekar M K, Deshmukh S S and Jadhav K M 2017 Deposition, characterization, magnetic and optical properties of Zn doped $CuFe_2O_4$ thin films *J. Alloys Compd.* **695** 1573–82

[23] O'Neill H S C and Navrotsky A 1983 Simple spinels; crystallographic parameters, cation radii, lattice energies, and cation distribution *Am. Mineral.* **68** 181–94

[24] Jamil M T, Ahmad J, Bukhari S H, Sultan T, Akhter M Y, Ahmad H and Murtaza G 2017 Effect on Structural and Optical Properties of Zn-Substituted Cobalt Ferrite $CoFe_2O_4$ *J. Ovonic Res. Vol* **13** 45–53

[25] Dagenais K, Chamberlin M and Constantin C 2011 Energy Band Gap Behavior as a Function of Optical Electronegativity for Semiconducting and Insulating Binary Oxides *APS Southeastern Section Meeting Abstracts*

[26] Bagbi Y, Sarswat A, Mohan D, Pandey A and Solanki P R 2016 Lead (Pb^{2+}) adsorption by monodispersed magnetite nanoparticles: Surface analysis and effects of solution chemistry *J. Environ. Chem. Eng.* **4** 4237–47

[27] Mathubala G, Manikandan A, Antony S A and Ramar P 2016 Photocatalytic degradation of methylene blue dye and magneto-optical studies of magnetically recyclable spinel $Ni_xMn_{1-x}Fe_2O_4$ ($x = 0.0 – 1.0$) nanoparticles *J. Mol. Struct.* **1113** 79–87

[28] Kittel C 2005 *Introduction to solid state physics* (Wiley)

[29] Taufiq A, Sunaryono, Hidayat N, Hidayat A, Putra E G R, Okazawa A, Watanabe I, Kojima N, Pratapa S and Darminto 2017 Studies on Nanostructure and Magnetic Behaviors of Mn-Doped Black Iron Oxide Magnetic Fluids Synthesized from Iron Sand *Nano* **12** 1750110

[30] Cullity B D and Graham C D 2008 *Introduction to Magnetic Materials* (Hoboken, N.J: Wiley-IEEE Press)

[31] Taufiq A, Sunaryono, Rachman Putra E G, Okazawa A, Watanabe I, Kojima N, Pratapa S and Darminto 2015 Nanoscale Clustering and Magnetic Properties of $Mn_xFe_{3-x}O_4$ Particles Prepared from Natural Magnetite *J. Supercond. Nov. Magn.* **28** 2855–63

[32] Li J, Gong X M, Lin Y Q, Liu X D, Chen L L, Li J M, Mao H and Li D C 2014 Investigation into loss in ferrofluid magnetization *AIP Adv.* **4** 077123

[33] Blanco-Mantecon M and O'Grady K 2006 Interaction and size effects in magnetic nanoparticles *J. Magn. Magn. Mater.* **296** 124–33

[34] Scherer C and Figueiredo Neto A M 2005 Ferrofluids: properties and applications *Braz. J. Phys.* **35** 718–27

IOP Conf. Series: Journal of Physics: Conf. Series **1091** (2018) 012030 doi:10.1088/1742-6596/1091/1/012030

[35] Taufiq A, Sunaryono, Rachman Putra E G, Pratapa S and Darminto 2015 Nano-Structural Studies on Fe_3O_4 Particles Dispersing in a Magnetic Fluid Using X-Ray Diffractometry and Small-Angle Neutron Scattering *Mater. Sci. Forum* **827** 213–8
[36] Lin L, Li J, Lin Y, Liu X, Chen L, Li J and Li D 2014 The extrinsic hysteresis behavior of dilute binary ferrofluids. *Eur. Phys. J. E--Soft Matter* **37**

Institute of Physics
Dirac House, Temple Back
Bristol BS1 6BE UK

CURRAN ASSOCIATES INC.
proceedings.com

ISSN: 1742-6588
ISBN 978-1-5108-7433-6

9 781510 874336